2

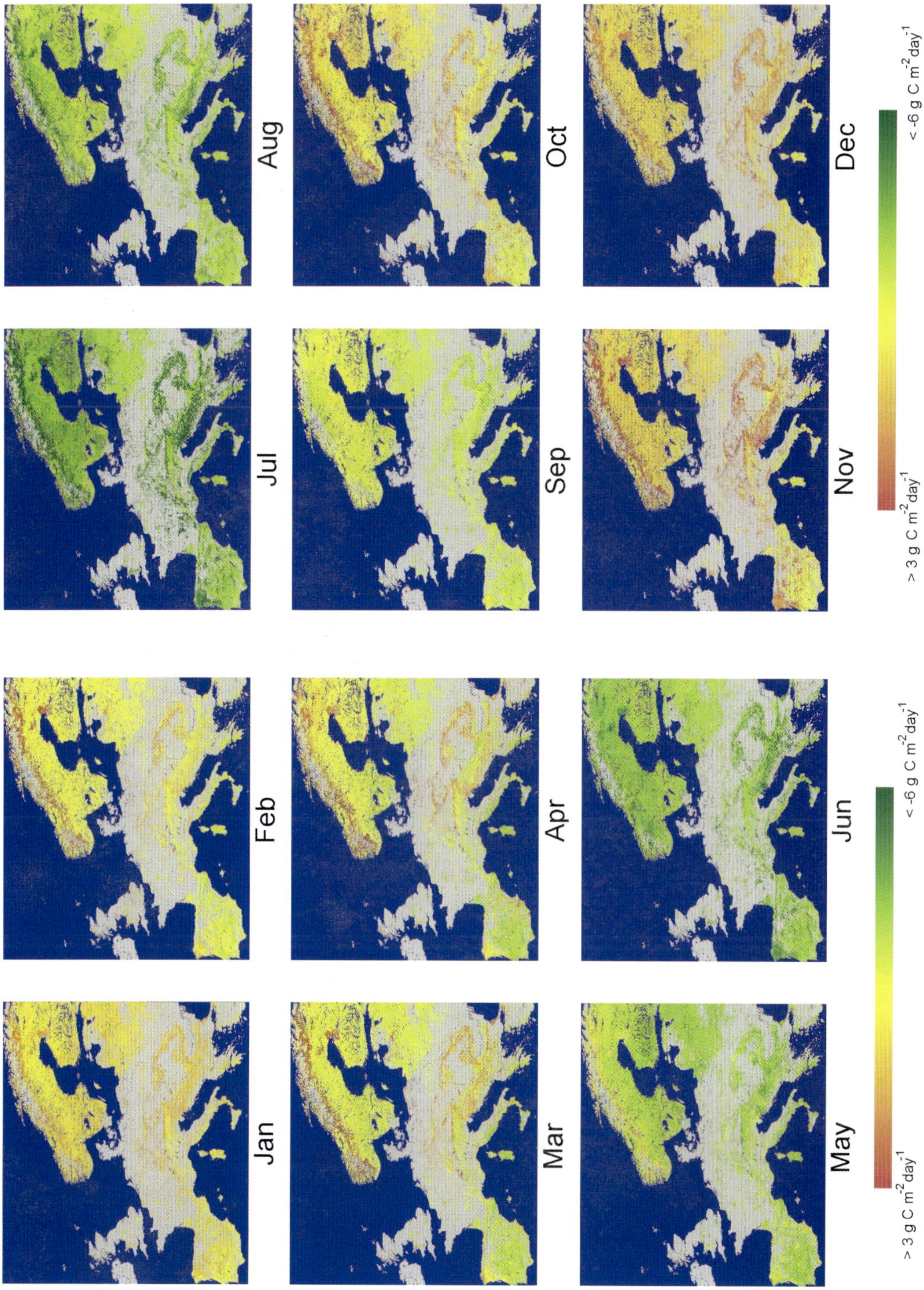

Fig. 14.17. Examples of NEE maps generated. Second week of each month.

Forests at the Land–Atmosphere Interface

Forests at the Land–Atmosphere Interface

Edited by

M. Mencuccini, J. Grace, J. Moncrieff and K.G. McNaughton
School of GeoSciences
University of Edinburgh
Darwin Building
Mayfield Road
Edinburgh EH9 3JU
Scotland, UK

CABI Publishing

CABI Publishing is a division of CAB International

CABI Publishing
CAB International
Wallingford
Oxon OX10 8DE
UK
Tel: +44 (0)1491 832111
Fax: +44 (0)1491 833508
E-mail: cabi@cabi.org
Website: www.cabi-publishing.org

CABI Publishing
875 Massachusetts Avenue
7th Floor
Cambridge, MA 02139
USA
Tel: +1 617 395 4056
Fax: +1 617 354 6875
E-mail: cabi-nao@cabi.org

A catalogue record for this book is available from the British Library, London, UK.

Library of Congress Cataloging-in-Publication Data
Forests at the land–atmosphere interface / edited by M. Mencuccini ... [et al.].
p. cm.
Proceedings of a conference held in Edinburgh, September 2001.
Includes bibliographical references and index.
ISBN 0-85199-677-9 (alk. paper)
1. Forest meteorology--Congresses. 2. Carbon sequestration--Congresses. 3. Stomata--Congresses. 4. Forest management--Congresses. I. Mencuccini, M., (Maurizio) II. Title.
SD390.5.F67 2004
634.9′01′5515--dc21
2003009468

ISBN 0 85199 677 9

Typeset in 10pt Souvenir by Columns Design Ltd, Reading
Printed and bound in the UK by Cromwell Press, Trowbridge

Contents

Contributors

Archer, N., *Division of Environmental and Applied Biology, Biological Sciences Institute, School of Life Sciences, University of Dundee, Dundee DD1 4HN, UK.*

Berninger, F., *Department of Forest Ecology, PO Box 27, 00014 University of Helsinki, Finland. E-mail: Frank.berninger@helsinki.fi*

Betts, A.K., *Atmospheric Research, Pittsford, VT 05763, USA. E-mail: akbetts@aol.com*

Brown, K.J., *Department of Environmental Biology, Ohio University, Athens, OH 45701-2979, USA.*

Cannell, M.G.R., *Centre for Ecology and Hydrology, Bush Estate, Penicuik, Midlothian EH26 0QB, UK. E-mail: mgrc@ceh.ac.uk*

Dolman, A.J., *Department of Hydrology and Geo-Environmental Sciences, Faculty of Earth and Life Sciences, Vrije Universiteit, De Boelelaan 1085, 1081 HV Amsterdam, The Netherlands. E-mail: a.j.dolman@alterra.wag-ur.nl*

Engel, V., *Department of Biology, Duke University, Box 90340, Durham, NC 27708-0340, USA.*

Grace, J.C., *New Zealand Forest Research Institute, Private Bag 3020, Rotorua, New Zealand. E-mail: jenny.grace@forestresearch.co.nz*

Griffin, K.L., *Lamont–Doherty Earth Observatory, Columbia University, Palisades, NY 10964-8000, USA.*

Hutjes, R.W.A., *Alterra, Green World Research, PO Box 47, 6700 AC Wageningen, The Netherlands.*

Isaac, P., *53 Brewer Road, Bentleigh, Victoria 3204, Australia.*

Jones, H.G., *Division of Environmental and Applied Biology, Biological Sciences Institute, School of Life Sciences, University of Dundee, Dundee DD1 4HN, UK. E-mail: h.g.jones@dundee.ac.uk*

Landsberg, J., *22 Mirning Crescent, Aranda, Canberra, ACT 2614, Australia. E-mail: joe.landsberg@landsberg.com.au*

Lee, X., *School of Forestry and Environmental Studies, Yale University, 370 Prospect Street, New Haven, CT 06511, USA. E-mail: xuhui.lee@yale.edu*

Leuning, R., *CSIRO Land and Water, FC Pye Laboratory, PO Box 1666, Canberra, ACT 2601, Australia. E-mail: ray.leuning@csiro.au*

Malcolm, D.C., *School of Forestry, Institute of Ecology and Resource Management, University of Edinburgh, Mayfield Road, Edinburgh EH9 3JU, UK. E-mail: d.malcolm@ed.ac.uk*

Mansfield, T.A., *Department of Biological Sciences, Lancaster University, Lancaster LA1 4YQ, UK.*

Medlyn, B., *School of Biological, Earth and Environmental Sciences, University of New South Wales, Sydney 2052, Australia. E-mail: b.medlyn@unsw.edu.au*

Overdieck, D., *TU-Berlin, Institut für Ökologie, FG Landschaftsökologie/ Ökologie der Gehölze, Königin-Luise-Str. 22, D-14195 Berlin, Germany. E-mail: dieter.overdieck@tu-berlin.de*

Papale, D., *University of Tuscia, Department of Forest Science and Environment, via C. de Lellis 01100 Viterbo, Italy. E-mail: darpap@infinito.it*

Perrier, A., *Environnement et Grandes Cultures, INRA-INA PG 78850 Thiverval Grignon, France.*

Rotenberg, E., *ESER, Weizmann Institute of Science, Rehovot 76100, Israel. E-mail: eyalr@wisemail.weizmann.ac.il*

Schuster, W.S.F., *Black Rock Forest Consortium, 129 Continental Road, Cornwall, NY 12519, USA.*

Stewart, J.B., *Department of Geography, University of Southampton, Highfield, Southampton SO17 1BJ, UK. E-mail: j.b.stewart@soton.ac.uk*

ter Maat, H.W., *Alterra, Green World Research, PO Box 47, 6700 AC Wageningen, The Netherlands.*

Tissue, D.T., *Department of Biological Sciences, Texas Tech University, Lubbock, TX 79409-3131, USA.*

Turnbull, M.H., *Department of Plant and Microbial Sciences, University of Canterbury, Private Bag 4800, Christchurch, New Zealand.*

Tuzet, A., *Environnement et Grandes Cultures, INRA–INA PG 78850 Thiverval Grignon, France.*

Valentini R., *University of Tuscia, Department of Forest Science and Environment, via C. de Lellis 01100 Viterbo, Italy. E-mail: rik@unitus.it*

van der Molen, M.K., *Department of Hydrology and Geo-Environmental Sciences, Faculty of Earth and Life Sciences, Vrije Universiteit, De Boelelaan 1085, 1081 HV Amsterdam, The Netherlands.*

Walcroft, A.S., *Landcare Research, Private Bag 11052, Palmerston North, New Zealand.*

Wang, Y.-P., *CSIRO Atmospheric Research, Private Bag No. 1, Aspendale, Victoria 3195, Australia. E-mail: yingping.wang@csiro.au*

Waring, R.H., *Oregon State University, College of Forestry, OR 7331, USA. E-mail: richard.waring@orst.edu*

Whitehead, D., *Landcare Research, PO Box 69, Lincoln 8152, New Zealand. E-mail: whiteheadD@LandcareResearch.co.nz*

Zhou, G., *South China Institute of Botany, Chinese Academy of Sciences, Guang Zhou, China.*

Preface

As we have grown more aware of the fragility of the earth's climate over the past two decades we have also become more aware of the interrelationships between its vegetation and its climate. This book contains chapters on the biophysical processes occurring at the interface between forest ecosystems and the atmosphere. It brings together topics such as the role of stomata in regulating water fluxes to the atmosphere, radiation penetration into forest canopies, forest meteorology, mesoscale modelling, remote sensing of canopy characteristics of forests, neural network predictions of carbon sequestration in Europe, and policy-making and forestry. The hope of the editors is that the book transmits a sense of scientific unity in a field that extends in scale from the physiology of intra-leaf organelles to the meteorology and carbon sequestration of whole forest ecosystems.

This volume also celebrates the achievements of Professor Paul Jarvis, a highly distinguished environmental biologist and forest physiologist, who devoted his career to understanding these interrelationships. The individual chapters are based on papers presented at the meeting 'Forests at the Land-Atmosphere Interface', which was held in his honour at the University of Edinburgh on 17–19 September 2001. The topics were chosen to reflect the range of Paul's interests and the scientists invited to present the papers were selected for their international expertise in each topic. It is a sign of the high esteem in which Professor Jarvis's is held that almost every invitation was accepted. The chapters printed here have been developed from a selection of the papers presented at the meeting. Our thanks go to the authors for their excellent efforts.

As fate would have it, our meeting was held just a week after the appalling tragedy of 11 September in New York. Air traffic across the Atlantic was disrupted and most of our American colleagues could not attend. We regret the absence of invited speakers such as John Norman, Piers Sellers, Tom Gower, Steve Wofsy, Joe Berry, Mike Unsworth and Forrest Hall from our meeting.

Gaps in the programme were filled by impromptu papers from volunteers recruited on the spot. The success of the meeting owes much to their contributions, as well as to the excellent quality of the papers from speakers who could attend.

Many people contributed to organizing the meeting and the final production of this book. We particularly wish to single out a few for their outstanding support: Sheila Wilson, who assisted during this whole process; and the producers from CAB International, particularly Rachel Robinson, Claire Gwilt and Rebecca Stubbs, who very patiently listened to the editors while they were making up their minds during the production process. We also thank our sponsors, Campbell Scientific, Delta-T Devices, Li-Cor, ADC BioScientific, PP Systems and The Royal Society of Edinburgh. Finally, a big vote of thanks to all the contributors to this volume and to all those who corresponded with us at several stages with useful information.

M. Mencuccini
J. Grace
J. Moncrieff
K.G. McNaughton

Foreword

Professor Paul Gordon Jarvis retired in April 2001 after 26 years as Professor of Forestry and Natural Resources at the University of Edinburgh, and after more than four decades as one of the world's most prominent plant ecophysiologists. His particular specialty is the exchanges of water and CO_2 between forest trees and the atmosphere. During his career Paul stayed at the forefront as this field changed from a minor topic on the margins of forestry and ecophysiology into a major focus of international research with implications for the future of the whole earth and its climate. For his achievements Paul Jarvis received one of science's highest honours when, in 1997, he was elected a Fellow of the Royal Society.

Paul was born on 23 May 1935 in Tunbridge Wells, Kent. His father had joined the Royal Flying Corp and flown Sopwith Camels in World War I. Later, in World War II, he was a founder member of the Royal Air Force Regiment, looking after aerodrome defences throughout the UK. When not serving in the RAF, Paul's father was a farmer and in the post-war period had a milk round, delivering milk from a pony and trap. Later he became an agricultural adviser, subsequently returning to farming in Hertfordshire. Paul's mother had been secretary to the geneticist Karl Pearson at University College, London. Paul and his two brothers Brion and Richard, therefore, had a rural upbringing – but with an academic slant to it: they built a raft to use on the farm pond and they became interested in the wildlife on the farm. Paul and his younger brother Richard attended Sir Anthony Brown's School in Brentwood, Essex. Paul was academically gifted and went to Oriel College, Oxford, where he read Botany and rowed.

Many careers are built on a number of important events, some planned and some arising by accident. For Paul, a very important event was undoubtedly his marriage to Margaret, in September 1957, just 3 months after both had graduated with BA degrees in Botany. They moved to Sheffield, and together began graduate studies at the university. Paul's research concerned the growth and regeneration of sessile oak, *Quercus petraea*, in nearby relict native woods, while Margaret investigated environmental limits to the geographical distribution of *Prunus padus* and other northern and southern species. They received their PhDs in 1961 and were ready to explore the wider world.

Two research careers required two research positions, so their next move was to the Institute of Plant Physiology at the University of Uppsala, for which both had been awarded NATO fellowships. Paul was to study the mycorrhizal requirements of oak and pine with Elias Melin while Margaret would work with Henry Rufelt on the water relations of tree seedlings and their relationships to environmental conditions. Paul soon decided that Margaret's line of work was the more interesting, so they teamed up. This brought Paul to the field that was to become his particular specialty. Together with Henry Rufelt they compared the growth and transpiration responses of tree seedlings to water stress. Significantly, they began using the leaf chambers developed by Olle Björkmann and Paul Holmgren to study the transport of CO_2 and water vapour to and from leaves. When the NATO fellowships ran out they continued, supporting themselves on local grants and a half-time research assistantship for Paul with Carl-Olof Tamm at the Royal College of Forestry in Stockholm. At the suggestion of Nils Fries, their very supportive Head of the Institute, Paul then submitted his Swedish work to Uppsala University, gaining a second doctorate (Fil. dr) and his 'competence' to teach at Swedish Universities, whereupon he became a senior lecturer in the Department of Plant Physiology at the Royal College of Agriculture in Uppsala. Together Paul and Margaret produced an impressive ten joint papers in these 4 years. Paul and Margaret's partnership entered a new phase when their son Eric was born in 1964. Margaret remained active in science as a scientific editor and translator, notably translating *Stålfelt's Plant Ecology*, with Paul's assistance, published in 1972.

In 1963 Paul attended two scientific meetings that significantly affected his career. The first was a Symposium on Water Stress in Plants, held in Prague. Michal Marek recalls the 'vigorous invasion' by a young fellow from the UK who took part in many discussions with much enthusiasm, and considerably contributed to the success of this meeting. Those who know Paul will understand exactly what Michal means. Paul subsequently joined the editorial board of *Photosynthetica*, and reviewed an endless stream of manuscripts for them. In this way Paul helped many young scientists from Eastern Europe to maintain international standards during the Soviet era of isolation. When the barriers were removed Paul helped them back into the mainstream of European science, working with Michal Marek to design a Czech project to study the effect of increasing CO_2 concentration on forests and to find EU funding for it. Paul also helped them to set up their large semi-open chambers and to carry out the long-term experiments. This unstinting support over many years reveals something of the man beneath Paul's 'vigorous' exterior. For his contributions to Czech plant physiology, Paul received the Gregorius J. Mendel medal of the Czech Academy of Sciences in October 2000. But such honours were for the distant future. More importantly for Paul's immediate future, at the Prague symposium he also impressed Paul Weatherly, then Professor of Botany at the University in Aberdeen, who offered him a lectureship at Aberdeen a few years later.

The second significant meeting in 1963 was a UNESCO Arid Zone Research symposium, held in Montpellier. There Paul met Ralph Slatyer, with whom he had corresponded. In those days the CSIRO Division of Land Research and Regional Survey in Canberra kept a full position for visiting scientists and this had just fallen vacant. Ralph Slatyer wasted no time in offering it to Paul, who accepted and moved there the following year. It was a good choice as their interests overlapped splendidly. Ralph Slatyer's research interests had evolved through ecoclimatology to a heavy involvement in micrometeorology and then to environmental plant physiology. By the early 1960s he was attempting to set plant/water relationships into a full environmental context and was using leaf chambers to gain crucial information. Paul's interest had evolved out of botanical ecology, but he had arrived at the same point with an equal desire to apply the exact methods of the physical sciences to the problems of understanding how plants interact with their environment. In that group, which included such notable scientists as Calvin Rose and John Begg, Paul began to measure stomatal conductance in the field using portable diffusion porometers, and learned how to measure fluxes of water and CO_2 using micrometeorological techniques. This combination of laboratory and field research was to become a feature of Paul's scientific research.

The Aberdeen Years, 1966–1975

A letter from Paul Weatherly, offering a lectureship in the then vibrant (now extinct) Botany Department at the University of Aberdeen brought Paul back to Britain. He arrived in 1966 with his family, now increased by the arrival of Kathryn in Australia. He was filled with enthusiasm to apply the Australian approach to ecophysiology to a Scottish forest: he wanted to understand its exchanges of water and CO_2, and so ultimately its growth, by building upwards from a detailed knowledge of the functioning of its needles. In particular he wanted to match up the layer-by-layer transpiration and CO_2 assimilation calculated from physiological models with the gross sources and sinks calculated from the humidity and CO_2 concentration profiles in the canopy air space. It was a bold plan and with his characteristic drive he won from NERC the largest research grant they had made up to that time to carry it out.

A field site was chosen in a stand of Sitka spruce at Fetteresso Forest, about 30 km from the University, where they worked from a caravan powered by a diesel generator. His university laboratory was in an old woolshed between the Botany and Forestry departments. Much time was spent travelling back and forth between the field and lab in a canvas-topped Land Rover, which made for chilly travel in winter. On the site, access to the upper canopy was by a plank between two ladders lashed to trees in adjacent rows. Rope handrails ran between the ladders a little higher up to give a sense of security, but even so this arrangement was a bit tricky when there was ice on the planks. Strong winds, lightning and mice added to the difficulties of maintaining the measurements. His lab in the woolshed became known as 'Jarvisland' by others in the Botany department. In it Paul built his tight-knit team from a progression of students, post docs, technicians and visitors. Merv Ludlow was his first physiologist and Joe Landsberg his first micrometeorologist. John Norman and Neil Turner soon followed, as did Lubosh Nátr from Czechoslovakia, Stan Tajchman from Germany, and many more.

Some of Paul's ambitions exceeded the technology and understanding available at that time. For example, use of the Bowen ratio approach to measure water and CO_2 fluxes over forest canopies was in its infancy, requiring accuracy of measurement to be at least an order of magnitude higher than over crops. Thus, most of the micrometeorological flux measurements were never published, but from it all Paul produced a landmark study and some enduring innovations. Diffusion porometers suitable for use on conifer twigs did not exist, so Paul, with Mike Beardsell and Ewan Neilson, developed the 'Dingbat' porometer. It introduced the null-balance principle that has become the standard today. His study with John Norman of the effect of canopy architecture on radiation penetration into the canopy was highly influential and laid the foundation for the later MAESTRO model. With Brian James and Joe Landsberg he made some of the earliest measurements of CO_2 fluxes over a forest, and from their work they learned enough to write a review that remains a standard work today. Little was known at the time about the physiology of Sitka spruce, despite its pre-eminent place in commercial forestry in the UK. In parallel with the micromet approach in the forest, Paul developed a strong programme of leaf, root and whole plant physiology of Sitka spruce in the laboratory. Together with the micromet work, this led ultimately to 30 publications and five Ph.D. theses on Sitka spruce.

One wonders how Paul fitted it all in. Aside from his research work, Paul developed his Czechoslovakian connection further, co-editing *A Manual of Photosynthetic Methods* with Zdanek Sestak and Jiri Catsky. In a personal tour de force, he co-authored six of its chapters. Despite these commitments, Paul maintained a strong presence in the lab., contributing good ideas of his own and, equally importantly, listening to the good ideas from his team, shaping them and backing them with resources. In discussion Paul was often challenging, but his team came to recognize this as one of his strongest traits. As John Norman put it 'We could both pour our best efforts into the discussions with great intensity, without any personal animosity and with a kind of enjoyment that goes with being able to concentrate on the problem in hand, unencumbered by protocol'. This could be daunting to students, but they came to understand Paul's underlying generosity of spirit and his concern for their success. It is a tribute to Paul's qualities as an educator that of the more than 30 PhD students that he supervised in his career, all completed their degrees successfully. His post docs were similarly fortunate. He taught them to ask significant questions, to be quantitative and rigorous in pursuit of answers and, when writing their papers or theses, to state the question at the beginning and answer it at the end. Not least, he taught them to write science using symbols and quantities in the SI-approved manner! Even when he was away for long periods the cooperative spirit and the close loyalties Paul had built up within the team carried them along, and the work went on as usual.

This part of Paul's career ended with a period of leave from the University, to spend time in Seattle as a consultant to the US Coniferous Forest Biome component of the International Biological Program. This 'integrative' period produced some of his most influential papers. With Ken Read, he developed a phenomenological model of photosynthesis in relation to environment, using an optimization approach. Based on that approach, he formulated the 'Jarvis model' of the response of stomatal conductance to leaf water potential, saturation deficit, light temperature and atmospheric CO_2 concentration. This became the standard phenomenological model of stomatal responses to environment. Also whilst in Seattle he wrote an influential review of hydraulic transport mechanisms in plants, extending the resistance model by adding a resistance law for conduction in the xylem. Together these made it possible to model water vapour and CO_2 exchanges from whole forests – a tremendous achievement in just 9 years.

The Edinburgh Years, 1975–2001

While still in Seattle Paul was recruited by the University of Edinburgh to become its Professor of Forestry and Natural Resources. He moved there in 1975. As Professor he automatically became Head of Department. Previously, he had had no significant administrative experience, except as the leader of a research group. The Department was medium-sized by the standards of the day, with a strong forestry tradition. However, what had been the Department of Forestry in 1968 had become the Department of Forestry and Natural Resources, and the forestry degree had become Ecological Science with Honours streams that also included Resource Management and Wildlife Management. It now taught courses in animal ecology, remote sensing and hydrology as well as plant science and forestry. Only two members of staff had ever met Paul, but his reputation preceded him, and the Edinburgh colleagues braced themselves, expecting him to deprecate some of the activities and especially the management-related teaching that had evolved over many years. Paul must have sensed the unease, and when asked what changes he would be making, replied 'I want to see how it all works first'. For that one sentence he gained huge respect.

However, he soon set about introducing rigour into teaching, and especially into practical classes. He had the departmental workshop constructing batches of Scholander pressure chambers, and second year students were required to make, and fully analyse, pressure–volume relationships of leaves. This was an excellent experience, involving some comparative ecology (leaves from sun and shade, different life forms, different treatments), manipulative skills, conceptual understanding, and data handling. Most students rose to the challenge and were able to infer the elastic modulus of the cell walls from pressure–volume curves, even though they were only second year students. In his Honours teaching, students would construct and calibrate thermocouples, and his first undergraduate lectures on the Omega factor preceded publication of his paper introducing Omega to the rest of the world. Amazingly, the students mostly took to Omega like ducks to water.

Paul was also enthusiastic about undergraduate field courses, and decided to take the second year field course to Bettyhill on the north coast of Scotland, staying in an old school house belonging to Aberdeen University, and sometimes taking his younger daughter to act as cook. It was very handy for visits to the Flow Country, a vast area of blanket bog with very controversial private plantations of conifers. The weather in late September was generally 'bracing' but the coastal scenery was quite magnificent, providing a habitat for rare plants such as *Primula scotica*. This was never blooming when the course ran, but it didn't seem to matter, as Paul's pride and joy was a stand of *huge* Sitka spruce trees grown on a local estate. Bettyhill was almost a day's journey from Edinburgh, and after several years was abandoned in favour of Firbush, the University's outdoor centre on Loch Tay. Paul's enthusiasm for field courses has persisted into his retirement and he still goes to the Firbush field course – always with a good idea for a student project on trees.

On their move to Edinburgh Paul and Margaret saw advertised a large stone house named Belmont, in Dalkeith just 5 miles south of the university. It was described as having 'fourteen rooms but no kitchen'. Belmont was in a poor state. Despite some misgivings, they bought it and began its repair and improvement – a project that lasted most of their 26 years in Edinburgh. They were very generous with this house, so that Belmont became a base for many of Paul's old colleagues and students on their visits to Edinburgh. One of its endearing features was the shower: a Victorian masterpiece of engineering with seven taps, by which one could direct water to any chosen region of the body. No-one had seen anything like it, not even in museums. The house had flower beds, a prolific vegetable patch, several *Eucalyptus* trees raised from seed brought from Australia (a symbol perhaps of the Aussie connection), and a large lawn where croquet was played. At the end of the academic year Paul would invite colleagues (with families) and Honours students to a party at Belmont. Wine flowed, dogs became excited, croquet was played, and speeches were made. A good time was had by all. He and Margaret also held an annual party on Boxing day, where relatives, neighbours, colleagues and visitors could meet each other. There was always a gigantic jigsaw puzzle in progress, so the more introverted could escape from the crowd and work on the sky.

Paul brought people, equipment and ideas with him to Edinburgh. This influx invigorated the Department and increased its size and output. Most of Paul's efforts in the Edinburgh years went into his research – he loved to be in the field, and he loved to receive visitors. A favourite annual event for members of his group and the visitors and their families was a trip to the highlands in the spring, staying in a Youth Hostel and walking the hills. Paul was always very fit and an extremely strong walker (he, Margaret and Eric ran marathons). His research interests in the early years included: the mechanism of stomatal function, the response of stomata to humidity, sap flow in trees, photosynthesis, light interception by foliage. He established a photosynthesis laboratory and a water relations laboratory; and the departmental workshop was much-excercised in making various pieces of equipment to satisfy Paul's desire for technical excellence. He extended his modelling work, developing the MAESTRO model of radiation penetration and absorption in plant canopies with Jenny Grace and Yingping Wang. He interacted with his research group through morning coffee and Friday evening sessions in Lesley's bar, and he successfully balanced the role of being Head of Department with that of being leader of the group, although there were inevitable conflicts of interest. One of his pleasures was to escape the tedium of administrative duties and join his team at their field base: the caravan this time being in Roseisle Forest near Elgin. Paul's review of plant hydraulics had proposed a theory of homeostasis in water-conducting properties as stands developed. Paul built a team to test this. David Whitehead and John Roberts, of the Institute of Hydrology, were the main researchers, and Dick and Doris Waring joined in during their year's visit. Sometimes being department head had its advantages, as when there was a rush to strip hundreds of bags of branches from the forest, which were starting to grow mouldy in storage. Paul not only joined in the task but commandeered research and administrative personnel to help as well!

After 6 years Paul decided he would rather not be Head of Department any more. He duly went to the Dean and made his position clear. To Paul's surprise the Dean agreed to release him from the position, instigating the scheme of 'rotating Headship', which caught on fast, and has prevailed throughout the university ever since.

To mark his renewed focus on research Paul took a sabbatical year, visiting old colleagues Joe Landsberg, at the time Chief of CSIRO Forestry Division, Neil Turner, then at CSIRO Division of Plant Industry in Canberra, and Jenny Grace and David Whitehead, at the Forest Research Institute in Rotorua, New Zealand. He also spent 3 months in Palmerston North where his brother, Brion, lectured in microbiology at Massey University. Paul had arranged a desk at DSIR Plant Physiology Division and Keith McNaughton was assigned to be his minder. No collaboration was planned but they hit it off immediately. Keith was a micrometeorologist with an interest in evaporation and the theory of advection. They wrote a review of evaporation from catchments, *for the first time* treating the ground vegetation and the boundary layer above as one interactive system. Keith McNaughton visited Edinburgh for a month in 1984 and the work was extended to a discussion of 'scaling-up'. This paper appeared just when the likelihood of man-induced climate change was becoming generally recognized and their work allowed ecophysiologists to see how information on single leaves, single trees and small stands of trees could project upwards onto the regional and global scales. The 'Omega factor' from those papers is now permanently linked with Paul's name. Through this work Paul led the way as the principal concerns of environmental physiology changed from the economic productivity of crops and forests in individual fields and stands to their ability to adapt to a world with higher atmospheric CO_2 concentrations. With Derek Eamus, he wrote the first comprehensive review of the effects of CO_2 fertilization on tree growth. This helped set the research agenda as the field grew dramatically.

With the improvement of eddy covariance for measuring fluxes over canopies, the science that he had started back in his Aberdeen days had become much easier. One could collect data continuously rather than in short campaigns, and this allowed understanding of seasonal changes in the control of gaseous exchanges between forests and the atmosphere, and subsequently total fluxes to be compared between years. Paul returned to Sitka spruce, selecting a stand near Aberfeldy where he also had a cottage (the first cottage was shared with his brother, the second cottage is at the edge of the forest). At that time NERC was starting a thematic programme TIGER (Terrestrial Initiative in Global Environmental Research), and he tapped NERC funds to participate in BOREAS. He loved participating in international projects – and there were many. He especially enjoyed BOREAS; it was a chance to 'get away from it all' and immerse himself in his research in the 'field'. The comradeship and contact with old friends was part of the appeal. He was also involved in the HAPEX-Sahel experiment and a whole string of European projects, and was a leading player in EUROFLUX, a project to measure fluxes of CO_2 and water vapour over forests throughout Europe.

At the same time, he was making a huge effort as co-ordinator of ECOCRAFT to evaluate the response of trees to elevated CO_2 by using open-top chambers, branch-bags and other facilities at almost every European site where forest research was active. In this period his capacity to aid, encourage and inspire younger scientists was especially evident. ECOCRAFT laid the foundation for modelling forest response to rising CO_2 concentration and temperature across Europe and was the forerunner for the subsequent suite of projects in CarboEurope. It provided the launching pad into climate change and carbon cycle research for many young scientists across Europe. Yes, he was often away, but when available he was uncommonly good value.

Paul also enjoyed editing. With Harry Smith and John Raven, he co-founded *Plant Cell and Environment*, which rapidly became one of the top journals in plant sciences, and still is. He would always exercise full 'Jarvisian rigour' in dealing with submitted papers, and sometimes his critical comments incurred the wrath of authors. Distinguished scientists would visit Paul to discuss their science and sometimes the matter of Paul's comments on their manuscripts would be raised. Not all authors could bear Paul's robust style of discussion: on one occasion a very senior Harvard professor required post-Paul therapy in John Grace's office. A stiff cup of tea in the Atholl Room was required to achieve full recovery. On other occasions, discussion frequently continued over 'a pint' in the local pub.

During his Edinburgh years Paul led substantial research teams working on a range of topics from the role of vegetation type in catchment water balances, through stomatal behaviour in the field and physiology of stomatal action, boundary layer properties of leaves and canopies, water transport in plants, agroforestry, carbon balance and growth of vegetation, nitrogen supply rate, growth, photosynthesis and carbon allocation, to scaling up CO_2 and water exchange processes from leaf to regional scale. This work involved collaborations with the Institute of Hydrology, the Institute of Terrestrial Ecology, the Forestry Commission and the Macaulay Land Use Research Institute, as well as laboratories in Belgium, France, Germany, Italy, the Czech Republic, Sweden and Finland. He was international co-ordinator for an IGBP/GCTE Core Project studying the impact of elevated CO_2 and temperature on European Forests (ECOCRAFT). He was involved in the BOREAS with over 200 scientists, largely from North America. With colleagues in the Institute of Ecology and Resource Management he set up the Edinburgh University Biosphere–Atmosphere Programme (EUBAP), which had three major components of impact studies, modelling and flux measurements. His most exciting recent development is a theoretical analysis of the role of spatial scale in exchange of water and carbon by plants, crops – natural and semi-natural vegetation. The analysis leads to far-reaching consequences with respect to our fundamental ideas about use of controlled environments, water use, vegetation types, experimental design and such questions as the likely long-term effects of rising atmospheric CO_2 on vegetation.

Some measure of Paul's success and prominence can be gauged by the formal honours he has received and the prestigious positions he has held. He was elected to Fellow of the Royal Society, the Royal Society of Edinburgh, the Royal Swedish Academy of Agriculture and Forestry, the Royal Science Society of Uppsala, the Institute of Chartered Foresters, and the Institute of Biology, and was awarded the Gregorius J. Mendel Medal of the Academy of Sciences of the Czech Republic. He has served as President and on the Council of the Society for Experimental Biology, on the Natural Environment Research Council Terrestrial and Freshwater Sciences Technology Board, as a Commissioner for the Countryside Commission in Scotland, on the AGBP/BAHC Scientific Steering Committee of the International Geosphere–Biosphere Programme, on the John Muir Trust, and on the committees of many other national bodies. He was co-founder and Sectional Editor of the journal *Plant Cell and Environment*, and served on the Review Boards of *Photosynthetica*, *Tree Physiology* and *Agricultural and Forest Meteorology*.

Retirement in Aberfeldy

Before he retired Paul and Margaret had bought a small farmhouse with 5 acres of land in the shadow of Griffin Forest, near Aberfeldy, and had begun to plant an arboretum and improve the house during weekend visits. On retirement they sold Belmont and moved there permanently. Paul's new interests include a practical project concerned with restoration of native woodland on severely over-browsed hill land. Paul and Margaret also win prizes at local horticultural shows for the fruit and vegetables from their garden. They are now grandparents and visit their younger daughter, Alice, and her family in Namibia as often as they can.

This does not mean that Paul has entirely adopted the lifestyle of a retired country gentleman. In 2000 Paul became a 'Director' of the Edinburgh Centre for Carbon Management, a small company located near the University, concerned in part with carbon sequestration by forests to meet 'Kyoto goals' and is now able to take a more active part in the business. He maintains an office at the university and comes in weekly, when not out of the country, attending seminars and discussing research. Paul still has research results to digest and papers to publish. His involvement with EU projects continued through 2003 as his last graduate students complete their degrees. Paul also continues on as Principal Investigator of a NERC project with John Moncrieff and Keith McNaughton. Also, with Griffin Forest on his doorstep, Paul is keen to see the next stage of the work there, the thinning experiment, go ahead and he is doing what he can to assist that.

So 'retirement' is a relative term for Paul. He has diverted some of his boundless energy into new directions, but most of his old interests remain. We will see his name in print again, and scientists throughout physiological ecology will continue to feel his influence for many years to come. We wish Paul and Margaret a long, happy and *productive* retirement.

Keith McNaughton
John Grace

1

Stomatal Control of Transpiration: a Major Dilemma 100 Years Ago

T.A. Mansfield

Department of Biological Sciences, Lancaster University, Lancaster LA1 4YQ, UK

The important role of stomata in facilitating the gas exchange of leaves had already been recognized in the second half of the 19th century, but nevertheless there remained a great deal of confusion about the extent of the control that they could exert. The very influential textbook by Strasburger *et al.* (1903) left the question of the degree of the control of transpiration unresolved:

> The cell walls of all living organisms are saturated with water, and, when the cuticle of the epidermis is not too strongly developed, water is constantly evaporated, even from uninjured cells... Evaporation is also promoted by the numerous Stomata (Air-pores) which penetrate the epidermis, and which give the air, saturated with watery vapour, an opportunity to escape from the intercellular spaces.

This clearly suggests that a significant proportion of transpiration is outside the control of the stomata, a conclusion that appeared to be supported by a lot of contemporary experimental evidence. Although we now have a clear picture of the significance of stomatal control, and we view stomata as being of central importance in the evolution of land plants (Edwards *et al.*, 1998), it is right to acknowledge the contribution of one plant scientist in particular, J.V.G. Loftfield, made over 80 years ago. It is salutary to recall that Einstein had introduced his special theory of relativity in 1905 and his general theory of relativity in 1915, and yet the full significance of the role of stomata – something so indispensable to our understanding of living systems on earth – remained unresolved until 1921.

At the Close of the 19th Century

One of the most influential early papers on stomatal physiology was that of Francis Darwin (1898), and it serves as a convenient point from which to start this brief historical survey. Darwin gave a comprehensive account of some earlier publications, most of which appeared after what he terms 'the reform in plant physiology of about the years 1879–80'. This was when 'a new era began, in which it was recognised that the reaction of plants to their environment is a phenomenon of irritability'. Darwin's 90-page paper is remarkable for the amount of valuable physiological information he managed to assemble using a very simple instrument employing the hygroscopic awn of the grass *Stipa* (Fig. 1.1). When the device was placed on the surface of a leaf, the awn bent into a new position, depending on the rate of transpiration, within 5–10 s. Darwin was able to calibrate it convincingly using gravimetric measurements of transpiration, and he also validated the data by comparing them with stomatal apertures determined microscopically. The response time of the hygrometer was rapid enough for him to describe in detail the temporary opening of the stomata after a leaf is excised (Fig. 1.2) – a phenomenon now known as the 'Iwanoff effect', named after that author's much later studies in the 1920s.

Darwin used his hygrometer to provide basic descriptions of stomatal responses to light and humidity, and he also came to the important conclusions that stomata remain open in light in the absence of CO_2 and that 'the closure of the stomata in darkness is not due to the starvation of the chloroplasts in the guard cells, but is an adaptive action of the same type as the other phenomena of irritability in plants'. The assumption that the measurements of transpiration by the hygrometer were a genuine indication of stomatal movements was, however, strongly disputed by later authors.

Important Progress, and Much Confusion, in the Early 1900s

The next very influential publication was the 142-page account of Lloyd's (1908) studies of stomata published by the Carnegie Institution of Washington, DC. Lloyd was much concerned to establish whether water loss from the plant is

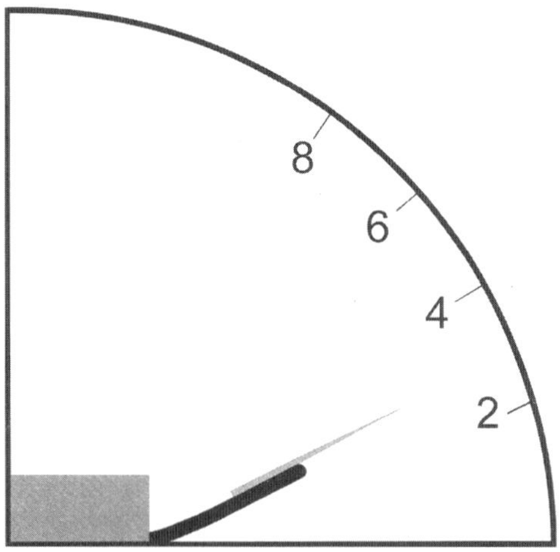

Fig. 1.1. The 'horn hygroscope' of Francis Darwin. The awn of *Stipa* (shown in black) bent in response to changes in humidity close to the leaf surface.

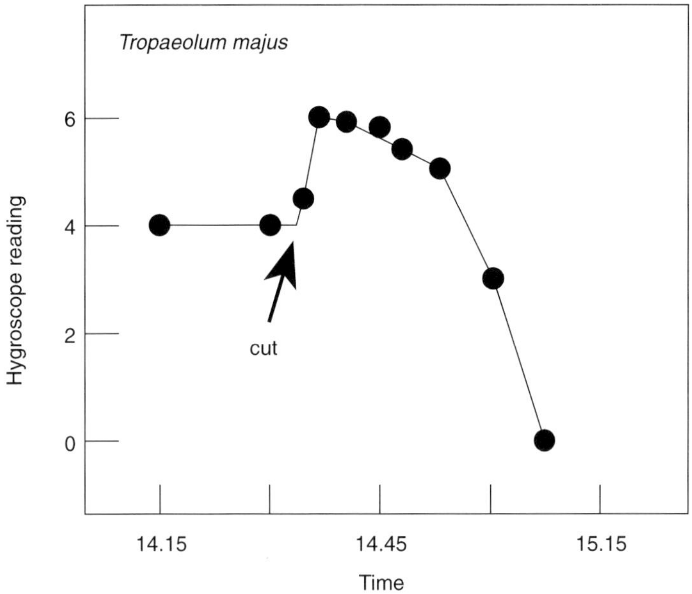

Fig. 1.2. Francis Darwin's demonstration of transient stomatal opening when a leaf was excised.

genuinely regulated by the stomata, for that issue had not then been satisfactorily resolved. He was not happy about using Darwin's hygrometer for his own studies because it was an indirect method of observation and it involved assuming what had still to be proved. He made use of leaves from which it was possible to remove strips of epidermis without obvious damage to the cells, and he plunged them immediately into absolute alcohol to 'fix' the openings of the stomata for subsequent measurement. Although later authors cast doubt on the assumption that the stomatal apertures remained unaltered, 'Lloyd's strips' became familiar to several generations of plant physiology students in practical classes until at least the 1950s. Lloyd is now remembered because he was the first to give detailed thought to the metabolism of the guard cells. After various experimental manipulations of light conditions and the supply of CO_2, he concluded:

> that photosynthesis, even in stomata, whose plastids are well supplied with chlorophyll, is connected only indirectly with their movements (cf. Francis Darwin, 1898), and that, therefore, there is a marked quantitative difference between the physiology of the typical chlorenchyma-cell and that of the guard-cell.

He reasoned that photosynthesis in the guard cells is not the driving force behind stomatal opening, and he presented detailed evidence that starch appears in the chloroplasts of the mesophyll during the day and in those of the guard cells at night. The account of the loss of starch from the guard cells as stomata opened in the morning led to what became known as the starch $\leftrightarrow$ sugar hypothesis: the breakdown/formation of starch to provide/remove osmotically active sugars to drive the turgor changes for stomatal opening and closing. This became a standard feature of plant physiology textbooks until the 1960s (a well-presented example can be found in Devlin (1966)), whereupon it rapidly faded from view when it became known that potassium ions are strongly involved in the turgor mechanism. Before Lloyd's paper was published in 1908, there was an important clue in the literature that potassium might be important in stomatal physiology, for Macallum (1905) introduced a new histochemical technique for determining potassium distribution in cells and found high concentrations in guard cells. Sixty-two years then elapsed before Fujino (1967) used Macallum's method to demonstrate very clearly the important role of potassium in the regulation of stomatal movements, and thereafter the ionic relations of guard cells became a major focus for many plant-cell physiologists. There are many excellent reviews of recent studies of stomatal mechanisms and it is not appropriate to attempt to deal with the more recent history of the subject here, for it was extensively covered in the book by Willmer and Fricker (1996).

Lloyd's Dilemma

Lloyd's own studies had added significantly to the basic understanding of the factors governing stomatal movements, but he was left with a huge physiological puzzle because he was unable to resolve the question: do stomatal movements regulate transpiration? His own work, and that of Darwin and others before him, had shown that stomata are highly 'sensible' of their environment and that they exhibit very clear diurnal patterns of behaviour. These qualities could presumably be attributed to features of the guard cells, which are qualitatively quite different from other leaf cells. But what was the purpose of stomatal movements and the controls exerted upon them by the environment? Lloyd wrote:

> The problem here considered has, as I conceive it, two aspects:
> **(1)** The degree of correlation between the normal rates of transpiration and the normal changes in the size of the stomatal openings. There is, however, a logical difficulty to be encountered in that, even should a close correlation be shown to occur, it would remain unproved that the movements of the stomata were immediately related, in the causal sense, to the rise and fall of transpiration rate. This must be fully reckoned with in attributing a regulatory function to the stomata, for, unless marked quantitative differences between the functions of the chloroplasts of the chlorenchyma and those of the stomata exist, it would be quite possible that the photosynthetic activity of the chlorenchyma and of the guard-cells would run parallel without a necessary causal relation between stomatal movements and changes in transpiration rate. That such a causal relation exists is to be doubted from the consideration that the stomata, in very many cases at least, have a greater capacity for the outward diffusion of water-vapor than actually has been found to occur (Brown & Escombe, 1900) and this may be true at any given diffusion capacity of the stomata, depending upon their dimensions.
> **(2)** Since, in the final analysis, light, above all other factors, influences stomatal movement, and since also appropriate illumination is the condition *par excellence* under which photosynthesis takes place, if the supposed causal relation under consideration exists, wide and sudden changes in the degree of illumination should be accompanied by changes in transpiration rate, and these, at the same time, by corresponding changes in the diffusion capacity of the stomatal openings.

In an attempt to resolve these questions, Lloyd made microscopic observations of the daily course of stomatal movements in *Verbena ciliata* and attempted to determine the corresponding rates of transpiration. He placed a branch in a potometer and covered it with a bell-jar, within which he could measure increases in humidity by means of a hygrometer. He found that the maximum apertures of the stomata occurred 2–3 h before the highest rates of transpiration, and also concluded that the rise in transpiration during the morning was not due in the main to stomatal movements because of 'the continued rise in the curve of transpiration after 8 a.m., the hour of maximum stomatal opening' (Fig. 1.3). The biggest discrepancy did, however, occur in the afternoon when substantial stomatal closure occurred while transpiration fell by only a small amount. He also conducted experiments on *Fouquieria splendens*, and found that only a small amplitude of stomatal movement 'maintained a daily

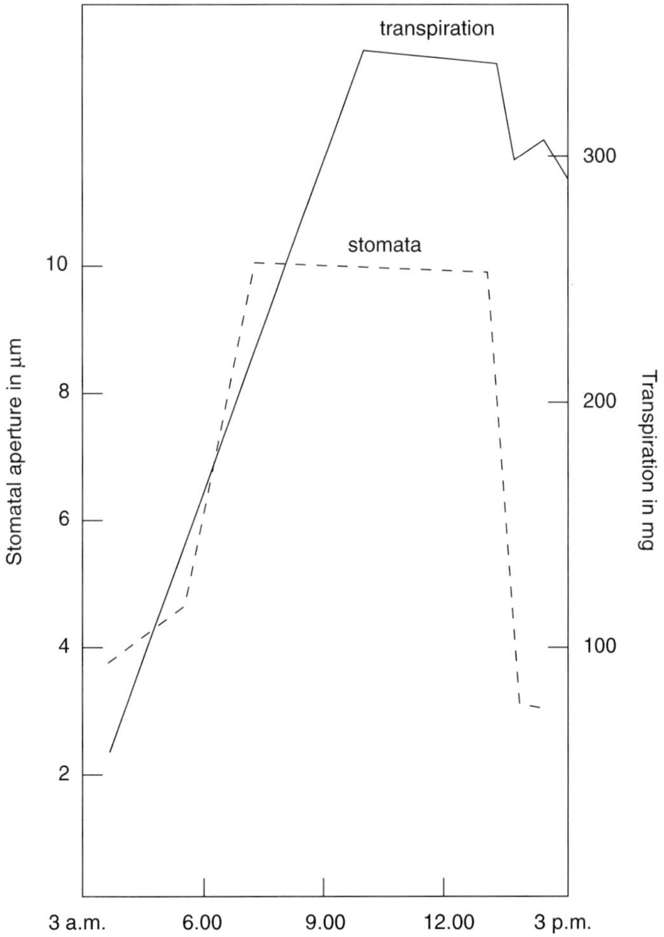

Fig. 1.3. Data such as these for *Verbena ciliata* convinced F.E. Lloyd that there was not a good correlation between stomatal opening and rate of transpiration (the time intervals for the measurements of transpiration were not given).

rise and fall of transpiration rate, apparently very nearly, if not wholly, normal', and he concluded that the functions of stomata are not regulatory because of the 'progress of transpiration when they are inactive'. Many further experiments in which changes in the dimensions of stomata were induced by alterations in illumination led to a final conclusion that 'stomatal regulation of transpiration does not occur, though, of course, conservation of the contained water follows on the complete closure of the stomata'. In attempting to offer an alternative explanation of a regulatory mechanism within the plant, he even suggested that the walls of the mesophyll cells might 'hold back the water' under dry conditions. While this is a possibility that still cannot be entirely ruled out, it has never been regarded as a factor to be realistically compared with the regulatory role of stomata.

Lloyd's paper was highly influential and, as subsequent authors followed his lead, more and more evidence was assembled suggesting that stomata had little control over the rate of transpiration. Loftfield (1921) cited work of Muenscher in 1915, which compared the rates of water loss from nine different species and did not demonstrate any relationship with the sizes of their stomata and the numbers per unit area of leaf. He also pointed to studies by Knight using a viscous-flow porometer, which had failed to show any correlation between 'relative transpiration' and 'aperture index', and to further porometer measurements by Trelease and Livingston, which also implied a lack of stomatal control.

The Crucial Contribution of J.V.G. Loftfield

The 104-page paper of Loftfield (1921), published by the Carnegie Institution of Washington, is undoubtedly one of the most remarkable in the field of stomatal physiology. He made very detailed observations of daily movements of stomata in a wide range of crop plants, and explored thoroughly the stomatal responses to light, temperature and plant water relations. Most important of all, however, was his resolution of the controversy concerning stomatal control of transpiration. There is a very real sense in which Loftfield's research laid the foundation for much of the work on stomatal regulation of transpiration of the last 80 years. His contribution has not, however, been generally recognized, and this volume provides an opportunity to view his work in an appropriate context.

The photographs of the equipment used show that Loftfield conceived his experiments on a scale that was exceptional for 1916 (plates 10 and 11 in his paper of 1921). He used a 'battery of potometers' to determine whether excised stems of lucerne could provide a reliable measure of the rates of water loss from rooted plants. He took many precautions, some of which must have appeared eccentric to observers: the entire apparatus was sterilized before the experiments, the stem of the plant was sterilized at the point of cutting with a sterile knife, and the water used was recently distilled and boiled just before it was used. He also ensured that the water pressure at the cut end of the stem was maintained constant. The outcome of these meticulous experiments was the firm conclusion that 'the transpiration of the cut stems as measured in the potometers is distinctly controlled by stomatal movement'. But he also went on to say:

> The stomatal movement of the field plants, watered and unwatered, did not in any manner resemble that of the cut stems, and the transpiration from the potometers in turn showed no relationship to the changes in stomatal opening of these plants. For both these reasons, in the case of lucerne at least, potometers cannot be used to represent the transpiration of rooted plants.

Similar studies with other species such as potato led to the same unequivocal conclusion, and he was able to argue strongly that previous workers had been mistaken in assuming that it was valid to determine water loss using potometers and to link the readings to concurrent measurements of stomatal aperture on intact rooted plants.

Although quite a lot of attention has been paid to the work of Loftfield in the literature dealing with stomatal mechanisms, his vital contribution to our understanding of stomatal control of transpiration has not been generally recognized. The subject was surrounded by contentious debate and muddled interpretations in the first 20 years of the 20th century. This clear statement by Loftfield (1921) provided a foundation for much of our present understanding, and its intrinsic insight is an appropriate foreword for the chapters in this volume:

> Although the factors concerned in evaporation have great influence upon transpiration, this influence is definitely controlled by the stomata. When the stomata are wide open or nearly wide open, transpiration is the result of the action of the factors of evaporation alone, since the stomata in nowise interfere with the action. As the stomata close, the influence of the factors is lessened, but until closure has reduced the apertures to 50 per cent or less, stomatal regulation is still largely overshadowed by the control exerted by them. When closure is almost complete, the regulation of water-loss by the stomata is very close and the effect of the factors overshadowed by the effect of even very small changes of the opening.

References

Brown, H.T. and Escombe, F. (1900) Static diffusion of gases in relation to the assimilation of carbon and translocation in plants. *Philosophical Transactions of the Royal Society, series B* 193, 223–291.

Darwin, F. (1898) Observations on stomata. *Philosophical Transactions of the Royal Society, series B* 190, 531–621.

Devlin, R.M. (1966) *Plant Physiology*. Reinhold, New York, 564 pp.

Edwards, D., Kerp, H. and Hass, H. (1998) Stomata in early land plants: an anatomical and ecophysiological approach. *Journal of Experimental Botany* 49, 255–278.

Fujino, M. (1967) Role of adenosinetriphosphate and adenosinetriphosphatase in stomatal movement. *Science Bulletin of the Faculty of Education, Nagasaki University* 18, 1–47.

Lloyd, F.E. (1908) *The Physiology of Stomata*. Carnegie Institution, Washington, DC, 142 pp.

Loftfield, J.V.G. (1921) *The Behavior of Stomata*. Carnegie Institution, Washington, DC, 104 pp.

Macallum, A.B. (1905) On the distribution of potassium in animal and plant cells. *Journal of Physiology* 32, 95–128.

Strasburger, E., Noll, F., Schenck, H. and Schimper, A.F.W. (1903) *A Textbook of Botany*, translated from the German by H.C. Porter. Macmillan, London, 671 pp.

Willmer, C. and Fricker, M. (1996) *Stomata*. Chapman & Hall, London, 375 pp.

2

Stomata as Part of the Soil–Plant–Atmosphere Continuum

Ray Leuning[1], Andree Tuzet[2] and Alain Perrier[2]
[1]*CSIRO Land and Water, FC Pye Laboratory, PO Box 1666, Canberra, ACT 2601, Australia;* [2]*Environnement et Grandes Cultures, INRA–INA PG 78850 Thiverval Grignon, France*

Introduction

Stomata are pivotal in regulating the exchanges of both water vapour and CO_2 between leaves and the atmosphere, and more than a century of scientific effort has gone into understanding the mechanism of stomatal responses to both physiological and environmental controls (Cowan, 1977; Buckley and Mott, 2002). Nature has not revealed its secrets easily because stomatal conductance at any instant is a resultant of complex, non-linear interactions between several factors such as light intensity, leaf temperature, water vapour pressure deficit, CO_2 concentration, and soil and leaf water status. From a mechanistic standpoint, stomatal conductance is a function of the relative turgor of the guard and epidermal cells (Wu *et al.*, 1985; Comstock and Mencuccini, 1998; Mencuccini *et al.*, 2000; Franks *et al.*, 2001), and turgor of these cells depends on the balance between loss of water through peristomatal transpiration (Cowan, 1977; Mott and Parkhurst, 1991; Maier-Maercker, 1999; Mott and Franks, 2001) and supply of water to the leaf from the soil via the xylem in roots and stem and along cell walls in the leaf. Stomatal regulation also requires metabolic energy from photosynthesis in either the guard cells or the mesophyll cells to regulate osmotic concentrations in the guard cells (Farquhar and Wong, 1984; Assmann, 1999; Blatt, 2000; Netting, 2000).

To our knowledge, no models have been developed that integrate the rapidly developing knowledge of stomatal function at the cellular level with descriptions of the coupled fluxes of water and CO_2 in the soil–plant–atmosphere continuum (SPAC), and all models of the latter use semi-empirical descriptions of stomatal responses to physiological and environmental factors. Professor Jarvis and his students have made important contributions to this field, and Jarvis (1976) presented an early empirical model to describe stomatal responses to light, temperature, atmospheric humidity deficit and soil moisture status. Noting the close coupling between maximum stomatal conductance

 Forests at the Land–Atmosphere Interface (eds M. Mencuccini, J. Grace, J. Moncrieff and K.G. McNaughton)

and photosynthetic capacity (Wong *et al.*, 1985), Ball *et al.* (1987) developed a relationship between stomatal conductance, assimilation rates and humidity, which was able to collapse a wide set of observations on to a single curve. Leuning (1990) coupled the original Ball *et al.* (1987) stomatal model with a model of photosynthesis (Farquhar *et al.*, 1980) to describe the responses of conductance, intercellular CO_2 concentrations and assimilation rates of well-watered leaves to a range of environmental factors. In a later development, Leuning (1995) replaced relative humidity in the Ball *et al.* stomatal model with a humidity deficit term, consistent with findings that stomatal conductance responds to transpiration rates rather than to relative humidity *per se* (Aphalo and Jarvis, 1991; Mott and Parkhurst, 1991; Dewar, 1995; Monteith, 1995).

Models of stomatal conductance that only account for physiological and environmental controls at the leaf level are incomplete because, while they describe stomatal response to atmospheric demand for water vapour, they do not consider supply of water to the leaves by transport of water within the soil and plant to maintain guard-cell and epidermal turgor. Wang and Leuning (1998) recognized this limitation and introduced an extra function to the coupled model of photosynthesis and stomatal conductance of Leuning (1995) to account for changes in soil water content. However, their function is unable to account for more dynamic responses of stomata to the competing demand and supply of water to leaves on a subdiurnal time-scale. To overcome this deficiency Tuzet *et al.* (2003) developed a model that describes the dynamics of stomatal conductance, transpiration and water transport through the plant and soil, which fully couples stomatal conductance, CO_2 assimilation and the leaf energy balance. Of course, this is not the only study to consider the full SPAC. Cowan (1977) was one of the first to show the importance of this approach, and many such models have been presented in the literature (e.g. McMurtrie *et al.*, 1992; Sellers *et al.*, 1992; Jensen *et al.*, 1993; Tardieu and Davies, 1993; Olioso *et al.*, 1996; Williams *et al.*, 1996; Baldocchi and Meyers, 1998; Wang and Leuning, 1998; Dewar, 2002; Whitehead *et al.*, Chapter 15, this volume). The model of Tuzet *et al.* (2003) differs from these studies in the way that stomatal conductance is parameterized and by use of dynamic solutions of the water-flow equations to calculate water uptake by the roots. Parameterization of the physiological controls of stomatal conductance is a variant of the Ball–Berry model (Ball *et al.*, 1987; Leuning, 1995), modified to include the responses of stomata to leaf water potential, which both responds to and controls supply and loss of water by the leaves. Thus this control mechanism depends on the transpiration rate, on resistances to water flow in the soil, roots, stem and leaf and on bulk soil water potential. Similar considerations of the model show that intercellular CO_2 controls stomatal conductance (Mott, 1988; Assmann, 1999) in response to supply of CO_2 by transport through the stomata and to biochemical demand for CO_2 by photosynthesis. The model of Tuzet *et al.* (2003) thus fully couples the physiological and environmental controls of the fluxes of water through the plant with exchanges of CO_2 by the leaves.

In this chapter we introduce the stomatal model and provide justification for the approach adopted, using findings in the literature, and describe briefly the equations for transport of water from the bulk soil to the roots and through the plant to the leaves. We then present some predictions of the model and compare these with experimental results reported in the literature.

The Model

Stomatal conductance and photosynthesis

A stylized diagram for the flow of water in the SPAC is shown in Fig. 2.1. Water flows from the soil, through the root and stem xylem and veins in the leaf to the moist surfaces of the mesophyll, epidermal and guard cells, where it evaporates into the substomatal cavity and diffuses through the stomata to the surrounding air, which is generally much drier than air within the leaf. The figure also shows the profile of soil pressure potential as a function of radial distance from the root surface to the bulk soil. At steady state the flux of water through the roots and stem per unit leaf area, J_w, equals the transpiration flux density, E.

Figure 2.2 presents the model for stomatal conductance and photosynthesis used by Tuzet *et al.* (2003). The model consists of three equations, which need to be solved simultaneously, for intercellular CO_2 concentration, c_i, stomatal conductance to CO_2, g_{CO_2}, and net assimilation rate, A. The rate of CO_2 supply to the mesophyll cells by transport through the stomata is given by Equation (1), while Equation (2) describes the biochemical demand for CO_2 by photosynthesis, which in turn is described by the photosynthesis model of Farquhar *et al.* (1980). Equation (3) relates g_{CO_2} to A and c_i, which is consistent with experimental observations by Mott (1988) showing that stomata respond to c_i, and with Bunce (1996) who found that stomatal sensitivity to atmospheric humidity diminishes as $c_i \rightarrow 0$, but contrasts with earlier models (Ball *et al.*, 1987; Leuning, 1995), which used CO_2 concentration at the leaf surface, c_s. Stomatal conductance is also modelled to depend on leaf water potential, ψ_v, through the function $f(\psi_v)$ shown graphically to the left of Fig. 2.2 and algebraically in Equation (4). The term s_f in this equation determines the sensitivity of stomatal conductance to ψ_v, while ψ_f is the 'threshold' potential at which conductance is reduced to half its maximum value.

What is the justification for using ψ_v when there are many experiments that show that g_{CO_2} can vary independently of ψ_v (e.g. Gollan *et al.*, 1986; Tardieu and Davies, 1993; Saliendra *et al.*, 1995; Fuchs and Livingston, 1996; Tardieu and Simonneau, 1998)? This has led to the hypothesis that conductance is controlled not by ψ_v, but by a chemical message, such as abscisic acid (ABA), which is transmitted from the roots to the shoot (Tardieu and Davies, 1993; Jones and Tardieu, 1998; Dewar, 2002). It is also well established that stomatal conductance depends on the difference in turgor pressure, and hence osmotic potential, between guard and epidermal cells, and not ψ_v itself (Wu *et al.*, 1985; Dewar, 1995; Franks *et al.*, 1998, 2001; Mencuccini *et al.*, 2000). Here we resolve these objections and show that many of the reported responses of stomatal conductance, transpiration and photosynthesis to environmental forcing can be explained without the need to invoke a chemical signal, provided the dynamics of water-supply from soil to roots is considered as part of a complete model.

How can we reconcile the fact that stomatal conductance depends on the difference in turgor pressure between guard and epidermal cells with the assumed dependence of conductance on leaf water potential? To do so we

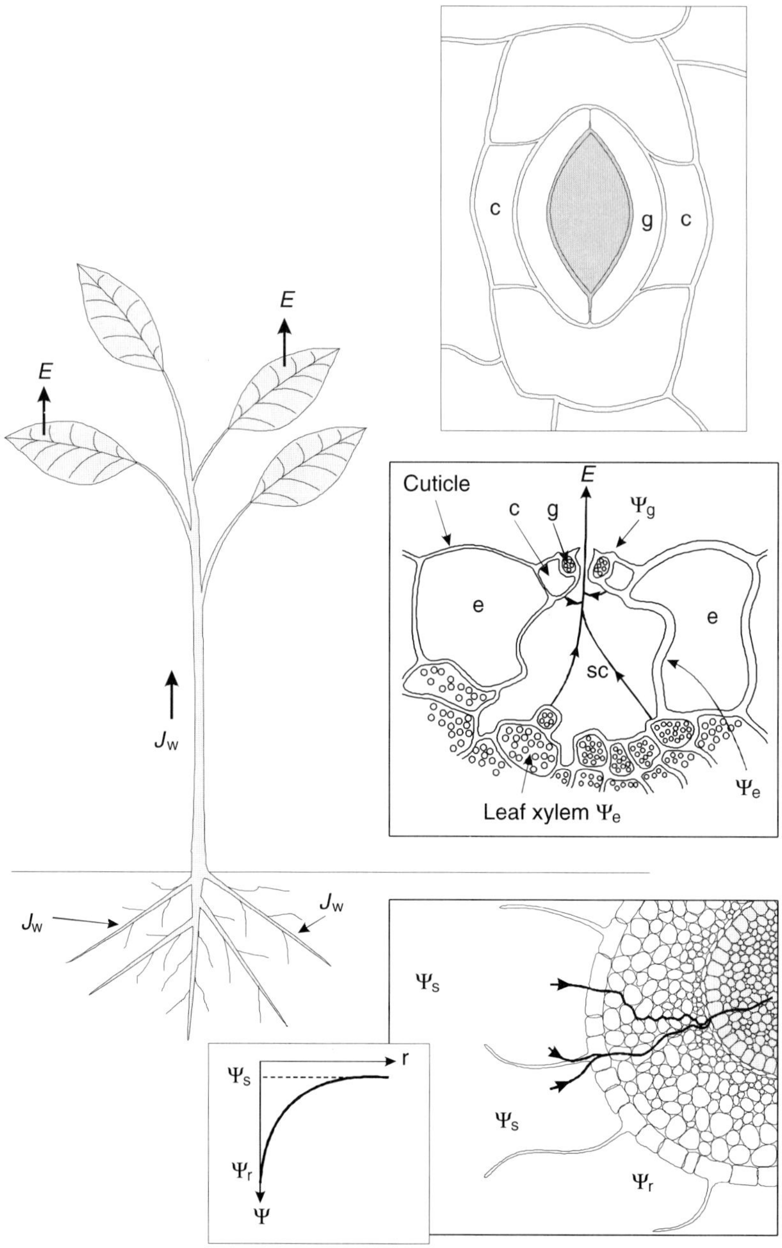

Fig. 2.1. Schematic diagram showing path of water from the soil to the atmosphere via the plant. Top panel on right shows a single stoma with guard cells (g), companion cells (c), surrounded by epidermal cells. J_w is the flux of water through the plant per unit leaf area and *E* is the transpiration flux density. These fluxes are equal at steady state. Middle panel on right shows cross-section through a leaf with pathways for water evaporated at walls of mesophyll cells in the substomatal cavity (sc). Lower panel on right shows pathway for transport of water from the bulk soil to the root surfaces and to the root xylem. The insert shows theoretical distribution in soil matrix potential between the bulk soil and root surface.

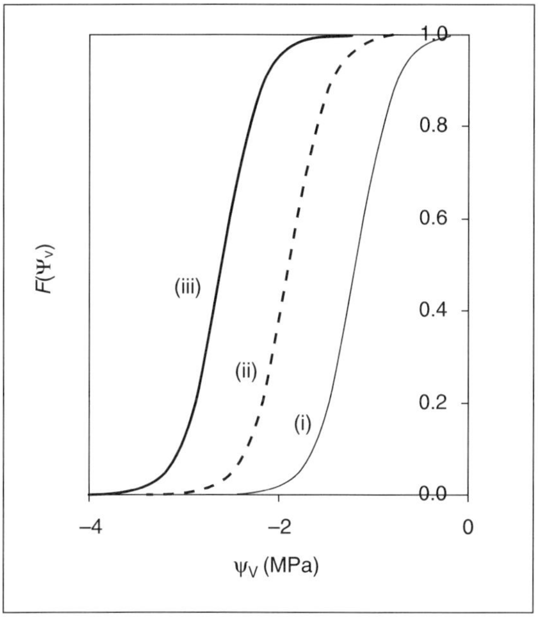

$$A = g_{CO_2}(C_S - C_i) \qquad (1)$$

$$A = \min(V_c, V_j) - R_d \qquad (2)$$

$$g_{CO_2} = g_0 + \frac{a\ A}{C_i - \Gamma} f(\Psi_v) \qquad (3)$$

where

$$f(\Psi_v) = \frac{1 + \exp\left[s_f \psi_f\right]}{1 + \exp\left[s_f(\psi_f - \psi_v)\right]} \qquad (4)$$

Fig. 2.2. Three coupled equations needed to solve simultaneously for intercellular CO_2 concentration, c_i, stomatal conductance to CO_2, g_{CO_2}, and net assimilation rate, A. Equation (1) describes the rate of CO_2 supply to the mesophyll cells by transport through the stomata, while Equation (2) describes the biochemical demand for CO_2 by photosynthesis. Equation (3) relates g_{CO_2} to A, c_i and leaf water potential, Ψ_v, through the function $f(\psi_v)$ shown graphically on the left and algebraically in Equation (4) (with parameter values: (i) $s_f = 4.9$, $\psi_f = -1.2$, (ii) $s_f = 3.2$, $\psi_f = -1.9$, (iii) $s_f = 2.3$, $\psi_f = -2.6$).

examine Fig. 2.3, first presented by Franks *et al.* (1998), which plots stomatal conductance as a function of guard-cell turgor pressure, P_g, for two values of epidermal turgor pressure, P_e (0 and 0.4 MPa). This figure shows that stomatal conductance has quite different values for a given P_g, depending on the turgor of the surrounding epidermal cells. We note from Fig. 2.5 below that there is a strong relationship between leaf turgor pressure and bulk leaf water potential under normal field conditions (Henson *et al.*, 1989), and hence stomatal conductance must also decline with leaf water potential as shown by the arrow in Fig. 2.3. The function $f(\psi_v)$ will not be linear because the epidermal- and guard-cell turgor pressures change at different rates with leaf water potential, and because the modulus of elasticity, χ, may also change with guard-cell turgor (see upper equation in Fig. 2.3; σ is the mechanical advantage).

Support for the role of ψ_v in regulating g_{CO_2} was presented by Jones and Sutherland (1991) and by Sperry (2000), who argued that stomata regulate water flow through the plant to avoid cavitation of xylem vessels and subsequent catastrophic loss of water-supply to the leaves. Sperry (2000) found that the minimum leaf water potential measured in the field for 73 species was always above the critical value needed to cause xylem cavitation. Considerable evidence is also accumulating on the apparent control of stomatal conductance by xylem cavita-

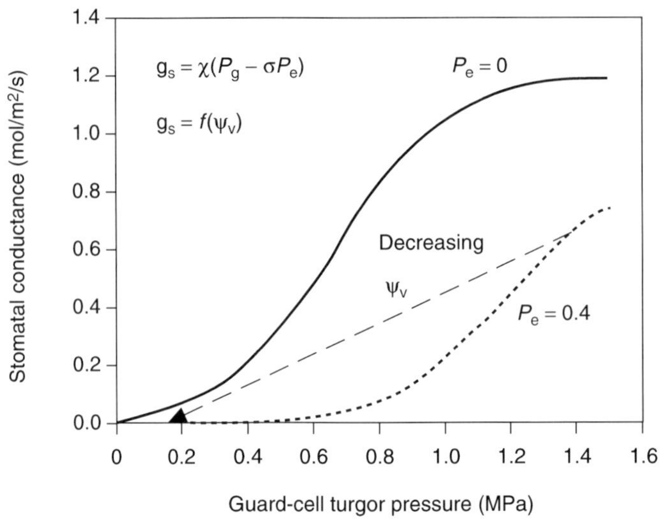

Fig. 2.3. Relationship between stomatal conductance, g_s, and guard-cell turgor potential, P_g, as a function of epidermal turgor pressure, P_e. Stomatal conductance is dependent on both turgor pressures through the modulus of elasticity, χ, and the mechanical advantage term, σ. Both P_g and P_e, and hence g_s, decrease with leaf water potential, ψ_v, but the relationship $f(\psi_v)$ will be non-linear. (After Franks *et al.*, 1998; Mencuccini *et al.*, 2000.)

tion (Sperry *et al.*, 1998; Hubbard *et al.*, 1999, 2001; Comstock, 2000; Aasamaa and Sober, 2001; Ewers *et al.*, 2001), with Nardini and Salleo (2000) postulating that the onset of cavitation may act as a signal for stomatal closure. Cavitation and consequent loss of hydraulic conductivity undoubtedly occur in the field (Tyree and Sperry, 1988, 1989) and the mechanism for repair of xylem embolisms recently proposed (Zwieniecki and Holbrook, 2000) suggests that cavitation is reversible even when adjacent vessels are under tension. Meinzer *et al.* (2001) concluded that xylem conductance varies continuously throughout the day and is the result of a dynamic balance between embolism formation and repair. Cavitation occurs through entry of air into vessels when the absolute potential (i.e. relative to free water) reaches a critical value, which varies between species, as shown in Fig. 2.4. Cottonwood is highly drought-susceptible and cavitation occurs at a small negative xylem potential and loss of hydraulic conductivity occurs over a narrow range of potentials. Loss of xylem conductance starts at a lower potential and drops over a wider range of potentials for Wyoming sagebrush than for cottonwood. For hoary leaf ceanothus, decreases in xylem conductance start at an even lower threshold than in the other two species.

Further support for a link between leaf water potential and stomatal conductance was obtained by measurements of Henson *et al.* (1989) on lupins and wheat grown in the field in identical soils and environmental conditions (Fig. 2.5). Plotted are stomatal conductance, assimilation rate and leaf turgor pressure as a function of bulk leaf water potential. The data show that both stomatal conductance and assimilation rates for wheat and lupins remained high and constant until a critical leaf water potential, followed by a rapid

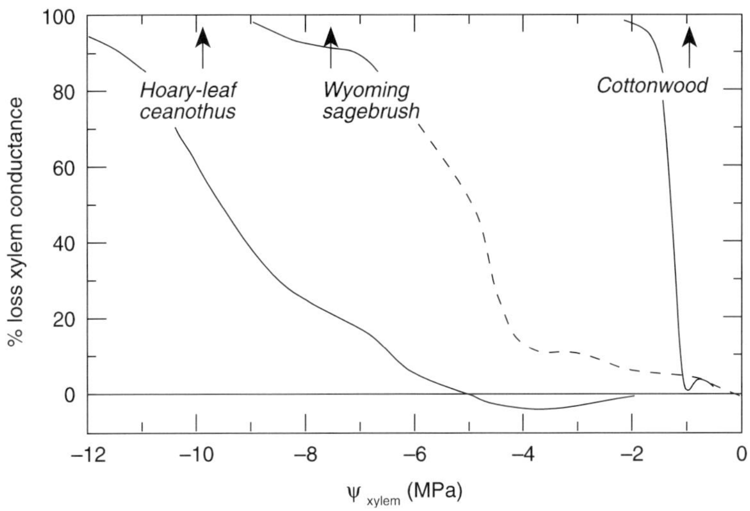

Fig. 2.4. Percentage loss in xylem conductance for three species as a function of xylem pressure potential (redrawn from Sperry, 2000, with permission of Elsevier Science B.V.).

decline thereafter. Compared with wheat, lupins had a higher stomatal conductance at high leaf water potentials, and they closed their stomata at a higher water potential. This is consistent with the cavitation hypothesis discussed previously, because stem hydraulic conductances and xylem vessels in lupins are greater than in wheat (Gallardo *et al.*, 1996) and hence will cavitate earlier. Similar results were presented by Schulze and Hall (1982), showing a marked decrease in stomatal conductance with decreasing leaf water status for plants subject to drought. Turner *et al.* (1984) also found a strong correlation between stomatal conductance and leaf water potential for ten species grown under controlled conditions. While these are only a few examples, there is growing evidence of close linkages between stomatal conductance, xylem hydraulic conductance and cavitation potential (Saliendra *et al.*, 1995; Fuchs and Livingston, 1996; Lovisolo and Schubert, 1998; Bond and Kavanagh, 1999; Comstock, 2000; Nardini and Salleo, 2000; Sperry, 2000; Aasamaa and Sober, 2001; Hubbard *et al.*, 2001). Such results have led Buckley and Mott (2002) to postulate that water potential at the evaporating sites within the leaf controls stomatal conductance through the balance between supply and demand for water at the leaf epidermis. Maier-Maercker (1998) and Bond and Kavanagh (1999) provide further support for this classical approach to modelling interactions between leaf water potential and stomatal conductance.

Water-supply and demand

Stomatal conductance in the model shown in Fig. 2.2 depends on leaf water potential, which in turn depends on the rate of transpiration from the leaves, the resistance to water-vapour transport across the stomata and leaf boundary layer

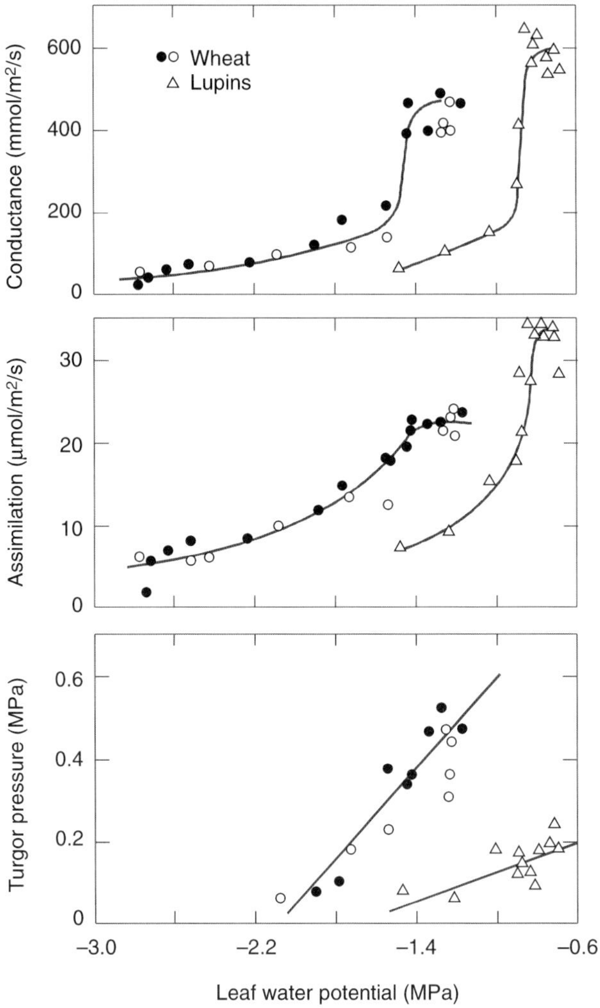

Fig. 2.5. Stomatal conductance, assimilation rate and leaf turgor pressure as a function of leaf water potential for lupins and wheat grown in the field under identical environmental conditions. Lupin has a higher maximum stomatal conductance and a higher leaf water-potential threshold than wheat. Turgor potential is linearly correlated with leaf water potential. (Redrawn from Henson *et al.*, 1989, with permission of CSIRO Publishing.)

and resistances to water flow through the plant and the soil. Tuzet *et al.* (2003) model the dynamics of the flow of water from the bulk soil to the root surfaces by solving the Richards' flow equation using the classical cylindrical geometry approach of Philip (1957) and Gardner (1960). Absorbing roots are taken to be uniformly distributed in the soil volume such that the associated soil cylinders occupy the entire soil volume. The roots are taken to be a long cylinder of uniform radius and water-absorbing properties, and water is assumed to move only in the radial direction. Complications due to root clumping (Tardieu *et al.*, 1992) or soil osmotica (Stirzaker and Passioura, 1996) are not considered in this model.

With the above assumptions, the flow equation is given by Richards' equation:

$$\frac{\partial\theta}{\partial t} = \frac{2\pi L z_r}{r}\frac{\partial}{\partial r}\left(rK_s\frac{\partial\psi_s}{\partial r}\right) \tag{5}$$

where θ is the soil water content; *t* is time; *r* is the radial distance from the axis of the root; *L* is root length per soil volume; z_r is root depth; and K_s is soil hydraulic conductivity and ψ_s soil water potential, both of which depend on soil water content, θ, according to the equations (Campbell, 1985):

$$\psi_s = \psi_a \cdot \frac{1}{\left(\theta/\theta_{sat}\right)^b} \tag{6a}$$

$$K = K_{sat} \cdot \left(\frac{\theta}{\theta_{sat}}\right)^{2b+3} \tag{6b}$$

where ψ_a is the air-entry water potential of the soil, θ_{sat} is the saturation water content, K_{sat} is the saturated conductivity and *b* is an empirical parameter. Equations (6a) and (6b) use the empirical parameters ψ_a, K_{sat}, θ_{sat} and *b*, which depend on soil physical characteristics. In calculations presented in this chapter, the initial water content is assumed to be uniform throughout the soil volume with some value θ_0, corresponding to a soil water potential ψ_0. At the half-average distance between neighbouring roots, the flow of water is zero and, at the boundary between the plant root and the soil, the rate of water uptake by the roots equals the flow of water through the plant.

Profiles of soil moisture content, θ, and soil-pore water potential, ψ_s, predicted by the model for noon of each day during the first 10 days of the dry-down cycle are shown as a function of radial distance in Fig. 2.6. Initially there is little variation in θ and ψ_s with distance, but as the soil dries both θ and ψ_s away from the roots decrease progressively and the local drawdown in both increases markedly near the roots. These profiles are similar to those presented by Cowan (1965).

Details of the daily dynamics of soil water potential on day 10 of the drying cycle are shown in Fig. 2.7. There is little variation in ψ_s with distance around dawn (06.00 h) but the potential well around the roots progressively increases in magnitude as evaporative demand by the atmosphere is transmitted through the plant to the roots and soil during the day. Water potential at the root surface reaches a minimum between 12.00 and 15.00 h. The ψ_s curves for mid-afternoon cross those for the morning as water in the root zone redistributes from the wetter regions away from the roots to regions closer in. By late evening the potential is approximately uniform with distance but is at a lower value than in the morning, commensurate with the extraction of soil water during the day. We note that the relationship between hydraulic conductivity and moisture content is highly non-linear (Equation 6b), and thus resistance to water flow in the soil increases rapidly as the soil dries, with the greatest changes in soil moisture content and potential occurring close to the root surfaces.

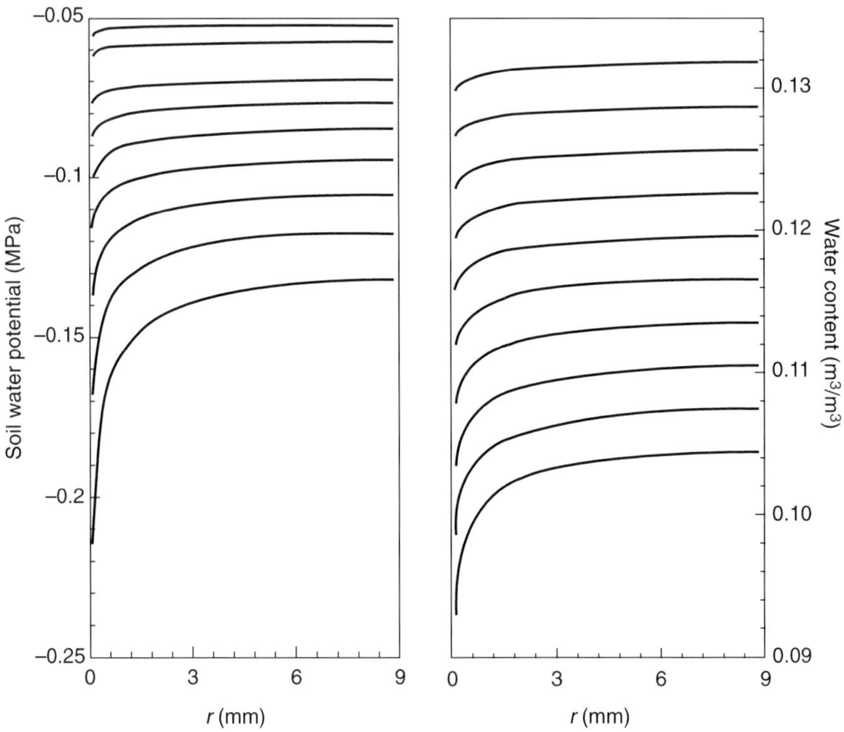

Fig. 2.6. The distribution of soil water potential (left) and soil water content (right) as a function of radial distance from the root surface. The figure represents the profiles of these quantities at noon over a 10-day drying cycle (top to bottom).

Under steady-state conditions, flow from the soil to the roots equals the flow through the plants to the atmosphere. This water flow, J_w, is expressed as:

$$J_w = \frac{\psi_r - \psi_v}{\chi_v} = \frac{D_s}{r_{H_2O} + r_{bv}} = \frac{M_v}{R} \frac{1}{r_{H_2O} + r_{bv}} \left(\frac{h_i P(T_{sv})}{T_{sv}} - \frac{P(T_{rv})}{T_{av}} \right) \tag{7}$$

where ψ_r is the water potential at the boundary between the plant roots and the soil and χ_v is the leaf-specific resistance to water flow through the plant from the roots to the stomata and is assumed constant in our calculations. In practice, it is extremely difficult to separate plant and soil resistances to water flow because the latter depends on microgradients around the root zone.

The second part of Equation (7) equates the steady-state flow of water through the plant with the transpiration by the leaves; D_s is the difference in humidity between the substomatal cavities and the air surrounding the leaves and $r_{H_2O} = 1/(1.6\ g_{CO_2})$ and r_{bv} are the resistances to water transport through the stomata and boundary layer, respectively. The factor 1.6 accounts for the ratio of molecular diffusivities for water and CO_2 in air. M_v is the molecular mass of water, R is the ideal gas constant, $h_i = \exp\ [M_v\ \psi_v/(\rho_v\ RT_{sv})]$ is the fractional relative humidity in the intercellular spaces and provides the link

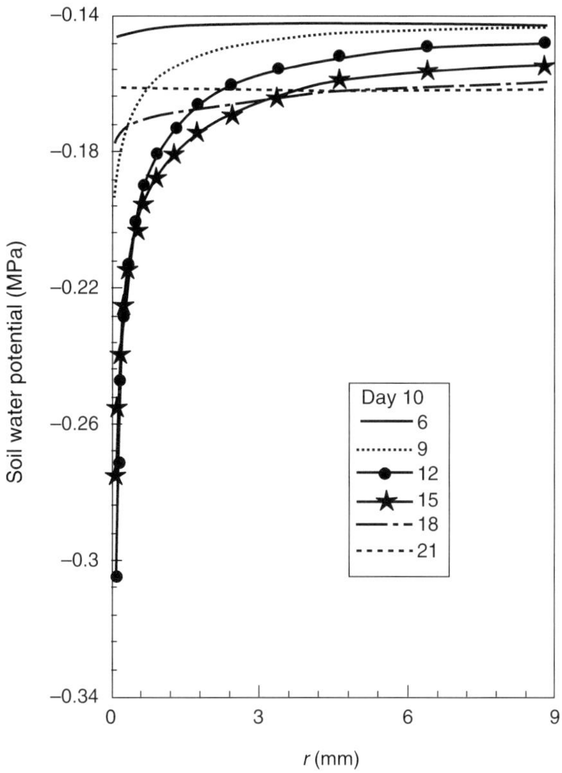

Fig. 2.7. The distribution of soil water potential as a function of radial distance from the root surface at 3-hourly intervals on day 10 of the drying cycle (6=06.00 h, etc.).

between the liquid and gas phases of water at the evaporation sites in the leaf. The function $P(T_{sv})$ is the saturation water-vapour pressure at foliage temperature T_{sv} and $P(T_{rv})$ is the corresponding function for the dew-point temperature of the air surrounding leaves T_{rv} while T_{av} is the air temperature.

Equations (1–4) in Figure 2.2 together with Equations (5–7) in the text link photosynthetic assimilation, stomatal conductance and water flow through the plant and soil. The full model combines these equations with expressions for plant energy balances and fluxes of heat, water vapour and CO_2 between the plants and atmosphere (see Tuzet *et al.* (2003) for full model details and parameter values used).

Contrary to the approach adopted by Leuning (1995), there is no explicit dependence of stomatal conductance on atmospheric humidity in the Tuzet *et al.* (2003) model. Equation (7) shows that ψ_v and D_s are linked, but the linkage is complicated by the dependence of stomatal conductance on leaf water potential. Using the relationship for $f(\psi_v)$ shown in Fig. 2.2, Equations (1) to (4) were solved to illustrate the form of the relationships expected between g_{CO_2} and D_s for two potential response functions $f(\psi_v)$ (Fig. 2.8a), when all other physiological and environmental conditions are non-limiting and when soil water potential is set to zero. Figure 2.8b shows that, after an initial constant region, g_{CO_2} is pre-

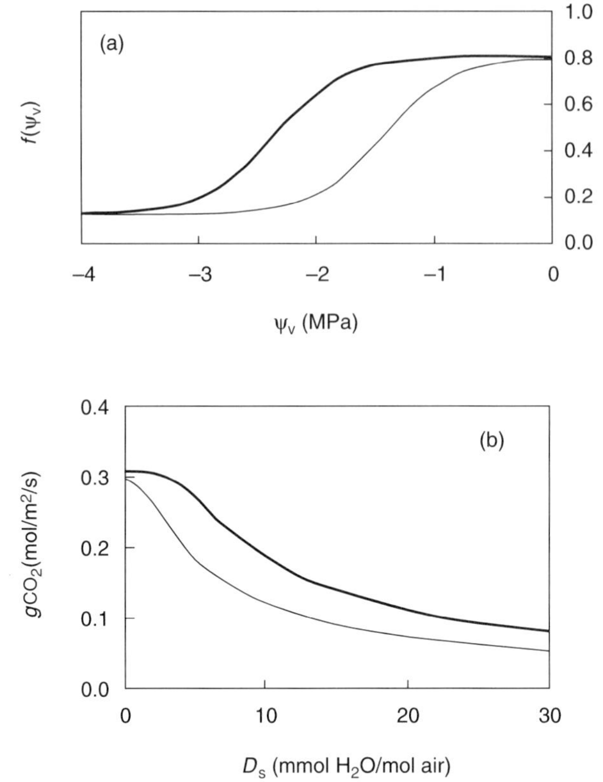

Fig. 2.8. (a) The functions $f(\psi_v)$ for two values of the threshold potential of −1.9 MPa (thick line) and –1.0 MPa (thin line); (b) stomatal conductance as a function of humidity difference between the substomatal cavity and air surrounding the leaf, D_s, for leaves with the functions $f(\psi_v)$ shown above.

dicted to decrease hyperbolically with D_s according to the relationship $g_{CO_2} = g_{cmax}/(1 + D_s/D_0)$. This is the form used by Leuning (1995) to describe the response of stomatal conductance to D_s, and is consistent with the fact that stomatal conductance must reach a maximum value determined by stomatal geometry when all physical and physiological limitations to conductance are eliminated. The major difference in the new formulation is that stomatal conductance is now a function of ψ_v, which depends on soil water potential, the rate of water flow through the plant, the hydraulic resistance to that flow and the threshold potential at which conductance begins to decline (Fig. 2.8).

Model Results and Discussion

We next examine some predictions of the full model. Variations in leaf and soil water potentials, stomatal conductance, transpiration and photosynthesis were calculated for a 30-day drying cycle, starting with an initial root zone uniformly wet ($\psi_s = -0.05$ MPa), and a repeated 24 h cycle of climatic data (Tuzet *et al.*, 2003). We present relationships between conductance and leaf water potential, water-vapour pressure deficit, latent heat flux and assimilation. Simulations are presented for daily cycles to allow assessment of responses of the SPAC to various factors, given that input climatic data and stomatal conductance may co-vary over the course of a day. The model has been applied to a homogeneous field crop with a leaf area index of 3, but the principles apply equally to individual plants.

Predicted diurnal variations in stomatal conductance, latent heat fluxes, assimilation and the ratio of intercellular to external CO_2 concentrations are shown in Fig. 2.9 for days 1, 11, 13 and 15 of the drying cycle. When soil water availability is non-limiting on day 1, g_{CO_2}, *A*, *LE* and c_i/c_s show a smooth and symmetrical variation during the day, but as soil moisture declines there is an increasingly asymmetrical pattern in g_{CO_2}, *A* and *LE*, with values always higher in the morning than in the afternoon. The patterns for g_{CO_2} and *A* are correlated, with both showing a midday depression that becomes more pro-

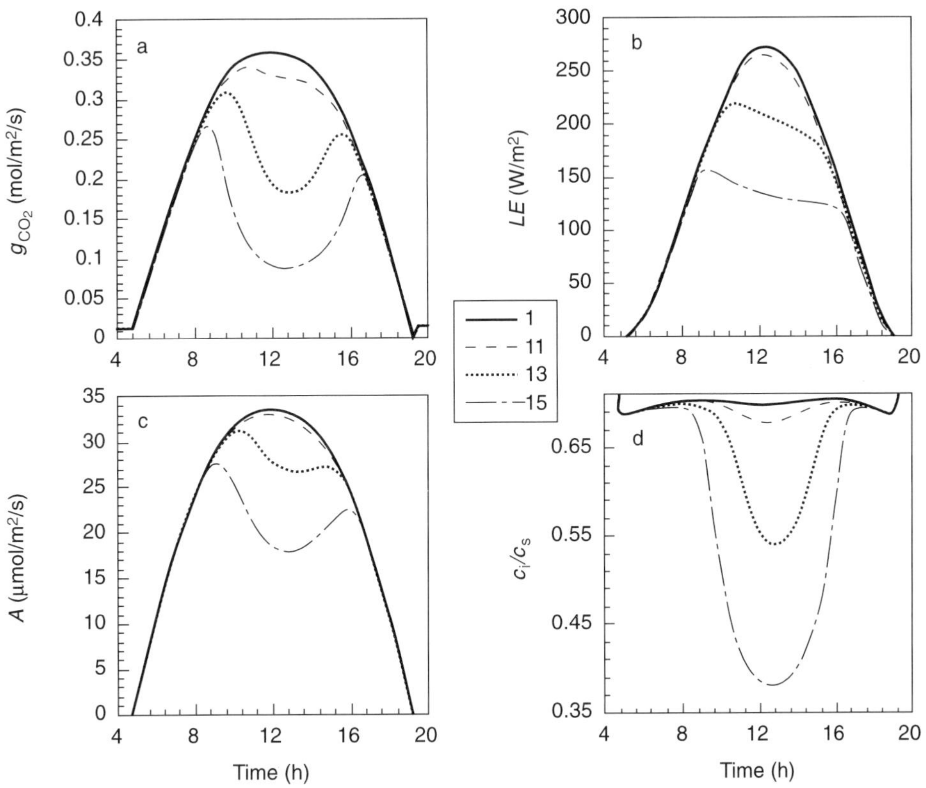

Fig. 2.9. Diurnal variations of: (a) CO_2 stomatal conductance, (b) latent heat flux, (c) assimilation rate, and (d) the ratio of the CO_2 concentration in the intercellular spaces and the leaf surface for days 1, 11, 13 and 15 of the drying cycle. The asymmetry in each quantity increases with the progression of the drying cycle and there is a marked midday depression predicted for g_{CO_2} and *A*.

nounced as soil drying continues. The pattern is not identical, however, with the model showing that g_{CO_2} decreases proportionally more rapidly than *A*, which is consistent with the results of Henson *et al.* (1989), presented in Fig. 2.5. The model predicts that the midday depression is less evident for *LE* than for *A*. The difference in the patterns arises because external CO_2 concentration is essentially constant during the day so variations in g_{CO_2} directly affect *A*, whereas humidity deficit increases steadily with rising temperature from morning to mid-afternoon, counteracting the decrease in midday stomatal conductance. This conforms to the analysis of McNaughton and Jarvis (1983), which showed that *LE* is insensitive to canopy conductance for dense canopies. The general patterns in g_{CO_2}, *A* and *LE* in response to soil drying shown in Fig. 2.9 are similar to reports in the literature as summarized by Schulze and Hall (1982), Olioso *et al.* (1996) and Maier-Maercker (1998). We note that, while earlier models (e.g. Leuning *et al.*, 1995; Baldocchi and Meyers, 1998) were able to reproduce much of the observed diurnal variation in g_{CO_2}, *A* and *LE*, they were unable to simulate the observed daytime asymmetry because the dynamics of water uptake by roots was not included.

Soil drying is also predicted to affect c_i/c_s, with small amplitude when there is ample soil water and a particularly strong drawdown later in the drying cycle (Fig. 2.9d). Such variation of c_i with increasing drought has been observed in the field (Henson *et al.*, 1989; Brodribb, 1996; Eamus *et al.*, 1999; Giorio *et al.*, 1999). We have assumed that the photosynthetic parameters are constant in these simulations, although there is some recent evidence showing that they may vary with leaf water potential (Tezara *et al.*, 1999; Nogues and Baker, 2000).

The model of Tuzet *et al.* (2003) can also be used to examine the expected responses of stomatal conductance to ψ_v, D_s, *A* and *LE*, as shown in Fig. 2.10 for days 1, 11, 13 and 15 of the drying cycle. Hysteresis in the relationship between g_{CO_2} and all the variables in Fig. 2.10 is predicted by the model, and the degree of hysteresis depends on the soil moisture content. This is despite the monotonic relationship between g_{CO_2} and ψ_v used in the stomatal conductance model (Fig. 2.2, Equations 3 and 4). Hysteresis occurs in the fully coupled model because conductance is also a function of light, temperature and intercellular CO_2 concentration (through the photosynthesis model, Fig. 2.2, Equation 2). Figure 2.10a shows that g_{CO_2} is higher in the morning than the afternoon at any given value of ψ_v. Similarly, there is no unique relationship between g_{CO_2} and D_s and the shape of the response curve depends on the soil moisture content, the transpiration rate and the hydraulic resistance of the plant and soil (Tuzet *et al.*, 2003).

The model predicts that in the early morning g_{CO_2} increases with increasing D_s and then decreases hyperbolically later in the morning and increases again late in the afternoon as evaporative demand diminishes (Fig. 2.10b). Such hysteresis has been observed by Pereira *et al.* (1987), Meinzer *et al.* (1997) and Takagi *et al.*(1998) for a variety of plant species when measurements were made in the field. The initial rise in g_{CO_2} occurs because both irradiance and D_s increase in parallel during the morning. Stomata begin to close around midday as leaf water potentials decrease and then begin to open again later in the afternoon before closing at

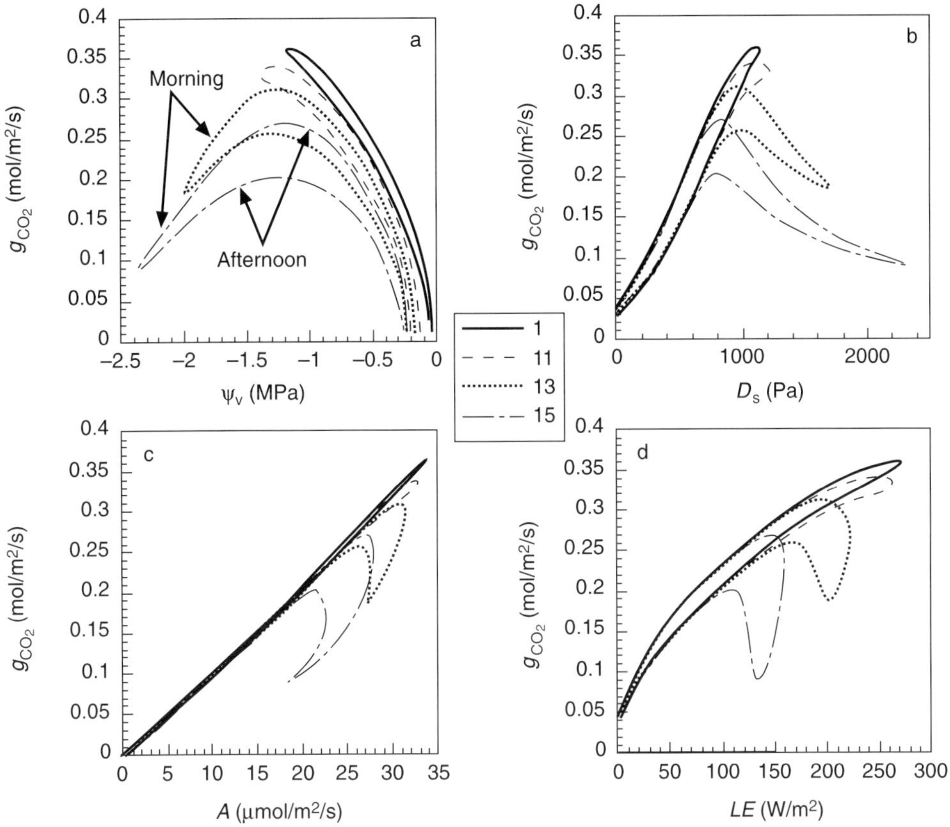

Fig. 2.10. Predicted relationships between stomatal conductance for CO_2 and: (a) leaf water potential, (b) water-vapour pressure deficit at the leaf surface, (c) assimilation rate, and (d) latent heat flux, for 4 selected days of the drying cycle. The model predicts hysteresis in all the relationships shown as a result of varying resistance to water flow in the soil during the day.

night. The hyperbolic relationship between g_{CO_2} and D_s used by many authors (e.g. Lohammer *et al.*, 1980; Leuning, 1995) is apparent only at high values of D_s when soil moisture starts to limit water-supply to the roots. Note that the maximum humidity deficit at the leaf surface, D_s, increases as the drying cycle progresses despite use of the same meteorological forcing for each day and hence the same maximum atmospheric humidity deficit at the reference height. This is because of the interaction between stomatal conductance, transpiration rate, leaf water potential, leaf temperature and D_s through the leaf energy balance.

A quasi-linear relationship between g_{CO_2} and A is predicted by the model when soil water is non-limiting (Fig. 2.10d), implying an approximately constant value of c_i (Fig. 2.9c). By day 11 the results show hysteresis, and the spread in the hysteresis curve increases as the soil dries out, similarly to results presented by Schulze and Hall (1982) and Pereira *et al.* (1987). Assimilation rates are predicted to be higher in the morning than in the afternoon for any given stomatal conductance (cf. Fig. 2.10c).

The relationship between g_{CO_2} and LE also shows hysteresis (Fig. 2.10d), very similar to results presented by Meinzer *et al.* (1997) for poplar leaves for daytime measurements. Stomatal conductance is predicted to decline while transpiration remains approximately constant during part of the hysteresis cycle for moderately dry soils (e.g. day 11, Fig. 2.10d). Farquhar (1978) argued that simple feedback between stomatal conductance and atmospheric humidity deficit was unable to account for this effect and suggested that stomata must sense water stress by mechanism(s) that involve more than peristomatal transpiration (so-called 'feedforward' response). Our model shows that it is unnecessary to invoke a 'feedforward' response of stomata to D_s, provided that the dynamics of the resistance to water transport from soil to roots is included in the model. Resistance to water movement in the soil is lower in the morning than in the afternoon when soil moisture in the zone immediately surrounding the roots is depleted (Fig. 2.6). Thus in the afternoon ψ_v (and hence g_{CO_2}, Fig. 2.2) continues to decline, even at a constant transpiration rate, because of increasing resistance to transport of water from the bulk soil to the roots. When transpiration ceases overnight there is a redistribution of soil moisture (Fig. 2.7), and hence resistance to water flow is lower next morning than the previous evening, even though bulk moisture content is the same.

Model predictions shown in Figs 2.9 and 2.10 illustrate the difficulty in developing simple, unambiguous relationships between stomatal conductance and controlling physiological and environmental variables. Stomatal conductance depends on the interaction of multiple factors and thus plots of g_{CO_2} versus single factors, such as leaf water potential or humidity deficit, will be confounded by the effects of the other controlling variables, such as light, temperature, intercellular CO_2 concentrations, plant and soil hydraulic properties, soil water potential and transport of water from soil to roots and from roots to leaves.

Conclusions

Understanding stomatal behaviour must involve a description of the coupled processes of leaf energy balance, transpiration, photosynthesis and the dynamics of water transport within the soil and plant. The model of Tuzet *et al.* (2003) captures many aspects of stomatal behaviour and leaf gas exchange reported in the literature. It successfully captures the daytime asymmetry in conductance, assimilation and transpiration under water-stress conditions and it describes the observed hysteresis in stomatal conductances versus leaf water potential, atmospheric humidity deficit, assimilation and transpiration rates. Predictions of the model represent a set of hypotheses to be tested under a range of laboratory and field conditions.

References

Aasamaa, K. and Sober, A. (2001) Hydraulic conductance and stomatal sensitivity to changes of leaf water status in six deciduous tree species. *Biologia Plantarum* 44, 65–73.

Aphalo, P.J. and Jarvis, P.G. (1991) Do stomata respond to relative humidity? *Plant, Cell and Environment* 14, 127–132.
Assmann, S.M. (1999) The cellular basis of guard cell sensing to rising CO_2. *Plant, Cell and Environment* 22, 629–637.
Baldocchi, D. and Meyers, T. (1998) On using eco-physiological, micrometeorological and biogeochemical theory to evaluate carbon dioxide, water vapour and trace gas fluxes over vegetation: a perspective. *Agricultural and Forest Meteorology* 90, 1–25.
Ball, J.T., Woodrow, I.E. and Berry, J.A. (1987) A model predicting stomatal conductance and its contribution to the control of photosynthesis under different environmental conditions. In: Biggens, I. (ed.) *Progress in Photosynthesis Research.* Martinus Nijhoff Publishers, The Netherlands, pp. 221–224.
Blatt, M.R. (2000) Cellular signaling and volume control in stomatal movements in plants. *Annual Review of Cell and Developmental Biology* 16, 221–241.
Bond, B. and Kavanagh, K.L. (1999) Stomatal behavior of four woody species in relation to leaf-specific hydraulic conductance and threshold water potential. *Tree Physiology* 19, 503–510.
Brodribb, T. (1996) Dynamics of changing intercellular CO_2 concentration (c_i) during drought and determination of minimum functional c_i. *Plant Physiology* 111, 179–185.
Buckley, T.N. and Mott, K.A. (2002) Stomatal water relations and the control of hydraulic supply and demand. *Progress in Botany* 63, 309–325.
Bunce, J.A. (1996) Does transpiration control stomatal responses to water vapour pressure deficit? *Plant, Cell and Environment* 19, 131–135.
Campbell, G.S. (1985) *Soil Physics with Basic: Transport Models for Soil–Plant Systems.* Elsevier Science Publishers, Amsterdam.
Comstock, J.P. (2000) Variation in hydraulic architecture and gas exchange in two desert subshrubs, *Hymenoclea salsola* (T. and G.) and *Ambrosia dumosa* (Payne). *Oecologia* 125, 1–10.
Comstock, J. and Mencuccini, M. (1998) Control of stomatal conductance by leaf water potential in *Hymenoclea salsola* (T. & G.), a desert subshrub. *Plant, Cell and Environment* 21, 1029–1038.
Cowan, I.R. (1965) Transport of water in the soil–plant–atmosphere continuum. *Journal of Applied Ecology* 2, 221–239.
Cowan, I.R. (1977) Stomatal behaviour and environment. *Advances in Botanical Research* 4, 117–128.
Dewar, R.C. (1995) Interpretation of an empirical model for stomatal conductance in terms of guard cell function. *Plant, Cell and Environment* 18, 365–372.
Dewar, R.C. (2002) The Ball–Berry–Leuning and Tardieu–Davies models: synthesis and extension within a spatially aggregated picture of guard cell function. *Plant, Cell and Environment* 25, 1383–1398.
Eamus, D., Myers, B., Duff, G. and Williams, D. (1999) Seasonal changes in photosynthesis of eight savanna tree species. *Tree Physiology* 19, 665–671.
Ewers, B.E., Oren, R., Phillips, N. and Stromgren, L.S. (2001) Mean canopy stomatal conductance responses to water and nutrient availabilities in *Picea abies* and *Pinus taeda. Tree Physiology* 21, 841–850.
Farquhar, G.D. (1978) Feedforward responses of stomata to humidity. *Australian Journal of Plant Physiology* 5, 787–800.
Farquhar, G.D. and Wong, S.C. (1984) An empirical model of stomatal conductance. *Australian Journal of Plant Physiology* 11, 191–210.
Farquhar, G.D., von Caemmerer, S. and Berry, J.A. (1980) A biochemical model of photosynthetic CO_2 assimilation in leaves of C3 species. *Planta* 149, 78–90.

Franks, P.J., Cowan, I.R. and Farquhar, G.D. (1998) A study of stomatal mechanics using the cell pressure probe. *Plant, Cell and Environment* 21, 94–100.

Franks, P.J., Buckley, T.N., Shope, J.C. and Mott, K.A. (2001) Guard cell volume and pressure measured concurrently by confocal microscopy and the cell pressure probe. *Plant Physiology* 125, 1577–1584.

Fuchs, E.E. and Livingston, N.J. (1996) Hydraulic control of stomatal conductance on Douglas fir [*Pseudotsuga menziesii* (Mirb.) Franco] and alder [*Alnus rubra* (Bong)] seedlings. *Plant, Cell and Environment* 19, 1091–1098.

Gallardo, M., Eastham J., Gregory, P.J. and Turner, N.C. (1996) A comparison of plant hydraulic conductances in wheat and lupins. *Journal of Experimental Botany* 47, 233–295.

Gardner, W.R. (1960) Dynamic aspects of water availability to plants. *Soil Science* 89, 63–73.

Giorio, P., Sorrentino, G. and d'Andria, R. (1999) Stomatal behaviour, leaf water status and photosynthetic response of field-grown olive trees under water deficit. *Environmental and Experimental Botany* 42, 95–104.

Gollan, T., Passioura, J.B. and Munns, R. (1986) Soil water status affects the stomatal conductance of fully turgid wheat and sunflower leaves. *Australian Journal of Plant Physiology* 13, 459–464.

Henson, I.E., Jensen, C.R. and Turner, N.C. (1989) Leaf gas exchange and water relations of lupins and wheat. I. Shoot response to soil water deficits. *Australian Journal of Plant Physiology* 16, 401–413.

Hubbard, R.M., Bond, B.J. and Ryan, M.G. (1999) Evidence that hydraulic conductance limits photosynthesis in old *Pinus ponderosa* trees. *Tree Physiology* 19, 165–172.

Hubbard, R.M., Ryan, M.G., Stiller, V. and Sperry, J.S. (2001) Stomatal conductance and photosynthesis vary linearly with plant hydraulic conductance in ponderosa pine. *Plant, Cell and Environment* 24, 113–121.

Jarvis, P.G. (1976) The interpretation of the variations in leaf water potential and stomatal conductance found in canopies in the field. *Philosophical Transactions of the Royal Society of London, Series B – Biological Sciences* 273, 593–610.

Jensen, C.R., Svendsen, H., Andersen, M.N. and Lösch, R. (1993) Use of the root contact concept, an empirical leaf conductance model and pressure–volume curves in simulating crop water relations. *Plant and Soil* 149, 1–26.

Jones, H.G. and Sutherland, R.A. (1991) Stomatal control of xylem embolism. *Plant, Cell and Environment* 14, 607–612.

Jones, H.G. and Tardieu, F. (1998) Modelling water relations of horticultural crops: a review. *Scientia Horticulturae* 74, 21–46.

Leuning, R. (1990) Modelling stomatal behaviour and photosynthesis of *Eucalyptus grandis*. *Australian Journal of Plant Physiology* 17, 159–175.

Leuning, R. (1995) A critical appraisal of a combined stomatal–photosynthesis model for C_3 plants. *Plant, Cell and Environment* 18, 339–355.

Leuning, R., Kelliher, F.M., DePury, D.G.G. and Schulze, E.-D. (1995) Leaf nitrogen, photosynthesis, conductance and transpiration: scaling from leaves to canopies. *Plant, Cell and Environment* 18, 1183–1200.

Lohammer, T., Larsson, S., Linder, S. and Falk, S.O. (1980) FAST – simulation models of gaseous exchange in Scots pine. *Ecology Bulletin* 32, 505–523.

Lovisolo, C. and Schubert, A. (1998) Effects of water stress on vessel size and xylem hydraulic conductivity in *Vitis vinifera* L. *Journal of Experimental Botany* 49, 693–700.

Maier-Maercker, U. (1998) Dynamics of change in stomatal response and water status of *Picea abies* during a persistent drought period: a contribution to the traditional view of plant water relations. *Tree Physiology* 18, 211–222.

Maier-Maercker, U. (1999) New light on the importance of peristomatal transpiration. *Australian Journal of Plant Physiology* 26, 9–16.

McMurtrie, R.E., Leuning, R., Thompson, W.A. and Wheeler, A.M. (1992) A model of canopy photosynthesis and water use incorporating a mechanistic formulation of leaf CO_2 exchange. *Forest Ecology and Management* 52, 261–278.

McNaughton, K.G. and Jarvis, P.G. (1983) Predicting effects of vegetation changes on transpiration and evaporation. In: Kozlowski, T.T. (ed.) *Water Deficits and Plant Growth*, Vol. 7. Academic Press, New York, pp. 1–47.

Meinzer, F.C., Hinckley, T.M. and Ceulemans, R. (1997) Apparent responses of stomata to transpiration and humidity in a hybrid poplar canopy. *Plant, Cell and Environment* 20, 1301–1308.

Meinzer, F.C., Clearwater, M.J. and Goldstein, G. (2001) Water transport in trees: current perspectives, new insights and some controversies. *Environmental and Experimental Botany* 45, 239–262.

Mencuccini, M., Mambelli, S. and Comstock, J. (2000) Stomatal responsiveness to leaf water status in common bean (*Phaseolus vulgaris* L.) is a function of time of day. *Plant, Cell and Environment* 23, 1109–1118.

Monteith, J.L. (1995) Stomatal responses to humidity: a reassessment. *Plant, Cell and Environment* 18, 357–364.

Mott, K.A. (1988) Do stomata respond to CO_2 concentrations other than intercellular? *Plant Physiology* 86, 200–203.

Mott, K.A. and Franks, P.J. (2001) The role of epidermal turgor in stomatal interactions following a local perturbation in humidity. *Plant, Cell and Environment* 24, 657–662.

Mott, K.A. and Parkhurst, D.F. (1991) Stomatal responses to humidity in air and helox. *Plant, Cell and Environment* 14, 509–515.

Nardini, A. and Salleo, S. (2000) Limitation of stomatal conductance by hydraulic traits: sensing or preventing xylem cavitation? *Trees* 15, 14–24.

Netting, A.G. (2000) pH, abscisic acid and the integration of metabolism in plants under stressed and non-stressed conditions: cellular responses to stress and their implication for plant water relations. *Journal of Experimental Botany* 51, 147–158.

Nogues, S. and Baker, N.R. (2000) Effects of drought on photosynthesis in Mediterranean plants grown under enhanced UV-B radiation. *Journal of Experimental Botany* 51, 1309–1317.

Olioso, A., Carlson, T.N. and Brisson, N. (1996) Simulation of diurnal transpiration and photosynthesis of a water stressed soybean crop. *Agricultural and Forest Meteorology* 81, 41–59.

Pereira, J.S., Tenhunen, J.D. and Lange, O.L. (1987) Stomatal control of photosynthesis of *Eucalyptus globulus* Labill: trees under field conditions in Portugal. *Journal of Experimental Botany* 38, 1678–1688.

Philip, J.R. (1957) Evaporation, and moisture and heat fields in the soil. *Journal of Meteorology* 14, 354–366.

Saliendra, N.Z., Sperry, J.S. and Comstock, J.P. (1995) Influence of leaf water status on stomatal response to humidity, hydraulic conductance, and soil drought in *Betula occidentalis*. *Planta* 196, 357–366.

Schulze, E.-D. and Hall, A.E. (1982) Stomatal responses, water loss and CO_2 assimilation rates of plants in contrasting environments. In: Lange, O.L., Nobel, P.S., Osmond, C.B. and Ziegler, H. (eds) *Physiological Plant Ecology II. Water Relations and Carbon Assimilation*. Springer-Verlag, Berlin, pp. 181–230.

Sellers, P.J., Berry, J.A., Collatz, G.J., Field, C.B. and Hall, F.G. (1992) Canopy reflectance, photosynthesis and transpiration. III. A reanalysis using improved leaf models and a new canopy integration scheme. *Remote Sensing of Environment* 42, 187–216.

Sperry, J.S. (2000) Hydraulic constraints on plant gas exchange. *Agricultural and Forest Meteorology* 104, 13–23.

Sperry, J.S., Adler, F.R., Campbell, G.S. and Comstock, J.P. (1998) Limitation of plant water use by rhizosphere and xylem conductance: results of a model. *Plant, Cell and Environment* 21, 347–359.

Stirzaker, R.J. and Passioura, J.B. (1996) The water relations of the root–soil interface. *Plant, Cell and Environment* 19, 201–208.

Takagi, K., Tsuboya, T. and Takahashi, H. (1998) Diurnal hystereses of stomatal and bulk surface conductances in relation to vapour pressure deficit in a cool-temperate wetland. *Agricultural and Forest Meteorology* 91, 177–191.

Tardieu, F. and Davies, W.J. (1993) Integration of hydraulic and chemical signaling in the control of stomatal conductance and water status of droughted plants. *Plant, Cell and Environment* 16, 341–349.

Tardieu, F. and Simonneau, T. (1998) Variability among species of stomatal control under fluctuating soil water status and evaporative demand: modeling isohydric and anisohydric behaviors. *Journal of Experimental Botany* 49, 419–432.

Tardieu, F., Bruckler, L. and Lafolie, F. (1992) Root clumping may affect the root water potential and the resistance to soil–root water transport. *Plant and Soil* 140, 291–301.

Tezara, W., Mitchell, V.J., Driscoll, S.D. and Lawlor, D.W. (1999) Water stress inhibits plant photosynthesis by decreasing coupling factor and ATP. *Nature* 401, 914–917.

Turner, N.C., Schulze, E.-D. and Gollan, T. (1984) The response of stomata and leaf gas exchange to vapour pressure deficits and soil water content. I. Species comparisons at high soil water contents. *Oecologia* 63, 338–342.

Tuzet, A., Perrier, A. and Leuning, R. (2003) Stomatal control of photosynthesis and transpiration: results from a soil–plant–atmosphere continuum model. *Plant, Cell and Environment* 26, 1097–1116.

Tyree, M.T. and Sperry, J.S. (1988) Do woody plants operate near the point of catastrophic xylem dysfunction caused by dynamic water stress? Answers from a model. *Plant Physiology* 88, 574–580.

Tyree, M.T. and Sperry, J.S. (1989) Vulnerability of xylem to cavitation and embolism. *Annual Review of Plant Physiology* 40, 19–38.

Wang, Y.P. and Leuning, R. (1998) A two-leaf model for canopy conductance, photosynthesis and partitioning of available energy. I. Model description. *Agricultural and Forest Meteorology* 91, 89–111.

Williams, M., Rastetter, E.B., Fernandes, D.N., Goulden, M.L., Wofsy, S.C., Shaver, G.R., Melillo, J.M., Munger, J.W., Fan, S.-M. and Nadelhoffer, K.J. (1996) Modelling the soil–plant–atmosphere continuum in a *Quercus–Acer* stand at Harvard Forest: the regulation of stomatal conductance by light, nitrogen and soil/plant hydraulic conductance. *Plant, Cell and Environment* 19, 911–927.

Wong, S.C., Cowan, I.R. and Farquhar, G.D. (1985) Leaf conductance in relation to rate of CO_2 assimilation. I. Influence of nitrogen nutrition, phosphorus nutrition, photon flux density, and ambient partial pressure of CO_2 during ontogeny. *Plant Physiology* 78, 821–825.

Wu, H.-I., Sharpe, J.H. and Spence, R.D. (1985) Stomatal mechanics III: geometric interpretation of the mechanical advantage. *Plant, Cell and Environment* 8, 269–274.

Zwieniecki, M.A. and Holbrook, N.M. (2000) Bordered pit structure and vessel wall surface properties: implications for embolism repair. *Plant Physiology* 123, 1015–1020.

3

Effects of Elevated CO_2 Concentration on Stomatal Conductance and Respiration of Beech Leaves in Darkness

D. Overdieck

TU-Berlin, Institut für Ökologie, FG Landschaftsökologie/Ökologie der Gehölze, Königin-Luise-Str. 22, D-14195 Berlin, Germany

Abstract

Beech saplings (*Fagus sylvatica* L.) were grown in 10 l pots with medium-fertile soil at ambient-air CO_2 concentration (control) and at ~700 µmol/mol in temperature- and humidity-controlled cabinets standing in the field. Respiration in darkness and transpiration of leaves in response to CO_2 concentration were measured with a mini-cuvette system. Stomatal conductance was calculated. Despite clearly decreasing stomatal conductance (~70%) and transpiration rate (~49%) with elevated CO_2 concentration from subambient (200 µmol/mol) to 800 µmol/mol, dark respiration rates did not decrease significantly (~3% on average). This suggests that stomata opened in response to a low CO_2 supply without increased CO_2 losses via respiration. This contradicts earlier suggestions in the literature assuming direct effects of ambient-air CO_2 concentration on respiration during darkness.

Stomatal conductance and transpiration rate of leaves grown under elevated CO_2 were greater than in the control and responded less to increased CO_2 concentrations. This indicates acclimatization of stomatal aperture to the permanently given CO_2 level.

Introduction

Concentration of CO_2 in the mesophyll of leaves has long been known to influence stomatal aperture (Akita and Moss, 1972; Morison, 1987) as much as other environmental factors, such as light photosynthetically active photon flux density (PPFD), leaf surface temperature, water shortage and air-to-leaf water-vapour pressure deficit (ALVPD). This is taken into appropriate account in many existing ecophysiological models (e.g. Farquhar *et al.*, 1980). Is ambient CO_2 concentration such a strong factor that it can determine stomatal aperture independently of other environmental factors or can even negate their effect?

 Forests at the Land–Atmosphere Interface (eds M. Mencuccini, J. Grace, J. Moncrieff and K.G. McNaughton)

In this study stomatal conductance was measured on leaves in darkness at varying ambient CO_2 concentrations under constant environmental conditions in order to contribute to the answer of this question.

Effects of CO_2 on net photosynthesis and transpiration as well as on many other growth parameters are relatively well known (among many others, see Overdieck and Forstreuter, 1994; Overdieck, 2001). However, until now, comparatively less has been known about the direct and indirect effects of CO_2 on dark respiration (cf. Mousseau, 1998). For instance, both enhancement and reduction of dark respiration at elevated CO_2 concentrations were found (Wullschleger *et al.*, 1994).

Materials and Methods

Beech saplings (*Fagus sylvatica* L., provenance north-west Germany, No. 81002) were grown in 10 l pots from the time they were 2 years old until the time of measurements at the end of their fourth year. The pots were filled with homogenized medium-fertile soil and placed in ten phytotron chambers in the field (eight trees per chamber). Air-temperature and relative-humidity regimes during the measurement period are described in Overdieck *et al.* (1995). Water status of the substrate in the pots was kept above field capacity by manual watering. Ten chambers were randomly selected to be included in one of two treatments: one set of five was continuously supplied with ambient air (i.e. about 385 μmol/mol CO_2) while the other set of five was continuously supplied with CO_2-enriched air (at about 700 μmol/mol CO_2). During the second year of CO_2 enrichment, one leaf per sapling (two trees per chamber) in the middle of the crown was selected for gas-exchange measurements. CO_2 exchange and transpiration were measured immediately after placing the leaves under darkness at 25°C, while ALVPD was kept at 1.3 ± 0.2 kPa by means of a WALZ minicuvette system (Walz, CMS-400).

Response curves of individual leaves to CO_2 were obtained by alternating two measurement protocols over two successive leaves. On the first leaf, CO_2 concentration was first elevated and then progressively lowered in the cuvette in a stepwise fashion (35 → 900 → 35 μmol/mol CO_2), whereas, on the subsequent leaf, measurements were started at the highest reachable CO_2 concentration, decreased and finally elevated again (900 → 35 → 900 μmol/mol CO_2). Thus each sapling grown at a different CO_2 concentration was measured at seven CO_2 steps twice (for a total of 20 leaves for each CO_2 treatment). Measurements took place during periods of darkness of about 2.5 h average length. Arithmetic means of each cuvette CO_2 concentration were calculated for both CO_2 growth conditions. Stomatal conductances to CO_2 were obtained by using the formulae given in the WALZ handbook. All values are relative to the area of one leaf surface.

In addition, the following growth parameters and chemical components were determined on the saplings: number of leaves per tree, leaf area, specific leaf area, length of the main axis, stem diameter at 2 cm above the substrate surface and N and C concentration of leaves (by means of a C/N analyser).

Results and Discussion

Growth parameters and C/N concentrations

Overall, growth parameters of beech saplings were significantly affected by the treatment and showed several effects associated with typical response patterns to CO_2 (Lee *et al.*, 1998). Results are given in Table 3.1.

Transpiration and stomatal conductance

All CO_2 response curves of transpiration rate during experimental darkness showed distinct increases at low CO_2 concentration levels (Fig. 3.1). Within a realistic range of ambient CO_2 concentrations, i.e. from 800 down to 200 µmol/mol CO_2, increases of transpiration amounted to ~49% in both treatments. At CO_2 concentrations above 400 µmol/mol, stable low levels of transpiration were reached. The mean difference between the CO_2 treatments amounted to ~33% on average.

Mean values of stomatal conductance showed similar patterns. Stomatal conductance was increased on average by ~69.8% between 800 and 200 µmol/mol CO_2. The mean difference between the two CO_2 treatments fluctuated around 32% (Fig. 3.2).

Table 3.1. Summary of data related to plant growth and to chemical composition of leaves of *Fagus sylvatica* saplings in response to an elevated CO_2 treatment.

	Elev. CO_2	Control	t-test	Diff. (%)
Leaf area (cm^2)	22.39	17.10	$P < 0.05$	31
SE	9.93	6.44		
Leaf dry mass (mg)	139.7	94.2	$P < 0.10$	48
SE	69.7	49.1		
SLA (cm^2/g)	166.11	192.17	$P = 0.01$	−14
SE	24.30	32.96		
Number of leaves per tree	46.8	37.2	$P < 0.10$	27
SE	11.3	18.7		
Height (cm)	43.2	38.3	NS	13
SE	11.3	16.0		
Diameter (mm)	5.9	5.4	NS	9
SE	1.6	1.1		
N (%) leaf	1.66	2.12	$P < 0.001$	−22
SE	0.2	0.4		
C/N leaf	29.1	23.5	$P < 0.001$	24
SE	4.5	4.7		

Elev. CO_2, elevated CO_2 treatment, ~700 µmol/mol CO_2; control, ~385 µmol/mol CO_2; SLA, specific leaf area; N leaf, leaf nitrogen concentration; C/N leaf, ratio of carbon to nitrogen in leaves; t-test, significance levels for a t-test; Diff., percentage difference between treatments; SE, standard error; NS, not significant.

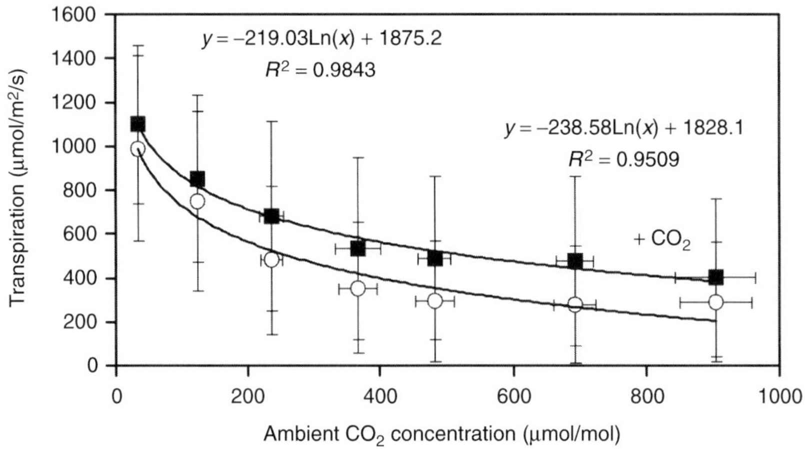

Fig. 3.1. Twenty *Fagus sylvatica* plants were grown either at ~700 (black squares) or at ~385 µmol/mol (empty circles) in phytotron chambers. Sample leaves were measured in June–July of their second year of CO_2 enrichment. For each leaf, ambient CO_2 concentrations were progressively changed until steady-state transpiration rates were obtained under complete darkness. More details on the protocol can be found in the text. Steady-state transpiration appeared to increase when ambient-air CO_2 concentration was progressively decreased from about 900 to about 50 µmol/mol. Regression equations are given in the figure for each treatment.

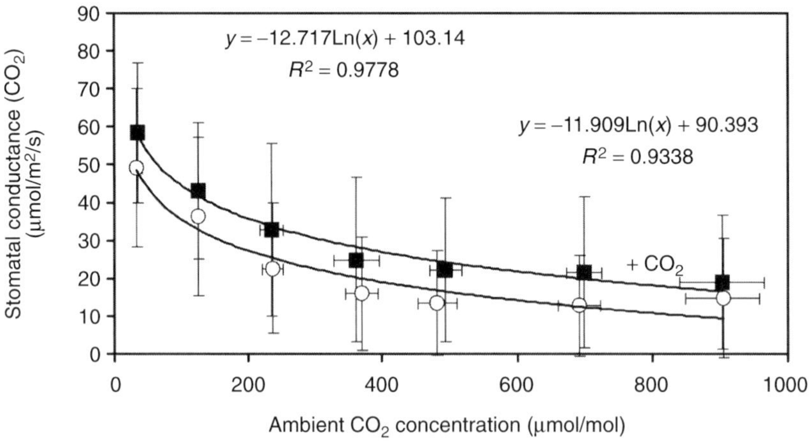

Fig. 3.2. Changes in stomatal conductance with ambient-air CO_2 concentration of *Fagus sylvatica* leaves. Details of the experimental protocol as in Fig. 3.2 and in the text. Black squares, plants grown at ~700 µmol/mol; empty circles, plants grown at ~385 µmol/mol. Measurements were carried out in complete darkness and at constant vapour pressure deficit (VPD).

The difference between the two CO_2 treatments in transpiration and stomatal conductance remained evident, although at a lower level (~22.5%), when results were expressed on a dry-mass basis (mg), again showing the same increase with decreasing CO_2 concentrations in the leaf cuvette. These results clearly indicate stomatal opening at low ambient CO_2 concentration during the respiratory phase in darkness.

Dark respiration

There was no clear response of dark respiration to changed CO_2 concentration levels in the leaf cuvette for either group of saplings (Fig. 3.3). However, a certain tendency appeared to a slight increase in dark respiration at the very lowest CO_2 concentrations, particularly in the leaves of plants grown at 700 µmol/mol CO_2. On a leaf-area basis, leaves grown under elevated CO_2 lost significantly more CO_2 by dark respiration. On average, within the range between 200 and 800 µmol/mol CO_2, the mean difference amounted to 14%. As shown before (Table 3.1), leaves grown under elevated CO_2 had on average greater values of dry mass per unit leaf area. Not surprisingly, when values are calculated on a dry-mass basis, the 14% difference disappears completely, and both response curves fall into a single one.

Discussion and conclusions

The assumption that stomata of beech leaves are always closed in darkness can be rejected. Their aperture clearly depends on the CO_2 concentration of the ambient air. The main consequence of this finding is that the 'CO_2 sensor' in *F. sylvatica* leaves reacts during darkness not only to the CO_2 concentration within the mesophyll but also, and very significantly, to the CO_2 level around the leaf surface. Using realistic ranges of CO_2 concentrations in a leaf cuvette, we demonstrated that changes in transpiration or stomatal conductance are quan-

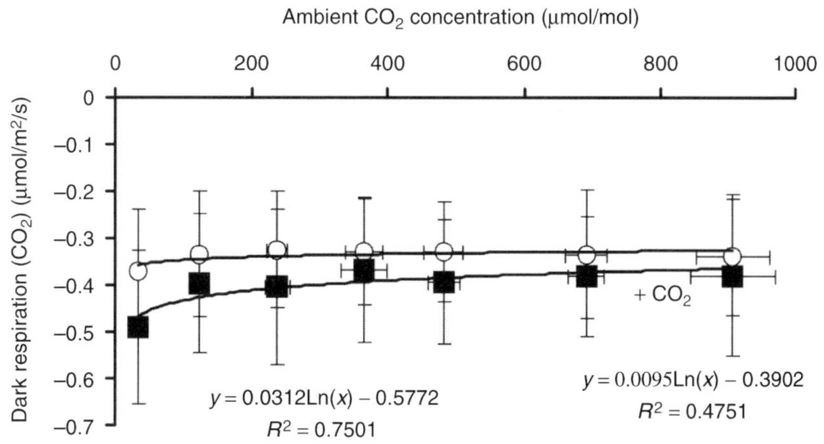

Fig. 3.3. Dark respiration versus ambient-air CO_2 concentration of *Fagus sylvatica* leaves grown at ~700 (black symbols) and ~385 µmol/mol in phytotron chambers, measured in June–July of the second year of a CO_2-enrichment experiment (n = 20).

titatively important in darkness. Particularly at dawn and at dusk, relatively low CO_2 concentrations can occur around plant canopies. Consequently, one might then expect that stomata will tend to stay open or open up again. During these periods, ALVPD can be very low, and therefore transpiration should also remain low under field conditions. However, if stomata increase their aperture in response to the decreased CO_2 concentration, as shown in this chapter, higher rates than expected based only on light and vapour pressure deficit (VPD) levels will be found. In other words, the findings reported in this chapter can also be interpreted to mean that transpiration rates around the light compensation point cannot be considered independently from concurrent changes in air CO_2 concentration.

Independently of whether leaf-area or dry-mass units were used and of the method of measuring CO_2 concentration level, transpiration rates were higher for the saplings grown under elevated CO_2. This finding indicates a certain degree of adaptation to higher CO_2 concentrations, i.e. stomata of leaves developed under a higher CO_2 level responded less sensitively to short-term increases in CO_2 concentration than stomata of leaves grown under ambient conditions. Consequently, an overall higher transpiration level was maintained.

Models that include a detailed representation of the physiological processes of CO_2 uptake and release in leaf cells and that link these processes to stomatal conductance should take into account the direct CO_2 effects on stomatal aperture that may occur at very low PPFD and in darkness. Despite changes in stomatal conductance in response to changes in ambient CO_2 levels, dark respiration rates did not change, or changed only slightly at the very lowest concentrations, suggesting that the biochemical processes influencing the amount of CO_2 released from leaves in darkness can be almost entirely stable regardless of stomatal closing or opening. Since dark respiration rates (expressed on a leaf dry-mass basis) did not differ between leaves of plants grown under elevated CO_2 concentrations and those grown under ambient air, one can conclude that CO_2 concentration has no direct (instantaneous) effect on dark respiration of beech leaves. Therefore, the higher dark respiration rates under elevated CO_2 (when data are calculated on a leaf-area basis) were entirely due to the direct effects of elevated CO_2 on leaf dry-mass per area.

Therefore, consideration of photorespiration as the main oxidative pathway of respiration during daytime will allow the construction of accurate carbon balances for European beech. However, further investigations are warranted to determine whether dark respiration rates are influenced by CO_2 concentration alone or by temperature in combination with CO_2 concentration.

Acknowledgement

This research was partly supported by the 5th Framework Programme of The European Commission through the MEFYQUE research contract (QIKS–CT–2001–2004).

References

Akita, S. and Moss, D.N. (1972) Differential stomatal response between C_3 and C_4 species to atmospheric CO_2 concentration and light. *Crop Science* 12, 789–793.

Farquhar, G.D., von Caemmerer, S. and Berry, J.A. (1980) A biochemical model of photosynthetic CO_2 assimilation in leaves of C_3 species. *Planta* 149, 78–90.

Lee, H.S.L., Overdieck, D. and Jarvis, P.G. (1998) Biomass, growth and carbon allocation. In: Jarvis, P.G. (ed.) *European Forests and Global Change. The Likely Impacts of Rising CO_2 and Temperature.* Cambridge University Press, Cambridge, UK, pp. 126–186.

Morison, J.I.L. (1987) Intercellular CO_2 concentration and stomatal response to CO_2. In: Zeiger, E., Farquhar, G.D. and Cowan, I.R. (eds) *Stomatal Function.* Stanford University Press, Stanford, California, pp. 229–252.

Mousseau, M. (1998) Is dark respiration rate changed? In: Jarvis, P.G. (ed.) *European Forests and Global Change. The Likely Impacts of Rising CO_2 and Temperature.* Cambridge University Press, Cambridge, UK, pp. 79–92.

Overdieck, D. (2001) Combinational effects of elevated CO_2 concentration and temperature on different tree species. *Verhandlungen der Gesellschaft für Ökologie* 31, 357.

Overdieck, D. and Forstreuter, M. (1994) Evapotranspiration of beech stands and transpiration of beech leaves subject to atmospheric CO_2 enrichment. *Tree Physiology* 14, 997–1003.

Overdieck, D., Stille, A. and Forstreuter, M. (1995) Temperaturerhöhung, Anstieg der CO_2-Konzentration und Länge der Vegetationsperiode (Beispiel: Buche und Bergahorn). *Schriftenreihe für Vegetationskunde* 27, 299–305.

Wullschleger, S.D., Ziska, L.H. and Bunce, J.A. (1994) Respiratory responses of higher plants to atmospheric CO_2 enrichment. *Physiologia Plantarum* 90, 221–229.

Top-down Models and Flux Measurements are Complementary Methods of Estimating Carbon Sequestration by Forests: Illustrations using the 3-PG Model

4

Joe Landsberg[1] and Richard H. Waring[2]
[1]22 Mirning Crescent, Aranda, Canberra, ACT 2614, Australia;
[2]Oregon State University, College of Forestry, OR 7331, USA

Abstract

Fluxes measured by eddy covariance and calculated using process-based ecosystem models must be consistent with integrated values of carbon sequestration obtained from biomass-inventory data. Ecosystem models provide the tools needed for spatial extrapolation from flux measurements and to evaluate the impacts of past, current and future alterations in the environment. This chapter illustrates the use of a simple process-based model (3-PG; Landsberg and Waring, 1997) as a tool to bridge the gap between flux and physiological measurements and inventory measurements. The structure and characteristics of the model are outlined and an example is provided of the accuracy with which it can be calibrated against biomass measurements made in an experiment in Australia where treatments resulted in widely differing nutrient status. The results illustrate the need to account for nutrition in estimating carbon sequestration. A comparison of several ecosystem models with flux measurements in Oregon is carried out, and the points emerging from the comparisons are discussed. The 3-PG model was used to simulate growth of a Sitka spruce stand in Scotland for which estimates of net ecosystem exchange (NEE) were available from integrated flux measurements. The implications of the assumptions that had to be made to estimate NEE from the model output are discussed.

We note progress in the use of measurements made from satellites to account for variability in conditions affecting rates of carbon exchange and sequestration.

Introduction

It is estimated that more than 50% of the carbon stored in terrestrial biomes is contained in forests (Dixon *et al.*, 1994). The size of these carbon sinks and the rate of exchange of CO_2 with the atmosphere are matters of immense importance in relation to the increasing concentrations of CO_2 in the atmosphere. On a global scale the size of the forest sink for CO_2 is subject to constant change as a result of forest clearing for agriculture, burning, logging, replanting and regeneration (see Malhi *et al.*, 1999). It is difficult to identify the total area of (relatively) undisturbed forest, but studies on carbon storage and exchange must be made on representative sites of established forest, whether natural or artificial (e.g. plantations). Two of the most important approaches to estimating the carbon balance of forest biomes are the micrometeorological measurement of surface fluxes – integration of eddy covariance measurements of CO_2 fluxes over diurnal, seasonal and annual cycles – and biomass inventories.

Eddy–flux covariance technology and the software needed to analyse the data have improved enormously in recent years (see, for example, Moncrieff *et al.*, 1997). The approach provides a measure of net CO_2 and water-vapour exchange. In the case of water vapour, the sources differ significantly: evaporation transfers water vapour from the surfaces of vegetation when wet, and from the litter and surface soil until dry; at the same time, plants exchange water through their stomata with the atmosphere (transpiration). In the case of CO_2, the measurements represent the differences between two very large fluxes: one negative (photosynthesis) into the system and one positive (respiration) out of the system.

Flux measurements routinely provide estimates of net ecosystem exchange (NEE) but, with instruments at different levels above and within a forest, they can also provide estimates of some of its components – soil/understorey respiration and the carbon stored in the system over any period during which conditions are favourable for measurements and analysis (Goulden *et al.*, 1996; Jarvis *et al.*, 1997; Valentini *et al.*, 2000). However, there may be considerable errors and uncertainties associated with flux measurements, which are affected by the current water balance of the forest biome (Goulden *et al.*, 1997), as well as by the nutrient status of the system (Valentini *et al.*, 2000). Despite this, the focus in many projects is on the collection and interpretation of the flux data alone;[1] there are relatively few sites where the flux measurements are supplemented by measurements of the various component fluxes and processes. Where possible these should include CO_2 exchange from soils, roots, litter, stems, branches and foliage, as well as frequently monitored changes in soil water content, the rates of liquid water movement through the vascular system of vegetation, and foliage measurements of gas exchange, foliar nutrition and plant water relations. These provide information useful in identifying changes in sources and sinks over monthly, annual and inter-annual cycles. To project forward or to assess past conditions, process-based ecosystem models that combine an adequate amount of structural complexity with sound functional relations between various pools and the environment must be employed (e.g. the BIOME-BGC model used by Thornton *et al.* (1997) and Law *et al.* (2001a) and the model of Liu *et al.* (1999)). Results from these can provide independent estimates of fluxes and carbon storage. With agreement between measured and process-modelled fluxes we would be in a better position to evaluate the impact of past, current and future alterations in the environment. For integration over long periods or large areas, relatively simple models, driven by readily available input data, are required.

In this chapter we draw attention to the progress made in comparisons between measured and process-modelled fluxes, and illustrate the potential of a simple process-based model (3-PG; Landsberg and Waring, 1997) as a tool for spatial and temporal extrapolation of carbon-balance estimates for forested areas. This model can be calibrated against biomass inventory data. It demonstrates the relationship between soil nutritional status and the efficiency with which forest canopies utilize radiant energy, provides some insight into the reasons for site-to-site variation in flux measurements and is a tool that can be used to explain variations in flux measurements and to extrapolate the results spatially and temporally.

The 3-PG Model: Structure and Characteristics

The 3-PG model, originally developed by Landsberg and Waring (1997) and updated by Sands and Landsberg (2002), is a deliberate attempt to bridge the gap between conventional, mensuration-based growth and yield and process-based carbon-balance models. It is a generalized stand model, applicable to plantations or even-aged, relatively homogeneous forests. 3-PG is not site- or species-specific, but may need to be parameterized for individual species. The output variables that it produces are those of interest to foresters, but, since it is a conservation-of-mass model, which produces information about – or takes account of – all the major processes involved in the carbon balance of forest stands, it has considerable potential as a tool to bridge the gap between flux and physiological measurements and inventory measurements. The model uses a time step of a month, considerably longer than the intervals that flux measurements deal with and longer than the daily time step employed by many process-based carbon-balance models (e.g. BIOME-BGC and the model of Liu *et al.*, 1999). There are advantages and disadvantages associated with this time step (see Coops *et al.* (2001) for a discussion), but one of the major advantages, in the context of the present discussion, is that 3-PG provides, in effect, integrated values of carbon fluxes and storage over monthly intervals. Since the model can be calibrated against standard, commonly available, measurements of forest growth, the estimates of carbon storage and loss are constrained by the measured values, and the non-measured components are therefore unlikely to be seriously in error.

3-PG consists of five simple submodels that deal with: (i) the assimilation of carbohydrates and calculation of net primary production (NPP); (ii) the distribution of biomass to roots – the amount depending on growing conditions – then to foliage and stems; (iii) the determination of stem number; (iv) the conversion of biomass values into variables of interest to foresters; and (v) the calculation of an ecosystem water balance. The state of the stand is updated at each monthly time step. The model can be run for any number of years, using actual weather data for each month or monthly averages for many years. Using averages is the normal procedure unless there is particular interest in specific events, such as droughts.

Biomass production is determined by applying an effective canopy quantum efficiency (QE, denoted α_c, mol C/mol photons) value to the photosynthetically active radiation (PAR) absorbed by a stand (APAR): total canopy photosynthesis GPP = α_c APAR. QE is determined from a maximum value (α_o) constrained by atmospheric vapour-pressure deficit (f_{VPD}), through its effects on stomatal conductance, and hence canopy conductance, as well as by air temperature, frost, water balance (f_{SW}) and nutrition. NPP is calculated from a simple ratio of NPP to GPP (0.47 ± SE 0.04), implying that autotrophic respiration for construction and maintenance is a fixed fraction of gross photosynthesis.

The assumption of a constant NPP/GPP emerged from ecosystem analysis of the productivity of temperate forests on annual time steps (Waring *et al.*, 1998). The finding has stimulated some controversy (Medlyn and Dewar, 1999) and some research aimed at testing it (Mäkelä and Valentine, 2001), but appears to be well founded (Landsberg and Gower, 1997: Fig. 5.7; Law *et al.*, 1999b, 2000b; Malhi *et al.*, 1999). The ratio does not appear to vary with the nutritional status of the vegetation, but nutrition affects effective canopy QE (see later).

Although 3-PG initially calculates GPP, it is in fact one of the family of epsilon (ε) models, which utilize the proportionality, first clearly demonstrated by Monteith (1977), between APAR and above-ground NPP (see Landsberg and Gower, 1997; Landsberg *et al.*, 1997; Gower *et al.*, 1999). Ruimy *et al.* (1994) and Gower *et al.* (1999) have provided extensive lists of the values of ε, characterized by considerable variation – from 0.2 to 2.5 g biomass/MJ APAR for forests. The advantage of using α instead of ε is that it allows direct comparison with detailed models or measurements of canopy photosynthesis (see Wang *et al.*, 1992; Raulier *et al.*, 1999) and also allows us to account for variations in above-ground productivity caused by differences in below-ground allocation of carbon that result from different growing conditions.

The maximum value of the canopy QE used in the original version of the model (Landsberg and Waring, 1997) was taken to be 0.03 mol C/mol photons, equivalent to 1.68 g C/MJ PAR, on the basis of work by Waring *et al.* (1995) at Harvard Forest, a deciduous forest in north-west Massachusetts. Law *et al.* (2000b) used 0.04 for *Pinus ponderosa* and Bond *et al.* (1999) reported values of 0.05 for *Pseudotsuga menziesii* and other highly productive conifers native to the Pacific North-west region of the USA. An analysis by Landsberg and Hingston (1996) indicated that *Eucalyptus* plantations in Western Australia were producing above-ground biomass at rates that indicated radiation conversion efficiency of up to 0.05 mol C/mol quanta or 2.73 g C/MJ APAR – approaching the value that seems to be about the norm for intensively managed field crops (see Landsberg *et al.*, 1997). Correcting for (estimated) below-ground production, converting the biomass to carbon and allowing for litter-fall indicated that values of $\alpha_c \geq 0.07$ mol C/mol quanta would not be unrealistic. This is taken as the upper allowable limit (α_o) for calibrating 3-PG (see Leverenz and Öquist, 1987; Sands and Landsberg, 2002). It is confirmed by recent experience, using 3-PG to analyse the growth of hybrid plantation eucalypt in Brazil (A. Almeida, Aracruz Celulose, 2001, personal communication). Based on these values we generally assume $\alpha_o = 0.07$ mol C/mol quanta.

Site nutritional status is represented by a fertility rating (FR) that takes values between 0 and 1. FR depends on soil fertility, although it is generally difficult to establish any quantitative relationship between soil chemical characteristics and tree growth: there is some dependence on expert knowledge about sites and regions to establish values for FR. Based on published information, reviewed by Ågren and Ingestad (1987), Landsberg and Gower (1997: p. 137) and Linder and Murray (1998), we have made α_c a linear function of FR. The equation used is

$$\alpha_c = \alpha_o f_N = \alpha_o (f_{N0} + (1 - f_{N0})\, FR) \qquad (1)$$

where $\alpha_o f_{N0}$ gives the minimum value of QE at very low fertility. The usual default value of f_{N0} is 0.5. It seems unlikely that the relationship is truly linear; we are currently collecting available information about canopy photosynthesis in relation to foliage nutrient status, with a view to establishing it more firmly.

Site water (SW) balance is calculated monthly using the Penman–Monteith equation to calculate transpiration. Only the smaller value of f_{VPD} and f_{SW} is used in the calculation of α_c.

Carbohydrate allocation is on a single-tree basis. The coefficient that determines allocation to roots is calculated first, with the proportion of NPP going to roots being influenced by moisture relations and soil nutrition. Allocation to stems and foliage relies on the ratio of the derivatives of the allometric equations describing leaf and stem mass in terms of stem diameter at 'breast' height (see Landsberg and Waring, 1997; Sands and Landsberg, 2002; Landsberg *et al.*, 2002). The carbon allocation procedure allows leaf area index (L*) to vary with growing conditions, but forces the allocation of carbon to follow patterns that lead to realistic structures. Initial tree populations are specified and changes in stem populations caused by natural mortality are calculated using the well-established −3/2 power law. Stem populations may also be changed by a thinning subroutine in the model.

Calibration of the 3-PG model

3-PG is calibrated by fitting to individual sets of observational data: in general, biomass production is determined by QE, FR, weather conditions and soil moisture-holding capacity at the site. With average monthly weather data relevant to the site and the best available estimates of soil water-holding capacity and fertility, specific leaf area (SLA, m^2/kg) and litter-fall as inputs, an iterative procedure is used to optimize parameter values and match model output to field observations. Time series of observations are preferable for calibration, but the model can be calibrated to reproduce a single set of measurements made at some specified age.

A number of parameter values can be varied to alter the output of the model – for example litter-fall rates, the relationship between carbohydrate allocation to roots and growing conditions – but the normal procedure is to use standard default or the best available empirical values for as many parameters as possible; if too much flexibility is allowed, it becomes very difficult to identify the factors that are determining growth and yield differences, little is learned and we have little guidance about parameter values to use in the model when predictive calculations are to be made. Successive adjustments of parameter values between model runs lead, in most cases, to good fits between observed and simulated variable values. The more observed/measured data available, the more precise will be the calibration. For example, if the time course of L* is available or, failing that, a reliable estimate of the maximum L* and the age at which it was produced, then clearly the maximum value of L* produced by 3-PG must correspond closely to this value. Similarly, the values of stem diameter, stem mass and stand volume produced by the model and the time course of change in the values of these variables must correspond to observation. In the rare case that root mass values are also available, these provide an additional constraint. Calibration of the soil water balance against measurements of soil water content provides the opportunity to adjust the parameter controlling the rate of change of stomatal conductance with VPD (see Ewers *et al.*, 2000).

Figure 4.1 shows results of calibrating 3-PG against two of the treatments in the biology of forest growth (BFG) experiment carried out on *Pinus radiata* in the Australian Capital Territory during the 1980s (see Special Issue, *Forest Ecology and Management* 52 (1–4), 1992). The 3-PG software includes allowance for specified thinning, for changes in FR at any time and for irrigation. α_o was 0.065 mol C/mol photons and in this case we used $f_{N0} = 0.6$. The initial stem population was 2500, reduced to 1200 after 5 years and to about 700 before treatments were applied in year 9. The FR values used in the calibration were 0.25 until the application of the treatments in 1983; then FR = 1 for the IL treatment, which was irrigated with liquid fertilizer to remove all constraints to growth by nutrients or water. FR for the control treatment was 0.5 after thinning – not 0.25; this was, presumably, because of the stimulus to the growth of the remaining trees from thinning. The changes in FR between the IL and control treatments resulted in $f_N\alpha_o = 0.052$ for the control. The other major cause of different growth rates was, of course, water stress.

The lines in Fig. 4.1 derived from the model provide a clear illustration not only of the accuracy with which 3-PG can be calibrated but, more relevant in this context, of the importance of nutrition in determining carbon sequestration by forest stands. Flux measurements would reflect such differences but could provide no explanation and no basis for extrapolation.

Comparisons of Process-based Simulations with Integrated Flux Measurements

Results from an Ameri-flux set of old and young ponderosa pine stands in Oregon provide examples of the insights obtainable when models and supplemental measurements are used to explain daily, seasonal and annual variation in eddy-covariance measurements. At the two sites, several models have been compared with direct measurements acquired over a 5-year period (Thornton *et al.*, 1997; Law *et al.*, 1999a,b, 2000a,b, 2001a,b). These comparisons have:

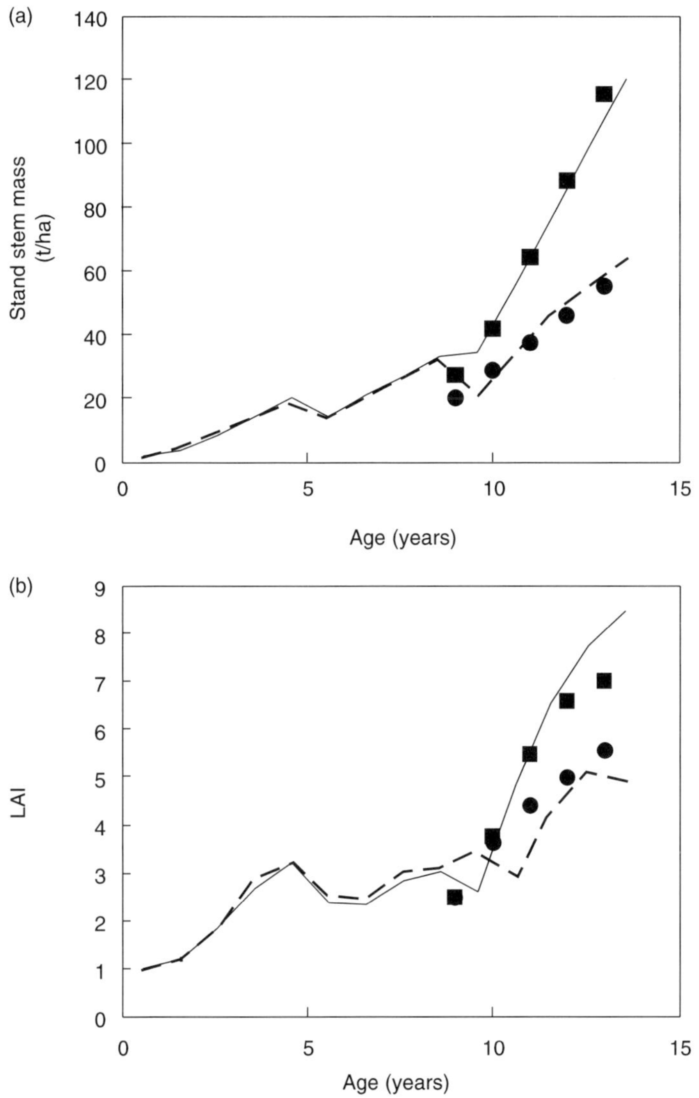

Fig. 4.1. (a) Stem dry mass in the plots irrigated with liquid fertilizer (IL; square symbols, upper line) and the control plots (spherical symbols, lower line) in the biology of forest growth (BFG) experiment at Pierce's Creek in the Australian Capital Territory (ACT). The points are values determined by destructive sampling; the lines are simulations using 3-PG after calibration against the IL treatment. (Data are from Snowdon and Benson, 1992.) (b) Leaf area index (LAI) in the IL and control plots in the BFG experiment. The lines are 3-PG output from the same calibration that produced (a).

- demonstrated that high-quality environmental data required to run process models can be generated from a grid of normal weather stations and topographic maps to predict daily temperatures within 1°C, daily radiation within 2.5 MJ/m^2/day and precipitation within 10–15%/year;

- confirmed the assumption that NPP/GPP approaches a constant ($\approx$ 0.5) that holds for young and old forests;
- defined how photosynthetic capacity changes with suboptimal temperatures and how stomatal conductance responds to frost, VPD and drought;
- explained how transpiration rates could remain constant while GEP varied by > 50%;
- indicated through simulations that a 20-year-old forest was unable to tap the water resources available to older forests with deeper root systems;
- identified strengths and weaknesses in model structure and functions;
- suggested specific times during the year when a few measurements would constrain estimates of soil rooting depth and provide an evaluation of how soil fertility affects photosynthetic capacity.

The 3-PG model was included in the tests at the Ameri-flux sites (Law *et al.*, 2000a), but here we illustrate its use in association with integrated flux measurements for Sitka spruce made at the Griffin site (Perthshire, Scotland; lat. 57°N), provided by Clement *et al.* (2001). They give annual net carbon fixed (net ecosystem exchange (NEE)) for 1997–2000, inclusive, as 6.38, 6.21, 6.15, 5.33 t C/ha/year (average = 6). NEE is given by: GPP − R_t, where total respiration, R_t, is the sum of autotrophic (R_a) and heterotrophic (R_h) respiration; i.e. $R_t = R_a + R_h$, where $R_a = R_{fol} + R_{wood} + R_{roots}$. We simulated the growth of Sitka spruce at the site, using information collected by Waring (2000) for the Griffin site. This is not a calibration, in the sense that we had no precise data for the site, but we matched the output of 3-PG to the information we had available (Table 4.1).

In the 3-PG calculations we assume R_a = 0.47 GPP, so R_a is calculated in the model, and NEE = 0.53 GPP − R_h. Using the data in Table 4.1, including branch + bark biomass in stems, and net root growth and assuming that R_h = 2 × litter-fall gives NEP ≈ 6 t C/ha/year, consistent with the flux measurements. The assumption that $R_h \approx 2 \times$ litter-fall in effect assumes that litter-fall and decomposition rates are in equilibrium and provide about half R_h, the balance coming from what Malhi *et al.* (1999) call below-ground detritus (fine-root turnover, root mortality and exudation), fed by transfer of carbon from foliage to the root systems (see Högberg *et al.*, 2001).

Table 4.1. Assumed stand characteristics and simulated growth of Sitka spruce at the Griffin flux site in Perthshire, Scotland, at 22–23 years old. Initial population was taken to be 2000 stems/ha, thinned to 1200 in year 20. The simulation was run using average monthly weather conditions for the site, α_o = 0.055 mol C/mol photons, FR ≈ 0.3 (so α_c= 0.04) and wood density = 300 kg/m^3.

Standing volume (m^3)	Stem biomass (t/ha)	Root biomass (t/ha)	LAI
327	126	48	8
MAI (m^3/ha/year)	Stem growth rate (t/ha/year)	Δ foliage mass (t/ha/year)	Litter-fall (t/ha/year)
16	13	0.8	1.5

LAI, leaf area index.

The exact correspondence – or lack of it – between the 3-PG outputs and flux measurements is not at issue in this exercise, which serves to illustrate that the 3-PG model, if calibrated against mensurational data from a site, can provide independent estimates of the carbon fluxes at a site. Given good correspondence between model output and integrated flux measurements, it also provides a tool with which to extend the results of flux measurements over wide areas, where it can account for periods of drought and differences in site nutrient status.

Scaling to Landscapes

If models are based on sound principles, they can be extrapolated spatially over large areas, although for this purpose model complexity must be reduced. It is at the landscape and regional level that land-use and climatic change can be best evaluated and coupled with global circulation models. Coops *et al.* (2001) showed that leaf area index could be estimated with useful accuracy using satellite measurements, and Means *et al.* (1999) have estimated growth and standing biomass using scanning airborne Lidar. Remote-sensing technology can provide independent estimates of the accuracy of model predictions of climate: Goward *et al.* (1994) and Prince and Goward (1995), using satellite-borne sensors, demonstrated the possibility of estimating solar radiation and ambient air temperatures with surprising accuracy. They also used satellites to quantify seasonal drought and variations in the canopy light absorption. Instruments are also available that measure canopy nitrogen content remotely (Matson *et al.*, 1994; Gitelsen and Merzlyak, 1997; Martin and Aber, 1997), but at present no satellite carries such sensors.

One of the goals in the development of 3-PG was that many of the driving variables, key parameters and predicted outputs should be assessable by remote-sensing technology. 3-PG has been used with monthly weather and satellite data to estimate the productive capacity of forest lands in Australia and New Zealand (Coops *et al.*, 1998a,b; White *et al.*, 2000) and the USA (Coops and Waring, 2001). By assuming that 3-PG can provide a reasonable estimate of the maximum potential productivity in a given region, and then reducing the fertility ranking to match above-ground estimates of periodic annual increment made using Lidar sensors (Lefsky *et al.*, 1997, 1999; Means *et al.*, 1999), it is possible that the model could be used to derive estimates of soil fertility across large areas.

Discussion

A number of points emerge from this evaluation of the use of ecosystem models, and particularly the 3-PG model, to provide comparisons with flux measurements.

It is well known that photosynthetic efficiency varies with the nutritional status of the foliage; incorporation of this knowledge into 3-PG leads to the need to link soil nutrient status to canopy QE – the efficiency with which the canopy utilizes PAR. Fitting the model to experimental data, with this constraint (Fig. 4.1), illustrates how site nutrition affects α_c and provides a partial explanation for the variation in ε values noted earlier. Gower *et al.* (1999) commented, in passing, that differences in soil nutrition cause changes in the proportion of NPP allocated below ground, but Ruimy *et al.* (1994) did not mention nutrition, and neither paper noted the influence of the soil water balance on ε. Clearly, nutrition and soil water affect NEE and should be considered explicitly in the analysis and interpretation of flux data. Using a model such as 3-PG, which includes allowance for the effects of these factors, calibrated for some flux-measurement sites, to analyse the growth patterns and responses to environmental factors of forest stands, will provide the insights needed for accurate interpretation and extrapolation of results from flux measurements.

A series of papers based on flux measurements at two ponderosa pine sites in the western USA, supported by a comprehensive range of supplemental measurements (Law *et al.*, 1999a,b, 2000a,b, 2001a,b), has shown how combination of the two approaches to forest carbon and water-balance estimation results in significantly improved ability to interpret flux measurements, as well as providing the data needed to test and refine process-based models. When the 3-PG model is 'tuned' to produce the biomass growth patterns observed at a particular site (Table 4.1), the model output can be interpreted in terms of the various components of NEE. Because the model includes parameters and procedures that account for litter-fall and fine-root turnover, it is possible, with some assumptions, to estimate R_h from surface soil layers and the rooting zone, and so calculate NEE (see also Waring and McDowell, 2002).

Flux measurements are made at sites that represent, as far as possible, large areas of particular vegetation types (biomes), so that the role of these biomes as carbon sinks and the rate at which they exchange CO_2 with the atmosphere can be determined. However, there is always a major problem arising from the fact that there is considerable variation across biomes, caused not only by disturbance but also by variations in topography, in soil fertility and water-holding capacity and in weather conditions. To account for this variation requires that extrapolation in time and space be done using process-based models and that these models be driven by variables measurable by satellites, which allow repeated coverage of large areas of the earth. We note progress in this area, not only in relation to the 3-PG model, but also in relation to the estimation of standing biomass and possible evaluation of soil fertility, using satellite measurements and process-based modelling.

Note

[1] We recognize that the reason for this is usually a matter of resource availability, including the availability of appropriate scientific expertise.

References

Ågren, G.I. and Ingestad, T. (1987) Root:shoot ratio as a balance between nitrogen productivity and photosynthesis. *Plant, Cell and Environment* 10, 579–586.

Bond, B., Farnsworth, B.T., Coulombe, R.A. and Winner, W.E. (1999) Foliage physiology and biochemistry in response to light gradients in conifers with varying shade tolerance. *Oecologia* 120, 183–192.

Clement, R., Moncrieff, J.B. and Jarvis, P.G. (2001) Long term C-sequestration by Sitka spruce in Scotland. Internal document, University of Edinburgh, Edinburgh, UK.

Coops, N.C. and Waring, R.H. (2001) Estimating forest productivity in the Eastern Siskiyou Mountains of southwestern Oregon using a satellite driven process model, 3-PGS. *Canadian Journal of Forest Research* 31, 143–154.

Coops, N.C., Waring, R.H. and Landsberg, J.J. (1998a) Assessing forest productivity in Australia and New Zealand using a physiologically-based model driven with averaged monthly weather data and satellite-derived estimates of canopy productivity. *Forest Ecology and Management* 104, 113–127.

Coops, N.C., Waring, R.H. and Landsberg, J.J. (1998b) The development of a physiological model (3-PGS) to predict forest productivity using satellite data. In: Nabuurs, G., Nuutinen, T., Bartelink, H. and Koorhonen, M. (eds) *Forest Scenario Modelling for Ecosystem Management at Landscape Level.* EFI Proceedings, Joensuu, Finland, pp. 173–191.

Coops, N.C., Waring, R.H., Brown, S.R. and Running, S.W. (2001) Comparisons of predictions of net primary production and seasonal patterns in water use derived with two forest growth models in South-West Oregon. *Ecological Modeling* 142, 61– 81.

Dixon, R.K., Brown, S., Houghton, R.A., Solomon, A.M., Trexler, M.C. and Wisniewski, J. (1994) Carbon pools and fluxes of global forest ecosystems. *Science* 263, 185–190.

Ewers, B.E., Oren, R., Johnsen, K.H. and Landsberg, J.J. (2000) Estimating maximum mean canopy stomatal conductance for use in models. *Canadian Journal of Forest Research* 31, 198–207.

Gitelsen, A.A. and Merzlyak, M.N. (1997) Remote sensing of chlorophyll concentration in higher plant leaves. *International Journal of Remote Sensing* 18, 2691–2697.

Goulden, M.L., Munger, J.W., Fan, S., Daube, B.C. and Wofsey, S.C. (1996) Exchange of carbon dioxide by a deciduous forest: responses to interannual climate variability. *Science* 271, 1576–1578.

Goulden, M.L., Daube, B.C., Fan, S., Sutton, D.J., Bazzaz, A., Munger, J.W. and Wofsy, S.C. (1997) Physiological responses of a black spruce forest to weather. *Journal of Geophysical Research* 102, 28987–28996.

Goward, S.N., Waring, R.H., Dye, D.G. and Yang, J. (1994) Ecological remote sensing at OTTER: satellite macroscale observations. *Ecological Applications* 4, 322–343.

Gower, S.T., Kucharik, C.J. and Norman, J.M. (1999) Direct and indirect estimation of leaf area index, FAPAR, and net primary production of terrestrial ecosystems. *Remote Sensing of Environment* 70, 29–51.

Högberg, P., Nordgren, A., Buchmann, N., Taylor, A.F.S., Ekblad, A., Högeberg, N., Nyberg, G., Ottosson-Löfvenius, M. and Read, D.J. (2001) Large-scale forest girdling shows that current photosynthesis drives soil respiration. *Nature* 411, 789–792.

Jarvis, P.G., Massheder, J.M., Hale, S.E., Moncrieff, J.B., Rayment, M. and Scott, S.L. (1997) Seasonal variation of carbon dioxide, water vapour, and energy exchanges of a boreal black spruce forest. *Journal of Geophysical Research* 102, 28953–28966.

Landsberg, J.J. and Gower, S.T. (1997) *Applications of Physiological Ecology to Forest Management.* Academic Press, San Diego, 354 pp.
Landsberg, J.J. and Hingston, F.J. (1996) Evaluating a simple radiation/dry matter conversion model using data from *Eucalyptus globulus* plantations in Western Australia. *Tree Physiology* 16, 801–808.
Landsberg, J.J. and Waring, R.H. (1997) A generalized model of forest productivity using simplified concepts of radiation-use efficiency, carbon balance and partitioning. *Forest Ecology and Management* 95, 209–228.
Landsberg, J.J., Prince, S.D., Jarvis, P.G., McMurtrie, R.E., Luxmoore, R. and Medlyn, B.E. (1997) Energy conversion and use in forests: an analysis of forest production in terms of radiation utilisation efficiency. In: Gholz, H.L., Nakane, K. and Shimoda, H. (eds) *The Use of Remote Sensing in the Modeling of Forest Productivity at Scales from the Stand to the Globe.* Kluwer Academic Publishers, Dordrecht, pp. 273–298.
Landsberg, J.J., Waring, R.H. and Coops, N. (2003) Performance of the forest productivity model 3-PG applied to a wide range of forest types. *Forest Ecology and Management* 172, 199–214.
Law, B.E., Baldocchi, D.D. and Anthoni, P.M. (1999a) Below-canopy and soil CO_2 fluxes in a ponderosa pine forest. *Agricultural and Forest Meteorology* 94, 171–188.
Law, B.E., Ryan, M.G. and Anthoni, P.M. (1999b) Seasonal and annual respiration of a ponderosa pine ecosystem. *Global Change Biology* 5, 169–182.
Law, B.E., Waring, R.H., Anthoni, P.M. and Aber, J.D. (2000a) Measurements of gross and net ecosystem productivity and water vapour exchange of a *Pinus ponderosa* ecosystem, and an evaluation of two generalized models. *Global Change Biology* 6, 155–168.
Law, B.E., Williams, M., Anthoni, P.M., Baldocchi, D.D. and Unsworth, M.H. (2000b) Measuring and modelling seasonal variation of carbon dioxide and water vapour exchange of a *Pinus ponderosa* forest subject to soil water deficit. *Global Change Biology* 6, 613–630.
Law, B.E., Thornton, P.E., Irvine, J., Anthoni, P.M. and Van Tuyl, S. (2001a) Carbon storage and fluxes in ponderosa pine forests at different developmental stages. *Global Change Biology* 7, 755–777.
Law, B.E., Goldstein, A.H., Anthoni, P.M., Unsworth, M.H., Panek, J.A., Bauer, M.R.J., Fracheboud, M. and Huiltman, N. (2001b) Carbon dioxide and water vapour exchange by young and old ponderosa pine ecosystems during a dry summer. *Tree Physiology* 21, 299–308.
Lefsky, M.A., Cohen, W.B., Acker, S.A., Spies, T.A., Parker, G.G. and Harding, D. (1997) LIDAR remote sensing of forest canopy structure and related biophysical parameters at the H.J. Andrews Experimental Forest, Oregon, USA. In: Greer, J.D. (ed.) *Natural Resources Management Using Remote Sensing and GIS.* ASPRS, Washington, DC.
Lefsky, M.A., Harding, D., Cohen, W.B., Parker, G. and Shugart, H.H. (1999) Surface lidar remote sensing of basal area and biomass in deciduous forests of eastern Maryland, USA. *Remote Sensing of Environment* 67, 83–98.
Leverenz, J. and Öquist, G. (1987) Quantum yields of photosynthesis at temperatures between −2°C and 25°C in a cold-tolerant C_3 plant (*Pinus sylvestris*) during the course of one year. *Plant, Cell and Environment* 10, 287–295.
Linder, S. and Murray, M. (1998) Do elevated CO_2 concentrations and nutrients interact? In: Jarvis, P.G. (ed.) *European Forests and Global Change.* Cambridge University Press, Cambridge, pp. 215–235.

Liu, J., Chen, J.M., Cihlar, J. and Chen, W. (1999) Net primary productivity distribution in the BOREAS region from a process model using satellite data and surface data. *Journal of Geophysical Research* 104, 27735–27754.

Mäkelä, A. and Valentine, H.T. (2001) The ratio of NPP to GPP: evidence of change over the course of stand development. *Tree Physiology* 21, 1015–1030.

Malhi, Y., Baldocchi, D. and Jarvis, P.G. (1999) The carbon balance of tropical, temperate and boreal forests. *Plant, Cell and Environment* 22, 715–740.

Martin, M.E. and Aber, J.D. (1997) High spectral resolution remote sensing of forest canopy lignin, nitrogen, and ecosystem processes. *Ecological Applications* 7, 431–443.

Matson, P., Johnson, L., Billow, C., Miller, J. and Pu, R. (1994) Seasonal patterns and remote spectral estimation of canopy chemistry across the Oregon transect. *Ecological Applications* 4, 280–298.

Means, J.E., Acker, S.A., Harding, D.J., Blair, J.B., Lefsky, M.A., Cohen, W.B., Harmon, M.E. and McKee, W.A. (1999) Use of large-footprint scanning airborne Lidar to estimate forest stand characteristics in the western Cascades of Oregon. *Remote Sensing of Environment* 67, 298–308.

Medlyn, B.E. and Dewar, R.C. (1999) Comment on the article by Waring, R.H., Landsberg, J.J. and Williams, M. relating net primary production to gross primary production. *Tree Physiology* 19, 137–138.

Moncrieff, J.B., Massheder, J.M., De Bruin, H., Elbers, J., Friborg, T., Heusinkveld, B., Kabat, P., Scott, S., Soegaard, H. and Verhoef, A. (1997) A system to measure surface fluxes of sensible heat, water vapour and carbon dioxide. *Journal of Hydrology* 188–189, 589–611.

Monteith, J.L. (1977) Climate and the efficiency of crop production in Britain. *Philosophical Transactions of the Royal Society* B281, 277–294.

Prince, S.D. and Goward, S.N. (1995) Global primary production: a remote sensing approach. *Journal of Biogeography* 22, 815–835.

Raulier, F., Bernier, P.Y. and Ung, C. (1999) Canopy photosynthesis of sugar maple (*Acer saccharum*): comparing big-leaf and multilayer extrapolations of leaf level measurements. *Tree Physiology* 19, 407–420.

Ruimy, A., Saugier, B. and Dedieu, G. (1994) Methodology for the estimation of terrestrial net primary production from remotely sensed data. *Journal of Geophysical Research* 99, 5263–5283.

Sands, P.J. and Landsberg, J.J. (2002) Parameterisation of 3-PG for plantation grown *Eucalyptus globulus*. *Forest Ecology and Management* 163, 273–292.

Snowdon, P. and Benson, M.L. (1992) Effects of combinations of irrigation and fertilisation on the growth and above-ground biomass production of *Pinus radiata*. *Forest Ecology and Management* 52, 87–116.

Thornton, P.E., Running, S.W. and White, M.A. (1997) Generating surfaces of daily meteorological variables over large regions of complex terrain. *Journal of Hydrology* 190, 214–251.

Valentini, R., Matteucci, G., Dolman, A.J., Schulze, E.-D., Rebmann, C., Moor, E.J., Granier, A., Gross, P., Jensen, N.O., Pilegaard, K., Lindroth, A., Greele, A., Bernhofer, C., Grunwald, T., Aubinet, M., Ceulemans, R., Kowalski, A.S., Vesala, T., Rannik, U., Berbigier, P., Loustau, D., Gudmundsson, J., Thorgeirsson, H., Ibrom, A., Morgenstern, K., Clement, R., Moncrieff, J., Montagnani, L., Minerbi, S. and Jarvis, P.G. (2000) Respiration as the main determinant of carbon balance in European forests. *Nature* 404, 861–865.

Wang, Y.-P., McMurtrie, R.E. and Landsberg, J.J. (1992) Modelling canopy photosynthetic productivity. In: Baker, N.R. and Thomas, H. (eds) *Crop Photosynthesis: Spatial and Temporal Determinants*. Elsevier Science Publishers, Amsterdam, pp. 43–67.

Waring, R.H. (2000) A process model analysis of environmental limitations on growth of Sitka spruce plantations in Great Britain. *Forestry* 73, 65–79.

Waring, R.H. and McDowell, N. (2002) Using a physiological process model with forestry yield tables to set limits on annual carbon balances. *Tree Physiology* 22, 179–188.

Waring, R.H., Law, B.E., Goulden, M.L., Bassow, S.L., McCreight, R.W., Wofsy, S.C. and Bazzaz, F.A. (1995) Scaling gross ecosystem production at Harvard Forest with remote sensing: a comparison of estimates from a constrained quantum-use efficiency model and eddy correlation. *Plant Cell and Environment* 18, 1201–1213.

Waring, R.H., Landsberg, J.J. and Williams, M. (1998) Net primary production of forests: a constant fraction of gross primary production. *Tree Physiology* 18, 129–134.

White, J.D., Coops, N.C. and Scott, N.A. (2000) Estimates of New Zealand forest and scrub biomass from the 3-PG model. *Ecological Modeling* 131, 175–190.

5

The Effects of Forests on Mesoscale Atmospheric Processes

A.J. Dolman,[1] M.K. van der Molen,[1] H.W. ter Maat[2] and R.W.A. Hutjes[2]

[1]*Department of Hydrology and Geo-Environmental Sciences, Faculty of Earth and Life Sciences, Vrije Universiteit, De Boelelaan 1085, 1081 HV Amsterdam, The Netherlands;* [2]*Alterra, Green World Research, PO Box 47, 6700 AC Wageningen, The Netherlands*

Introduction

There is a large body of evidence, primarily from modelling studies, which suggests that forests have a noticeable impact on the atmospheric boundary layer and precipitation patterns. Pitman *et al.* (1999) collected evidence for effects of land-surface properties on the atmosphere at a range of temporal and spatial scales, and concluded that most of this evidence is from modelling studies at global scales. Indeed, most of the evidence of forest–atmosphere interaction comes from global circulation model (GCM) modelling studies, in which one type of vegetation, for instance, grassland, is replaced by another, for instance, forest.

The most extreme cases are those where, globally, the land surface is transformed from one single type to another. Fraedrich *et al.* (1999) performed two 10-year integrations of a GCM where in one simulation the global land surface was bare (desert) and in the other the surface was completely covered by forests. This arguably extreme sensitivity study caused a doubling of continental precipitation, a three times as high evaporation from the land and a 25% increase of freshwater flow from the continents into the ocean (runoff). This was accompanied by a decrease in surface temperature for the forested run and an increase in tropospheric warming as a consequence of enhanced convective activity in the higher layers of the atmosphere. Although these results probably present the most extreme case of land-use change, they clearly suggest important potential for the role of forest in the global climate and hydrological cycle.

More recently, Betts (2000) described a feedback of forest albedo on climate that could potentially offset the beneficial effect of forests as terrestrial sinks for CO_2. Betts found that planting forests in the boreal zone could lead to significant warming due to their low albedo and snow-masking effect in high-latitude zones. Such a warming could offset the cooling effects associated with reducing atmospheric CO_2 levels.

Forests have a number of physical characteristics that are different from those of vegetation of lower density and height. Useful reviews can be found in Jarvis *et al.* (1976, 1989), Monteith (1976), and, more recently, in Dolman *et al.* (2003). In general, forests have high leaf-area densities, typically more than 3. They are also tall, and this may be related to the ability to carry more leaf area per unit ground surface. These two primary characteristics, the tallness and leaf-area density, generate aerodynamic and radiative properties of forests that are radically different from lower vegetations, such as grassland and agricultural crops. The distribution of leaf area with height and the spectral characteristics and orientation of leaves all affect the transmission and reflectivity of forests. In other words, they are more effective in capturing short-wave radiation than lower vegetation, as is shown by a comparison of albedos of forest and other vegetation types (e.g. Oke, 1978).

An important effect of forest on the atmosphere is enhanced recycling of water through evaporation, i.e. the atmospheric moisture–precipitation chain. These are processes that occur at an intermediate, regional, spatial scale, so considerable attention has been given to the effects of forest on regional-scale (mesoscale) atmospheric processes. Indeed, the mesoscale studies in particular highlight the primary feedbacks and responses of forest–atmosphere interaction very well.

The effects of forests on mesoscale atmospheric processes are again primarily documented by modelling studies, although more recently quantitative analysis of remote-sensing images has helped to elucidate some of the primary mechanisms (e.g. Cutrim *et al.*, 1995; Rabin and Martin, 1996). Cutrim *et al.* (1995) and Rabin and Martin (1996) found increased cumulus convection and cloudiness over cleared areas along the main road, the BR-364 in Rondônia, west Amazonia, later during the day, indicating that the changes in radiation and energy balances associated with deforestation had a marked effect on the atmosphere. Pielke *et al.* (1998), in an extensive review of land–surface–atmosphere interactions, list other studies.

Mesoscale atmospheric models have improved considerably over the last two decades so that they now incorporate realistic descriptions of land-surface processes and the atmospheric boundary layer (e.g. Pielke, 1984; André *et al.*, 1989). They are based on the standard equations of mass and energy conservation and dynamics of flow that form the core of all weather prediction and climate models (e.g. Pielke, 1984). Mesoscale modelling studies have been particularly useful in studying the effect of regional-scale land-surface changes on the atmosphere. They provide additional insight into this interaction at the scale at which most of the land-surface changes are happening. Unlike global studies, which often simulate rather ideal, extreme conditions, they can be used to study the real changes that occur in human-made landscapes. A further advantage is that local-scale flux and boundary-layer measurements can be used to validate the performance. This is arguably more complicated at the global scale.

In the remainder of this chapter we shall concentrate on the mesoscale effects of forest on the atmosphere, in both dry and wet conditions. First we shall outline some of the main feedback mechanisms. In principle, one-dimensional studies, such as those by De Bruin and Jacobs (1989), provide first-order evidence of the impact of forest on the atmosphere, so the chapter then briefly reviews such studies. They present the simplest case. Two- and three-dimensional studies provide further realism by incorporating lateral advection. We shall first give an overview of studies investigating the feedback in simplified forms and then show some more realistic case-studies. The chapter will end by highlighting a novel use of mesoscale models in the context of atmospheric carbon dioxide modelling and we set a research agenda for mesoscale models.

Mesoscale Experimental and Modelling Evidence

Feedbacks and processes

Figure 5.1 shows a by now classic example of the effects of forests on mesoscale cloud formation. The picture was taken on 16 June 1986, over the area in south-west France that included the HAPEX-MOBILHY experiment (André *et al.*, 1986). Clearly, over the forested area of Les Landes in the west, preferential cloud development can be seen to occur. This is related to the

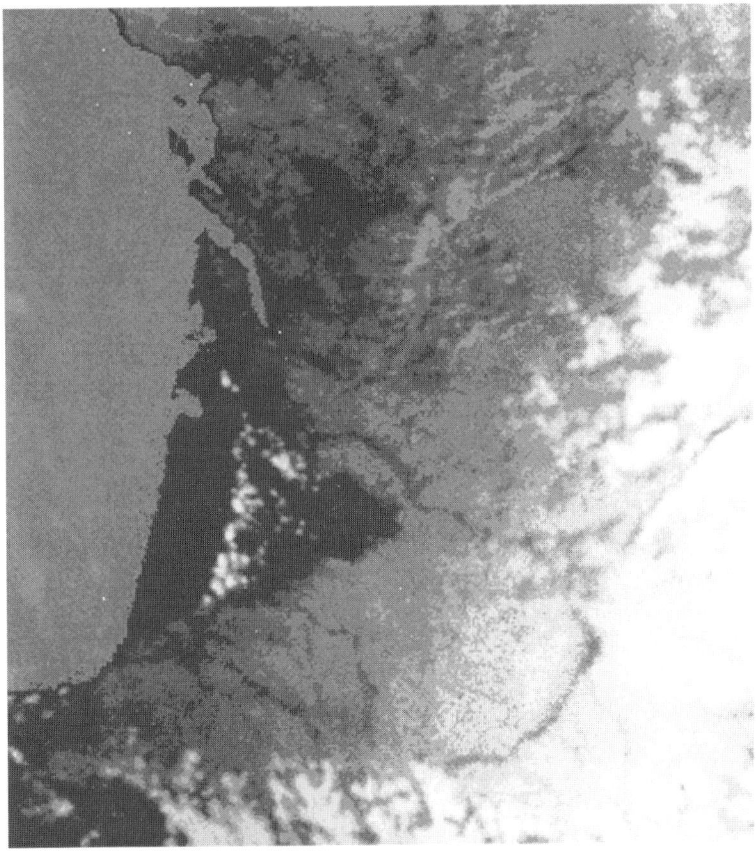

Fig. 5.1. Landsat image taken on 16 June 1986 over France showing preferential development of cloud over the Les Landes forest.

increased sensible-heat flux over the forest, which facilitates early cloud development over the forested area through higher vertical velocities. Modelling attempts have reproduced this effect (André *et al.*, 1989).

Figure 5.2 depicts schematically the most relevant feedbacks between forests and the atmosphere; the following numbers in parentheses refer to the numbers in the figure (after WBGU, 2001). Forest cover modifies the energy and water balances by means of two main effects (Fig. 5.2): (1) plants and forests, compared with land surfaces with no vegetation, have higher transpiration since they draw on soil water and groundwater and (2) they absorb more solar radiation because of their lower albedo. Both mechanisms work together and (3) tend to increase evaporation at the earth's surface. The direct consequences of this are (4) a reduction in surface temperatures and (5) an increase in humidity in the atmosphere. Three feedback effects from the water and radiation balance of the atmosphere change the direct impact of forests: (6) the hydrological cycle is increased through greater humidity in the atmosphere. This favours cloud formation and precipitation and can therefore lead to more evaporation (7). An increase in humidity and clouds also increases the long-wave radiation of the atmosphere and thus the water-vapour-induced greenhouse effect. (8) The radiation available at the earth's surface also changes: superficial cooling is caused by lower solar radiation penetrating the clouds and by higher evaporation. The cooler surface lowers the long-wave radiation of the soil and thus substantially attenuates the water-vapour greenhouse effect.

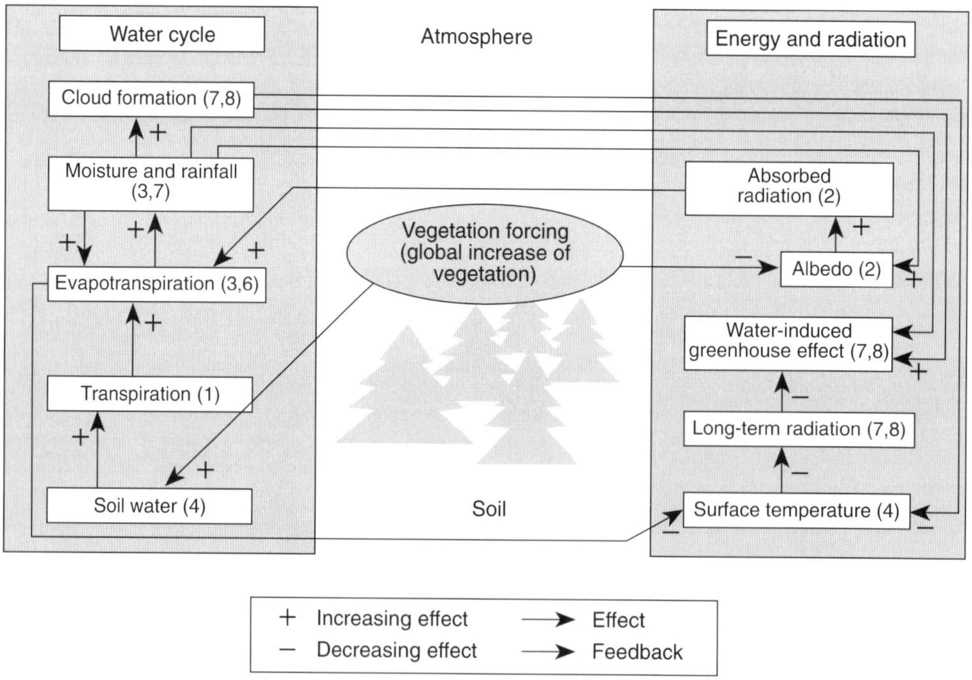

Fig. 5.2. Schematic representation of various interactions between land surface and atmosphere.

The two feedback effects work contrary to one another but, as a whole, vegetation reduces the greenhouse effect.

Often, the overall effect of forests on, for example, rainfall is a trade-off between the different feedbacks illustrated here. The subtle balance between mechanisms acting in opposing directions may introduce additional levels of non-linearity and may increase uncertainties when making predictions.

It is also important to distinguish between the effect of forest on the atmosphere in dry (non-rainy) and wet (rainy) conditions. Under dry conditions, i.e. conditions that are not favourable to the development of precipitation, the effect of forests is generally to increase the heating of the boundary layer through a tight stomatal control. In contrast, under wet conditions, for instance, at the large scale, increased precipitation of forested areas is due to increased low-level convergence of moisture. Convergence is enhanced lateral flow originating as compensation for increases in vertical flow. This is required to conserve mass. Enhanced vertical flow over forests is primarily due to the enhanced aerodynamic roughness of the forested surface compared with, for instance, grassland.

One-dimensional studies

De Bruin and Jacobs (1989) showed that linking a land-surface model to an atmospheric boundary-layer model produced results comparable to observations of the boundary layer. They showed that under wet canopy conditions the forest was able to use sensible heat from the overlying air to sustain high evaporation rates. This phenomenon, first determined experimentally by Stewart (1976) over a pine forest in the UK, explains why forests can sustain relatively high evaporation rates from a wet canopy, even in the absence of positive net radiation. Entrainment of dry air at the top of the boundary layer appeared to be important in the De Bruin and Jacobs (1989) study. Furthermore they showed that the atmospheric boundary layer over forested land grows more rapidly in the summer, as a result of increased heat input due to the lower surface conductance of the forest compared with grassland (both fully supplied with moisture). These results are readily explained by realizing that the growth of the cloudless convective atmospheric boundary layer is, in first-order approximation, equal (simple encroachment) to the square root of the sensible-heat input over time (see Garrat, 1992). The lower surface conductance of the forest changes the energy partitioning in such a way that less energy is diverted into latent heat while more goes into sensible heat and thus warms the atmospheric boundary layer.

Two-dimensional studies

Two-dimensional studies allow the interaction between land-cover types to be simulated more realistically. André *et al.* (1989) described a study investigating the interaction between a forest and an adjacent patch of agricultural land.

Note that, in these two-dimensional studies, infinite stretches of the land-surface are assumed. Differences in the land-surface characteristics are effectively integrated in the values of albedo, aerodynamic and surface resistances. So, although differences in rooting depth may occur, these are not explicitly modelled, as they play no role on the diurnal time-scales of interest. Figure 5.3 shows their main result. The simulation consisted of a cereal and forest patch over a moist sandy soil. Evaporation over the forest and cereal was almost equal (Fig. 5.3a), while the sensible-heat flux over the forest was higher. In this case, the available energy over the forest was higher than over the cereal crop, due to a decreased albedo and a smaller ground-heat flux. This excess available energy was not translated into evaporation but appeared as an increase in sensible heat, because the low surface conductance of the forest restricted the use of the energy for evaporation. This differential heating of the boundary layer creates a 'forest breeze' with ascending air over the forest (Fig. 5.3c) and descending air over the crop, with a low-level lateral flow from the crop to the forest at levels up to 1 km, which returns at higher levels (Fig. 5.3b).[1] The strength of the breeze was about one-third that of a typical sea breeze. The dif-

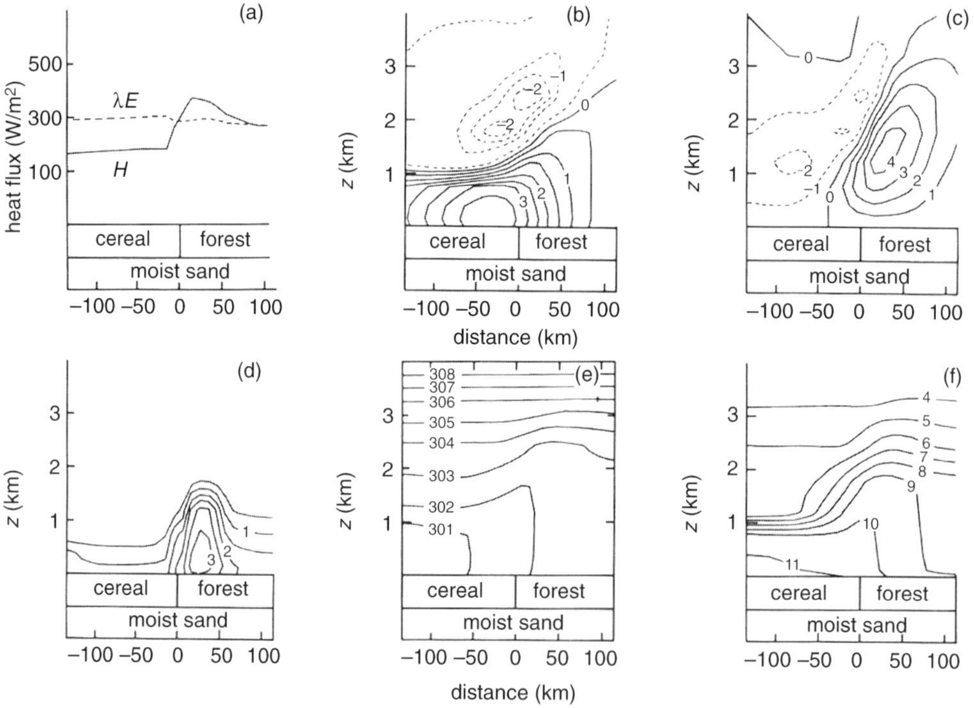

Fig. 5.3. Two-dimensional simulation of a cereal-crop–forest transition. (a) Sensible- and latent-heat flux; (b) cross-section of horizontal wind velocity (m/s); broken lines indicate wind from east to west, solid lines the reverse; (c) cross-section of vertical velocity (cm/s); broken lines indicate descending wind, solid lines, upward movement; (d) cross-section of eddy kinetic energy (m^2/s^2); (e) cross-section of potential temperature (K); and (f) cross-section of water-vapour mixing ratio (g/kg). H, sensible heat; LE, latent heat of evaporation; *z*, height. (Redrawn after André *et al.*, 1989.)

ferent input of sensible heat created a different structure of the boundary layer, as can be seen in Fig. 5.3d and e, with the air above the forest being more turbulent and warmer. The increased turbulence was primarily located around the edges of the transition and originated from the smooth–rough transition of the crop to the forest. Although evaporation rates were similar (Fig. 5.3a), the amount of moisture injected in the boundary layer over the forest was diluted over a larger volume, and lower values of specific humidity resulted. This mechanism of enhanced moist convection may contribute to preferred cloud formation over the forest, as observed in Fig. 5.1. These arguably simple two-dimensional simulations show the basic characteristics of forest–atmosphere interaction under dry conditions: the aerodynamically rough vegetation yields increased turbulence, differential heating leads to different boundary-layer growth and the result may be a mesoscale circulation ('forest breeze'). If mesoscale circulations appear within a heterogeneous domain, simple averaging linear rules no longer apply (e.g. Dolman and Blyth, 1996).

Three-dimensional studies

There have been numerous cases where mesoscale models have been used to study the effects of land use and land-use change on the atmosphere, in particular precipitation (Pielke *et al.*, 1998). Detailed three-dimensional mesoscale models incorporating the right level of complexity of dynamic and physical processes have become available since the early 1970s. These models have been applied to study the effects and generation of sea breezes (Pielke, 1974; Pielke *et al.*, 1992; Tijm *et al.*, 1999), the effects of different forms of land use on the meteorological flow (André *et al.*, 1989; Blyth *et al.*, 1993) and the effects of tropical deforestation on boundary-layer growth and rainfall (Silva Dias and Regnier, 1996; Dolman *et al.*, 1999; Lawton *et al.*, 2001; Van der Molen, 2002). We concentrate here on a review of studies dealing with temperate and tropical forest.

Temperate forests: the HAPEX-MOBILHY experience

André *et al.* (1989) give a review of mesoscale modelling that was executed around the HAPEX-MOBILHY experiment in south-west France (André *et al.*, 1986). Figure 5.1 gives an overview of the area that includes the experimental area. The work is described in a series of seminal papers (Bougeault *et al.*, 1991; Noilhan *et al.*, 1991). These describe, among other things, the implementation of a new land-surface model (Noilhan and Planton, 1989) in a mesoscale model. The model was initialized for a sunny day in 1986, 16 June. To be able to run the model, detailed maps of land use and soil type were prepared using remote-sensing imagery and soil maps. The land-surface model was calibrated against observations of the energy-balance partitioning over agricultural and forested land.

Figure 5.4 shows a cross-section of some of the results obtained during the field campaign with a flux aircraft flying across the domain. Indicated in the graph is the area that is covered by forest. Following the observations of sensi-

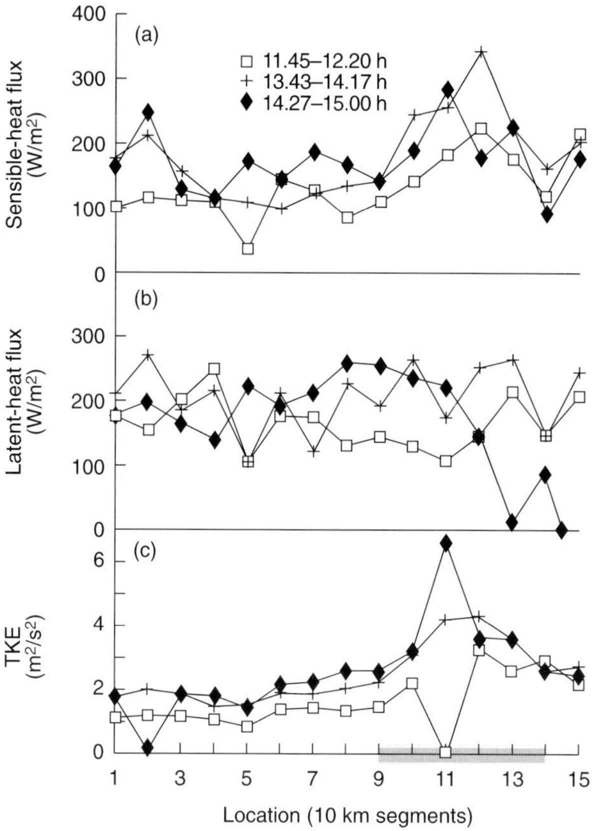

Fig. 5.4. Horizontal variation of sensible heat (a), latent heat (b) and turbulent kinetic energy (TKE) (c) measured over the HAPEX-MOBILHY area in south-west France on 16 June 1986. Each measurement is a 10 km average. Also indicated (hatched area) is the area of the forest (see also Fig. 5.1). (Redrawn after André *et al.*, 1989.)

ble heat, a definite increase in sensible-heat flux over the forested part of the flight path could be observed. Differences between the forest and the surrounding agricultural land were up to 100 W/m^2 for all three flights shown in the graph. Differences in latent-heat flux were less pronounced, with the flights early in the morning showing less difference with the agricultural areas, but in the late afternoon (14.30 h) differences were apparent, with forest evaporating less than the agricultural land. This was probably due to the tighter stomatal control of the pine forest limiting evaporation when atmospheric humidity deficits became large. The larger sensible-heat flux over the forest was also consistent with a roughness-induced enhanced turbulence, as seen in the observations of turbulent kinetic energy (see also Fig. 5.3c).

This situation was subsequently modelled with the French PERIDOT meso-beta-scale model (Bougeault *et al.*, 1991; Noilhan *et al.*, 1991). After calibration of the land-surface scheme, the basic pattern of low sensible-heat flux over the agricultural area compared with the forest was reproduced in the model. André *et al.* (1989) described a series of sensitivity experiments with the same model, in which they changed the roughness of the forest and decreased the stomatal conductance of the forest. A decrease in the forest's roughness decreased the turbulence over the forest and increased the surface temperature, since sensible heat was not as easily transported as before with a higher roughness. The increase in surface temperature over the forest was of the order of 2°C. This did not lead to any noticeable effects in the partitioning of energy between sensible and latent heat or on the development of the boundary layer and cloud formation over the forest. Decreasing the stomatal conductance, on the contrary, did change the energy partitioning, but hardly showed any effect on surface temperature. The resulting decrease in sensible heat over the forest led to considerably less buoyant energy and the formation of clouds was suppressed in a boundary layer with considerably smaller depth. Interestingly, the adequate simulation of the preferentially arising cloud above the forest (Fig. 5.1) was not possible with low conductances. Conductance and, to a lesser extent, surface roughness thus played a key role in the formation of different surface energy balances and resulting boundary layers above forests.

These simulations addressed the capability of the model to simulate dry, fair-weather conditions. Of considerable interest, however, is the interaction of wet forest with the atmosphere. Using the same set-up as before, Blyth *et al.* (1993) simulated a day with a frontal intrusion in south-west France. The model was initialized with data from the large-scale forecast of 5 June 1986. The model was able to reproduce the general features of the situation with warm, moist conditions before the front and colder and drier conditions behind the front. Also, the general distribution and amount of rainfall were in good agreement with the observations, although, due to initialization problems, the front was slightly misaligned in the model compared with the observations.

Figures 5.5 and 5.6 showed the observed and modelled profiles of temperature, humidity and wind speed and direction in the boundary layer from this case. Both the modelled and observed profiles of potential temperature showed a stable boundary layer. The wet forest, acting like a well-ventilated wet bulb, tended to cool the lower layers of overlying air. The heat flux associated with this was negative. Thus a downward heat flux was used to drive the evaporation of intercepted water in the absence of sufficient amounts of radiative energy. The evaporation rates of intercepted water in the model were similar to those observed by Stewart (1976) and by De Bruin and Jacobs (1989). The supply of warm air to the evaporating surface was maintained by downward entrainment of warm air into the boundary layer. Downward entrainment of this heat was sustained by the turbulent kinetic energy, generated by strong wind shear over the forest. It is interesting to investigate this phenomenon of evaporation of a wet canopy sustained by a downward heat flux. Unfortunately, few measurements are available that show this process in real circumstances. Stewart (1976) was the first to show this by taking very detailed and high-resolution measurements of temperature humidity above a forest. He showed that this heat transport could sustain evaporation rates of the order of 150 W/m^2, well in excess of the available net radiative energy.

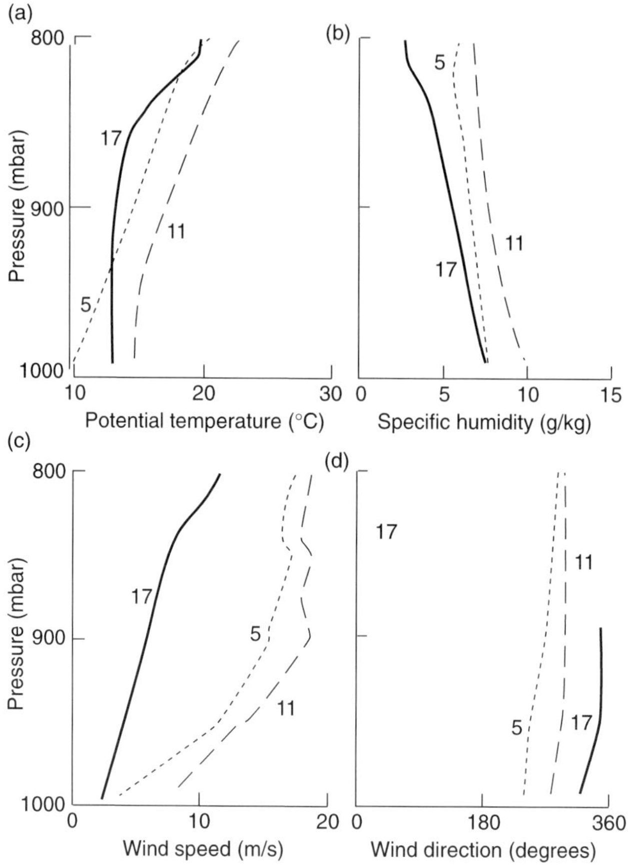

Fig. 5.5. Observed profiles of potential temperature (a), humidity mixing ratio (b), wind speed (c) and wind direction (d) over the central site located in the Les Landes forest (redrawn after Blyth *et al.*, 1993).

Blyth *et al.* (1993) performed a similar sensitivity experiment to that of André *et al.* (1989) and replaced the forest interception capacity and roughness length by values representative of bare soil. Running a model simulation with a forest roughness and zero interception capacity allowed separate determination of the effects of roughness and interception on rainfall. Table 5.1 shows the results of these simulations. The run with complete forest cover generated the largest amount of precipitation in the domain, followed by the run in which a forest with no interception capacity was used. This indicates that roughness played an important part, primarily because it generates an increase in low-level moisture convergence in the area. The evaporation rates of the two forests were rather different, as would be expected by the lack of intercepted evaporation in the forest with no interception capacity. There is little difference between the changes in soil moisture simulated for these two forested runs. This suggests that the increased atmospheric moisture due to rapid evaporation of intercepted rainfall is merely recycled within the boundary layer and does not reach

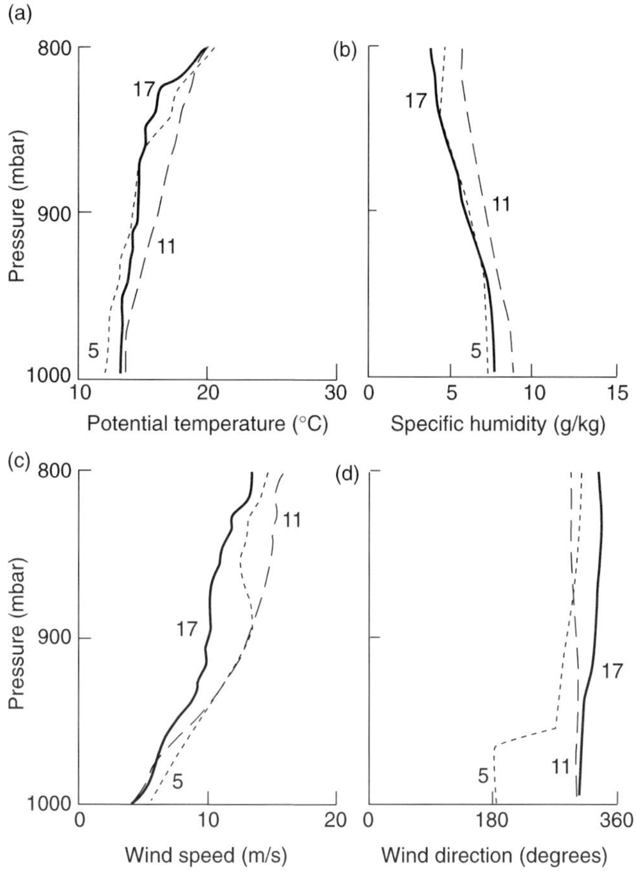

Fig. 5.6. Simulated profiles of potential temperature (a), humidity mixing ratio (b), wind speed (c) and wind direction (d) over the central site located in the Les Landes forest (redrawn after Blyth *et al.*, 1993).

Table 5.1. Sensitivity experiments for the wet case of 5 June 1986 in HAPEX-MOBILHY.

Experiment	z_0	w_{max}	P	Δm	E_{tot}
Control	0.28	0.17	3.3	1.42	1.89
Forest	1.00	0.46	3.8	1.93	1.83
Bare soil	0.01	0.00	2.9	1.22	1.68
Forest (no interception capacity)	1.00	0.00	3.4	1.94	1.46

z_0, aerodynamic roughness length (m); w_{max}, maximum water storage (mm); P, precipitation (mm); Δm, change in soil moisture storage; E_{tot}, total evaporation (mm).

the ground. From these results it may be concluded that, from the 30% increase in rainfall between the forested run and the bare-soil run, about 50% of the rainfall was recycled quickly and never reached the soil moisture store. The other 50%, which was due to the increased low-level moisture convergence as a result of the enhanced roughness of the forest compared with bare soil, represented a real increase of available water on the land. This is shown by the difference between the forest with no interception and the bare-soil run.

It is interesting to note that the modelling experiment with the forest set to zero interception capacity did not lead to enhanced rainfall. This could have been expected because of the potentially greater atmospheric instability that might have resulted from a forest that put more energy into sensible than latent heat. The implication may be that, in cases like the current one, where turbulence is primarily mechanically generated, atmospheric recycling of moisture is the dominant mechanism for enhancing rainfall. In cases where this is not the dominant mechanism producing turbulence, recycling may not be so crucial and atmospheric stability effects through enhanced convection may become more important. It is to these cases that we now turn our attention.

Tropical rain forest: the Amazon and lowland forest in Puerto Rico

Silva Dias and Regnier (1996) and Dolman *et al.* (1999) used mesoscale models to simulate the effects of deforestation on mesoscale flow in the Amazon. In particular, they found a complex interaction between topography and mesoscale circulation induced by different energy partitioning of the forest and pasture areas. In contrast to the previously mentioned studies in temperate regions, in the tropics the weather patterns and mesoscale flows are determined more by local-scale phenomena, due to the absence of large-scale weather disturbances and the minor importance of the Coriolis force at low latitudes. Silva Dias and Regnier (1996) suggested that preferential development of convective activity may occur due to the complex interactions induced by topography and mesoscale circulations. Silva Dias *et al.* (2003) used experimental evidence during the wet season to investigate these effects in more detail. Deforestation seemed to be of secondary importance during the wet season as compared with the dry season. However, there were differences between forest and pasture in the radiation budget and in the surface fluxes. There was a complex interaction between cloudiness and surface fluxes during the wet season, leading to different times of initiation of first clouds. Mixed layer heights grew faster over forest and reached about the same maximum values in the afternoon as over pasture in the wet season. Shallow clouds formed first over forest and, on average, 30 min later over pasture. These results suggest a complex interaction at several scales, from the synoptic to the local, which are important in generating convection. One important result of the experiments described by Silva Dias *et al.* (2003) is also that biogenic aerosols produced by the forest play an important role as cloud condensation nuclei in generating convection (Roberts *et al.*, 2001). Thus, not only the physical characteristics of the forest appear to play a role in forest–atmosphere interaction, but also the purely biological phenomena such as the emission of biogenic aerosols.

Lawton *et al.* (2001) and Van de Molen (2002) investigated the effect of deforestation of the lowland of Costa Rica and Puerto Rico. In the Caribbean basin, cloud forests occur where mountains force trade winds to rise above the lifting condensation level, the point of orographic cloud formation. Lawton *et al.* (2001) performed a simulation of lowland deforestation in which they assumed that parameter values of the Costa Rican forest and pasture were similar to those found during the ABRACOS experiments in Brazil (Gash *et al.*, 1996). The indication from their work is that deforestation in the lowlands increases the lifting condensation level (the height at which the ambient temperature and humidity are such that saturation starts to occur and subsequent cloud formation), and thereby reduces the environmental quality of the higher-altitude cloud forest. It simply becomes more difficult to generate low-level cloud in the deforested runs. However, it is questionable whether the transfer of Amazon parameters to a maritime climate is justified.

In contrast to using parameter values from Amazonian forest and pasture, Van der Molen (2002) took detailed micrometeorological measurements of Puerto Rican lowland wetland forest and pasture. He found that the forest evaporation was lower than that of pasture (3.3 mm/day versus 3.8 mm/day). Sensible-heat flux decreased to 50% of its initial value. This contrasts with findings of continental Amazonian rain forest, where deforestation was associated with a loss in evaporation and an increase in sensible heat. Van der Molen then used these experiments to calibrate a land-surface model that was part of the mesoscale model regional atmospheric modelling system (RAMS). Using the full three-dimensional set-up, he investigated the effects of deforestation of the lowlands, similarly to Lawton *et al.* (2001). The typical surface partitioning of energy generally led to warmer and slightly drier boundary layers over the forested area. However, on clear calm days, the sea breeze over the island tended to dominate the flow and tended to bring moisture to higher levels in the atmosphere, thereby increasing the likelihood of producing rainfall. As differential heating generates it over the ocean and the land, the sea breeze was also stronger in the forested runs, as here the simulated thermal contrast was greater. Thus, enhanced convergence at the sea-breeze front, rather than differential heating between forest and pasture (e.g. Lawton *et al.*, 2001), generated higher cloud water contents.

Figure 5.7 shows an example of this work. The top panels show a warmer boundary layer over a forested island than over a deforested and pastured island. As a result of the larger difference in heating between ocean and land, the sea breeze becomes stronger over a forested island (middle top). Stronger sea-breeze convergence consequently leads to stronger up draughts (middle bottom), which transport more moisture to higher atmospheric levels (bottom). This explains why fewer and smaller clouds occur over a deforested island. At the same time, the lifting condensation level is clearly lower over pasture, because the surface is cooler and moister. Apparently the lowering of the lifting condensation due to deforestation is of minor importance when compared with the reduction of sea-breeze convergence. This example clearly shows that care must be exercised when using mesoscale models with uncalibrated land-surface parameters or parameter values obtained at different locations (e.g. Lawton *et al.*, 2001). This may produce physically credible results, but not necessarily for the correct reason. As always, careful experimentation must go hand in hand with modelling.

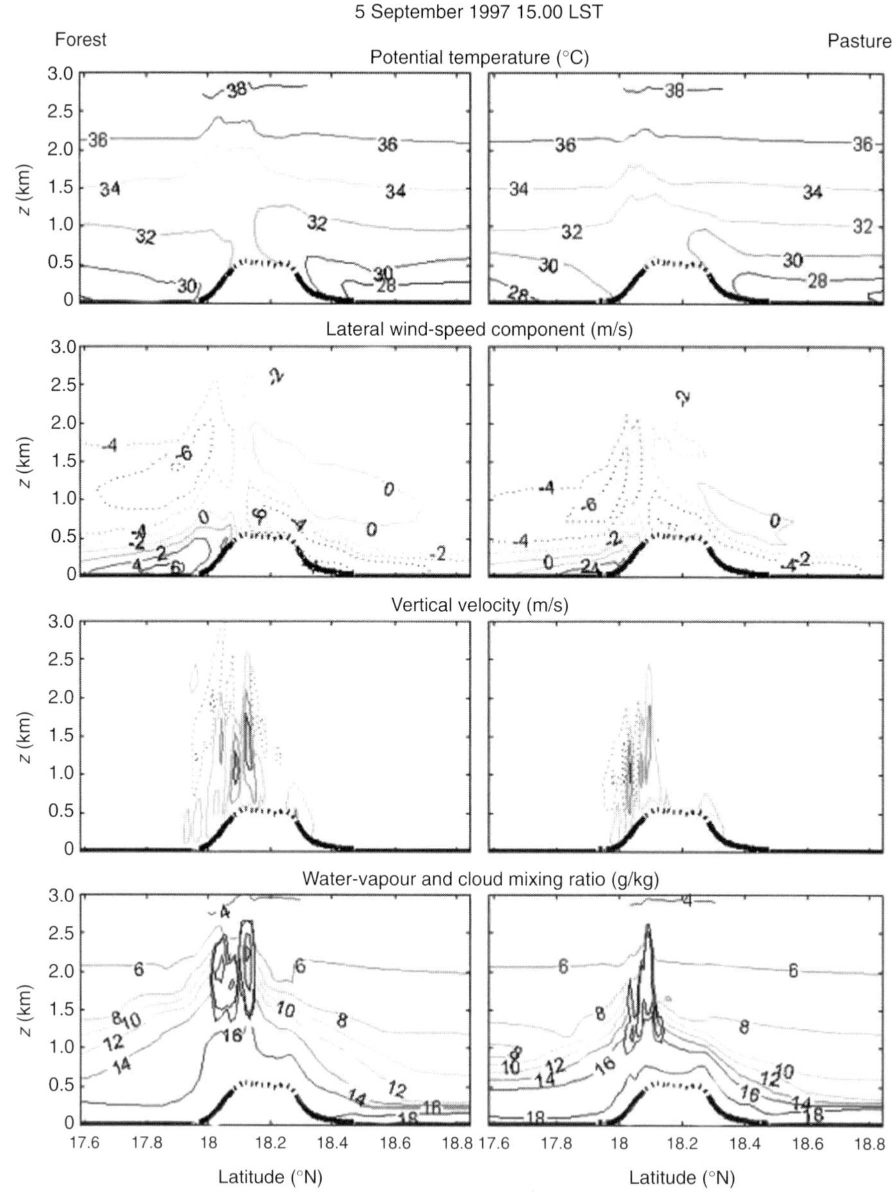

Fig. 5.7. Cross-sections of potential temperature (top), lateral wind speed (middle top), vertical velocity (middle bottom) and water-vapour and cloud-water mixing ratio (bottom) in the forest (left) and pasture run (right) at 15.00 h local solar time (LST) over the island Puerto Rico. All negative contours are dotted. Vertical-velocity contours are drawn every 0.25 m/s, starting from 0.125 m/s. The cloud-water contour lines are drawn at 0.001, 0.01, 0.1, 0.2,...g/kg. The atmospheric fields of the finest grid are averaged over the longitudinal extent of the domain (20 km). The thick solid line indicates the location of the island with lowland forest or pasture vegetation and the thick dotted line gives the location of the upland forest. It is important to realize that the cloud-water mixing-ratio contours represent values averaged over the longitudinal dimension of the grid. *z*, height.

Large-scale sensitivity studies in the Sahel

Mesoscale models can now give good accounts of scale atmospheric processes. It is relevant to ask if this capability can be used to investigate the relative importance of large-scale versus local (land-surface)-scale forcing. At their boundaries mesoscale models are driven by the large-scale flow; this forcing can easily be changed for experimental purposes. In particular, in the Sahel, where large-scale forcing through sea-surface temperatures is believed to determine rainfall, mesoscale modelling studies may be used to illustrate the strength of these forcings.

This issue was investigated by performing a simulation run (control) with a mesoscale model, RAMS (Pielke *et al.*, 1992), at 80 km grid resolution with two sea-surface temperature forcings (SSTs): one representing a dry year (1985) and one a wet year (1963). Additional to this, an artificial greening experiment was performed by changing the land-surface typology to the next denser vegetated class (green run), so that a desert became a semi-desert and a semi-desert became a forest and so on.

The effect of a difference in SSTs suggests a strong response in rainfall. In this case, rainfall reacts markedly, by a more than 40% increase in rainfall for the wet SST forcing (1963), in a very wide latitudinal band (Fig. 5.8). Also the rainfall season is prolonged. This difference was associated with significant differences in large-scale flow, especially a stronger south-westerly monsoon. Such differences in large-scale flow were not found in the greening experiments. In both the dry and wet years the imposed greening increased seasonal rainfall by 20% (Fig. 5.8). In the wet year of 1963 greening produced 120 mm more rainfall, whereas in the dry year of 1985 the increase in precipitation was only 40 mm. Interestingly, the area of maximum increase in precipitation is displaced northwards of the area of maximum increase in vegetation (greening), suggesting northward transport of recycled moisture (see below).

Increased seasonal rainfall from 'wet' SSTs results from both increased evaporation and moisture convergence (in a roughly 60/40 ratio). However, not all increases in evaporation lead to increases in precipitation. The enhanced rainfall simulated in response to greener vegetation was associated with an even more enhanced evaporation, but at the same time convergence substantially decreased too (Fig. 5.9). Thus, farther north in this simulation,

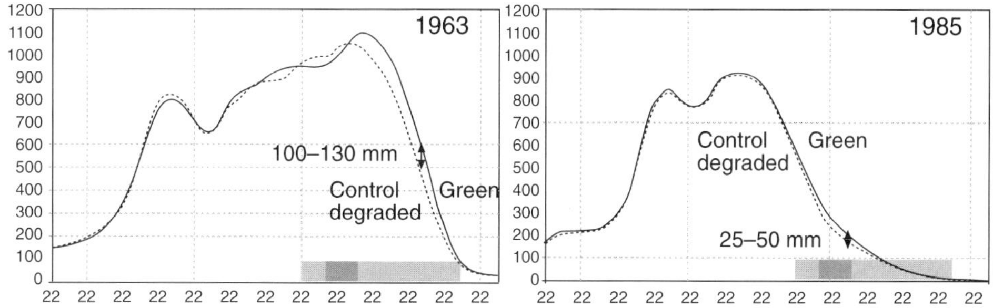

Fig. 5.8. Zonal precipitation (averaged over longitudes 5°W–7°E) of green and control runs for 1963 and 1985. The shaded box at the bottom indicates the zone with the most marked vegetation changes.

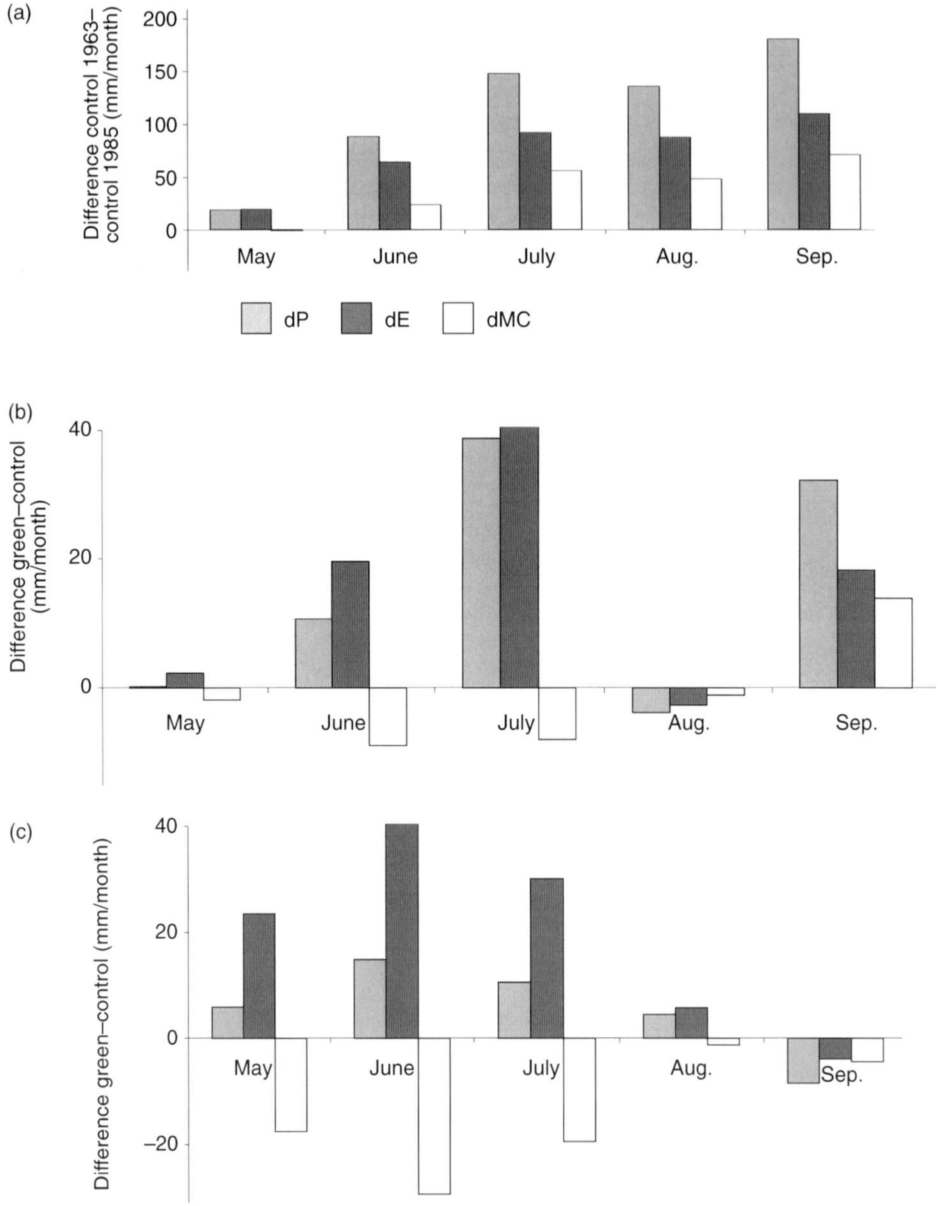

Fig. 5.9. Areal differences in precipitation (dP – light grey), evaporation (dE – dark grey) and moisture convergence (dMC – white) between 1963 and 1985 and the 5 months simulated. Units: mm per month. Area is from 13 to 17°N and 5°W to 7°E. Vertical axis intervals: 20 mm. (a) Control 1963–control 1985; (b) green–degraded for 1963 for an area of 17–21°N; (c) green–degraded (1985).

more rainfall was originating from recycled (evaporated) water instead of from water transported to the region by large-scale advection. In the middle Sahel (13–17°N) the difference between the 1963 and 1985 effects of greening appeared large, particularly in July, August and September. This is because in 1963 the enhanced rainfall from the greening effect occurred mostly north of 17°N, not in the 13–17°N band (Fig. 5.9b).

The current simulation results seem to favour the explanation that sea-surface temperatures play an overriding role in determining rainfall in the Sahel. SST anomalies appear to have a larger effect on Sahelian rainfall than do modest and perhaps realistic vegetation anomalies. Land-surface feedbacks affect rainfall only within the overall constraints set by the large-scale circulation as modulated by SST patterns.

Conclusions and Outlook

What have we learned from mesoscale models?

Mesoscale modelling studies have been particularly useful in studying the effect of regional-scale land-surface changes on the atmosphere. They clearly provide additional insight into this interaction at the scale at which most of the land-surface changes are happening. Unlike global studies, which often simulate rather ideal, extreme conditions, mesoscale models can be used to study the real changes that occur in the human-made landscape. A further advantage is that local-scale flux and boundary-layer measurements can be used to validate the performance. It is clear from this review that forests can affect the properties of the planetary boundary layer. They can do this by changing the partitioning of the available net energy into sensible and latent heat, but also by increasing convergence through their enhanced surface roughness. The development of forest breezes may be a final result, if the surrounding land use generates sufficient thermal contrast with the forest. Furthermore, evidence is mounting that forest can influence rainfall at the regional scale. The mechanisms are similar to those found in large-scale studies, i.e. increased convergence through enhanced roughness and potentially enhanced convection through a warmer and more unstable boundary layer as a result of differences in energy partitioning between forest and, for instance, grass. However, care must be exercised when generalizing these results, as the study of Van der Molen (2002) shows.

What can we still learn from mesoscale models?

In the mesoscale modelling as used, for instance, by Noilhan *et al.* (1991), the model is mainly used as a prototype assimilation tool. This allows the use of local- and regional-scale flux and boundary-layer measurements to be integrated in a physically consistent manner. Next to the investigation of primary feedback processes, this is arguably the second most important use of mesoscale models. It is likely that, with the advance of computing power, more high-resolution mesoscale models will be used as regional assimilation schemes.

The importance of forests as CO_2 absorbers has also generated an interest in the use of mesoscale models to study the effects of regional-scale sinks and sources of CO_2. There is increasing evidence that the mixing properties of the convective boundary layer may be used to infer regional-scale sinks and sources of CO_2 (Lloyd *et al.*, 2001). As an example we show some recent results in which the RAMS model was extended with a CO_2 carrying capacity, both in the land-surface model and in the atmosphere. Figure 5.10 shows an example of such a simulation for CO_2 fluxes. In Fig. 5.10 a snapshot is taken from a simulation of 27 June at 12.00 UTC. Clear differences in net ecosystem exchange of CO_2 can be seen at midday between the southern and middle parts of Europe. Also, the impact of large-scale weather systems is visible. The net exchange of CO_2 is smaller for the area of Austria/Bavaria (south Germany), where a system with considerable amounts of rain and clouds is present.

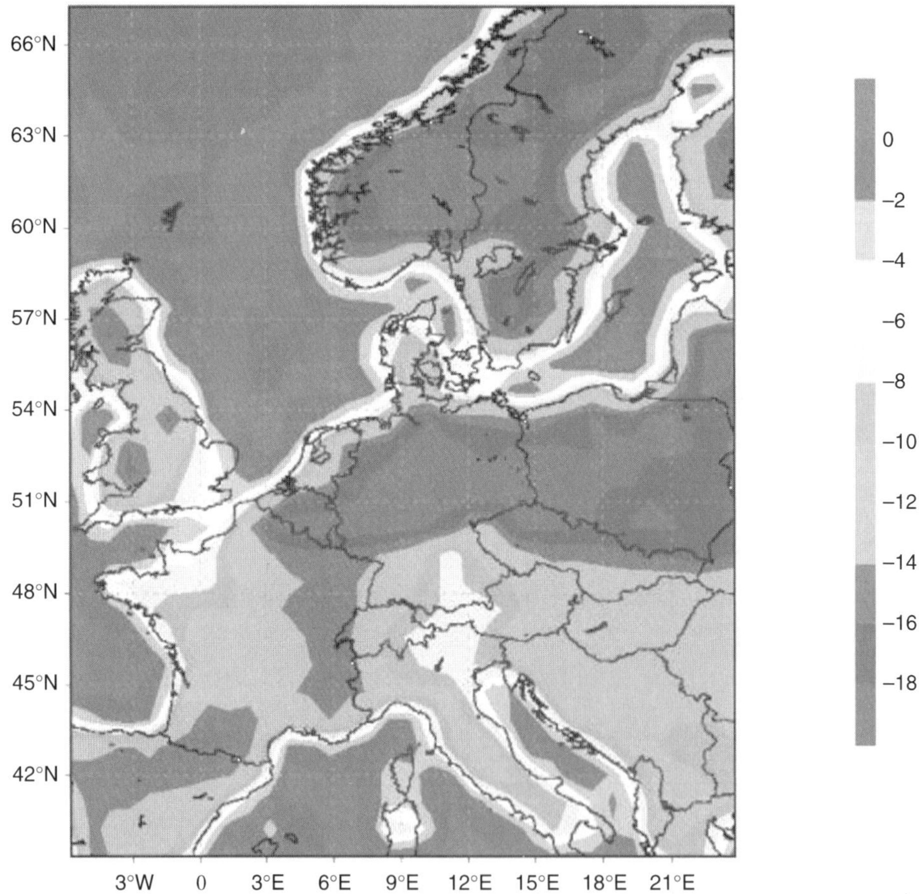

Fig. 5.10. Daytime CO_2 flux (also known as net ecosystem exchange, μmol/m²/s) at 27 June 12.00 UTC for a RAMS simulation over Europe.

A key result of this study was that, when leaf area index was reduced by 50%, the sensible-heat flux resulting from this artificial degradation declined substantially for large parts of Europe. This was primarily attributed to a reduction in albedo by 0.04%. Betts (2000) also found that large-scale planting (the reverse of the current degradation study) could lead to enhanced warming. This suggests that planting of forest for carbon uptake may have side-effects on climate that appear as warming through physical mechanisms. Only modelling studies where the essential parts of the system can be validated, i.e. mesoscale models, can realistically simulate these effects. This, then, is what we can still learn from mesoscale models: how do the land surface, biogeochemical cycles and atmosphere interact?

The simulation of the interactions of biogeochemical cycles with the land surface and atmosphere requires a refocusing of the time-scale commonly used in mesoscale modelling studies. As most of the nutrients, and certainly carbon, rest in pools that have relatively slow decay times compared with those of atmospheric transport, these simulation studies need to go beyond the typical 1–5-day simulations of current mesoscale work. This also holds, *inter alia*, for studies that aim to find statistically meaningful relationships between land-surface changes and, for instance, precipitation. There is a clear need to cover at least a seasonal cycle to achieve a reliable level of significance.

Simulation of atmospheric transport of CO_2 requires the interaction of a land surface with the atmosphere to be adequately modelled. Several examples of this coupling now exist. Chevillard *et al.* (2002) used the regional atmospheric model regional climate model (REMO) to investigate the temporal and spatial distribution of CO_2 over Europe and western Siberia and found good agreement between observed and simulated patterns. However, key to successful regional transport modelling of these atmospheric tracers are the boundary conditions applied at the edges of the domain, which need to include oceanic, atmospheric and other terrestrial sources of CO_2. Similarly, Eastman *et al.* (2001) used RAMS coupled to a sophisticated land-surface package to investigate the effects of a doubling of CO_2 on regional weather patterns in the USA. They found that it was necessary to include the feedback of plants on increased atmospheric CO_2 to realistically simulate changes in future weather patterns. Mesoscale models allow this interaction to be studied at a scale where the impact of human actions is most pronounced and at which important land-use management decisions are generally taken: the regional scale. This is where their greatest potential is to be found for the next decade.

Note

[1] Note that the use of the term 'forest breeze' in analogy with the sea breeze is incorrect; as the wind blows from the sea to the land in the case of a sea breeze, the correct name for the 'forest breeze' should be 'cereal breeze'. However, we stick to the common usage of the term 'forest breeze' in this chapter.

References

André, J.-C., Goutorbe, J.-P. and Perrier, A. (1986) HAPEX-MOBILHY: a hydrologic atmospheric experiment for the study of water budget and evaporation flux at the climatic scale. *Bulletin of the American Meteorological Society* 67, 138–144.

André, J.-C., Bougeault, P., Mahfouf, J.-F., Mascart, P., Noilhan, J. and Pinty, J.-P. (1989) Impact of forests on mesoscale meteorology. *Philosophical Transactions of the Royal Society, Series B* 324, 407–422.

Betts, R.A. (2000) Offset of the potential carbon sink from boreal forestation by decreases in albedo. *Nature* 408, 187–190.

Blyth, E.M., Noilhan, J. and Dolman, A.J. (1993) The effects of forests on mesoscale rainfall: an example from HAPEX-MOBILHY. *Journal of Applied Meteorology* 33, 445–454.

Bougeault, P., Noilhan, J., Laccarrè, P. and Mascart, P. (1991) An experiment with an advanced surface parameterization in a meso beta model. Part I. Implementation. *Monthly Weather Review* 119, 2358–2373.

Chevillard, A., Karstens, U., Ciais, P., Lafont, S. and Heimann, M. (2002) Simulation of atmospheric CO_2 over Europe and Siberia using the regional scale model REMO. *Tellus, B* 54(5), 872–894.

Cutrim, E., Martin, D.W. and Rabin, R. (1995) Enhancement of cumulus clouds over deforested lands in Amazonia. *Bulletin of the American Meteorological Society* 76, 1801–1805.

De Bruin, H.A.R. and Jacobs, C.M. (1989) Forests and regional scale processes. *Philosophical Transactions of the Royal Society, Series B* 324, 393–406.

Dolman, A.J. and Blyth, E.M. (1996) Patch scale aggregation of heterogeneous land surface cover for mesoscale meteorological models. *Journal of Hydrology* 190, 252–268.

Dolman, A.J., Silva Dias, M.A., Calvet, J.-C., Ashby, M., Tahara, A.S., Delire, C., Kabat, P., Fisch, G. and Nobre, C.A. (1999) Mesoscale effects of deforestation in Amazonia: preparatory LBA modelling studies. *Annales Geophysicae* 17, 1095–1110.

Dolman, A.J., Moors, E.J., Grunwald, T. and Berbigier, P. (2003) Factors controlling forest atmosphere exchange of water, energy and carbon. In: Valentini, R. and Matteucci, G. (eds) *Fluxes of Carbon, Water and Energy of European Forests*. Ecological Studies, Vol. 163. Springer Verlag, New York, pp. 207–223.

Eastman, J.L., Coughenour, M.B. and Pielke, R.A. (2001) The regional effects of CO_2 and landscape change using a coupled plant and meteorological model. *Global Change Biology* 7, 797–815.

Fraedrich, K., Kleidon, A. and Lunkeit, F. (1999) A green planet versus a desert world: estimating the effect of vegetation extremes on the atmosphere. *Journal of Climate* 6, 419–449.

Garrat, J.R. (1992) *The Atmospheric Boundary Layer.* Cambridge University Press, Cambridge.

Gash, J.H.C., Nobre, C.A., Roberts, J.M. and Victoria, R.L. (eds) (1996) *Amazonian Deforestation and Climate*. McWiley, London.

Jarvis, P.G., James, G.B. and Landsberg, J.J. (1976) Coniferous forests. In: Monteith, J.L. (ed.) *Vegetation and the Atmosphere*, Vol. 2. *Case Studies*. Academic Press, New York, pp. 171–240.

Jarvis, P.G., Monteith, J.L., Shuttleworth, W.J. and Unsworth, M.H. (eds) (1989) Forests, weather and climate. *Philosophical Transactions of the Royal Society, Series B* 324, 173–436.

Lawton, R.O., Nair, U.S., Pielke, R.A. and Welch, R.M. (2001) Climatic impact of tropical deforestation on nearby montane cloud forests. *Science* 294, 584–587.

Lloyd, J., Francey, R., Mollicone, D., Raupach, M.R., Sogachev, A., Arneth, A., Byers, J.N., Kelliher, F.M., Rebmann, C., Valentini, R., Chin-Wong, S., Bauer, G. and Schulze, E.-D. (2001) Vertical profiles, boundary layer budgets, and regional flux estimates for CO_2 and its $^{13}C/^{12}C$ ratio and for water vapour above a forest/bog mosaic in central Siberia. *Global Biogeochemical Cycles* 15, 267–284.

Monteith, J.L. (ed.) (1976) *Vegetation and the Atmosphere*. Academic Press, New York.

Noilhan, J. and Planton, S. (1989) A simple parameterisation of land surface processes for meteorological models. *Monthly Weather Review* 117, 536–549.

Noilhan, J., Laccarré, P. and Bougeault, P. (1991) An experiment with an advanced surface parameterization in a meso beta model. Part III. Comparison with the HAPEX-MOBILHY dataset. *Monthly Weather Review* 119, 2393–2413.

Oke, T.R. (1978) *Boundary Layer Climates*. Methuen, London.

Pielke, R.A. (1974) A three dimensional numerical model of the sea breeze over South Florida. *Monthly Weather Review* 102, 115–139.

Pielke, R.A. (1984) *Mesoscale Meteorological Modelling*. Academic Press, New York.

Pielke R.A., Cotton, W.R., Walko, R.L., Tremback, C.J., Lyons, W.A., Grasso, L.D., Nicholis, M.E., Moran, M.D., Wesley, D.A., Lee, T.J. and Copeland, J.H. (1992) A comprehensive meteorological modelling system – RAMS. *Meteorology and Atmospheric Physics* 49, 69–91.

Pielke, R.A., Avissar, R., Raupach, M.R., Dolman, A.J., Zeng, X. and Denning, S. (1998) Interactions between the atmosphere and terrestrial ecosystems: influence on weather and climate. *Global Change Biology* 4, 461–475.

Pitman, A.J., Pielke, R.A., Avissar, R., Claussen, M., Gash, J.H.C. and Dolman, A.J. (1999) The role of the land surface in weather and climate: does the land surface matter? *IGBP Global Change Newsletter* 39, 4–11.

Rabin, R.M. and Martin, D.W. (1996) Satellite observations of shallow cumulus coverage over the central United States: an exploration of land use impact on cloud cover. *Journal of Geophysical Research* 1010, 7149–7156.

Roberts, G.C., Andrea, M.O., Zou, J. and Artaxo, P. (2001) Cloud condensation nuclei over the Amazon basin: marine conditions over a continent? *Geophysical Research Letters* 28, 2807–2810.

Silva Dias, M.A.F. and Regnier, P. (1996) Simulation of mesoscale circulations in a deforested area of Rondonia in the dry season. In: Gash, J.H.C., Nobre, C.A., Roberts, J.M. and Victoria, R.L. (eds) *Amazonian Deforestation and Climate*. John Wiley & Sons, Chichester, pp. 531–547.

Silva Dias, M.A.F., Rutledge, S., Kabat, P., Silva Dias, P.L., Nobre, C., Fisch, G., Dolman, A.J., Zipser, E., Garstang, M., Manzi, A., Fuentes, J.D., Rocha, H., Marengo, J., Plana-Fattori, A., Sá, L., Alvalá, R., Andreae, M.O., Artaxo, P., Gielow, R. and Gatti, L. (2003) Clouds and rain processes in a biosphere atmosphere interaction context in the Amazon Region. *Journal of Geophysical Research* 107(D20), art. no. 8072.

Stewart, J. (1976) Evaporation from the wet canopy of a pine forest. *Water Resources Research* 13, 915–921.

Tijm, A.B.C., Holtslag, A.A.M. and van Delden, A.J. (1999) Observations and modelling of the sea breeze with the return current. *Monthly Weather Review* 127, 625–640.

Van der Molen, M.K. (2002) Meteorological impacts of land use change in the maritime tropics. PhD thesis, Vrije Universiteit, Amsterdam.

WBGU (German Advisory Council on Global Change) (2001) *World in Transition: Conservation and Sustainable Use of the Biosphere*. Earthscan Publications, London and Sterling, Virginia, 451 pp.

The Diurnal Cycle over Land

6

Alan K. Betts
Atmospheric Research, Pittsford, VT 05763, USA

Introduction

This chapter reviews, initially from an observational perspective, the nature of the diurnal cycle over land, using illustrations from high latitudes to the tropics. Understanding the coupling between different processes at the land surface is of fundamental importance, because, in global models, many processes are parameterized and are poorly constrained by routine observational inputs. Yet the diurnal cycle is observed synoptically, so it is an excellent indicator of whether the surface processes and their interaction with the boundary layer (BL) are modelled correctly. In addition, the diurnal range of surface temperature and humidity are important prognostic variables for society.

First some basic concepts are reviewed, with illustrative examples, and then additional controls on the diurnal cycle at high latitudes are discussed. Finally, a tropical example (over Rondônia in the southern Amazon basin) will be used to illustrate the difficulty in getting the diurnal cycle of precipitation right in a forecast model, because of the interaction of many processes.

Near the earth's surface, the incoming solar radiation drives a characteristic diurnal cycle for many variables. The atmosphere is relatively transparent to the short-wave radiation from the sun and relatively opaque to the thermal radiation from the earth. As a result, the surface is warmed by a positive net radiation balance in the daytime and cooled by a negative radiation balance at night. The surface temperature oscillates between a minimum at sunrise and a maximum in the afternoon. In warm seasons, the daily net radiation balance is positive, and the daily mean temperature is determined by the daily mean surface energy balance, which involves not only the short- and long-wave radiation components but also heat transfers to the atmosphere. The magnitude of this diurnal range of temperature is determined by many factors. The most important are the nature of the underlying surface, whether land or water, and the coupling to the atmosphere above. The phase change of water, particularly evaporation and condensation, plays an important role in moderating the diurnal range of temperature, because of the large latent heat of vaporization (in cold climates the freezing and thawing of the soil is also important on the seasonal time-scale).

Over the ocean (and large lakes), the diurnal temperature range is small, because the incoming solar energy is mixed downward into an ocean 'mixed layer', which is usually tens of metres deep. One day of solar heating will warm a layer of water 50 m deep by less than 0.1 K, because of its large thermal capacity. Only in light winds, when the downward mixing is small, does the diurnal range of sea-surface temperature reach 1 K. On time-scales longer than the diurnal, evaporation of water primarily balances the surface net radiation budget.

Over land, only a small fraction (< 20%) of the net radiation at the surface is conducted downward in the daytime, stored by warming trees on the surface, or partly used for photosynthesis. As a result, the surface temperature rises rapidly after sunrise, until near-balance is achieved between the net radiation and the direct transport of heat to the atmosphere (referred to as the sensible-heat flux) and evaporation of water (or transpiration from plants), referred to as the latent-heat flux. If the surface is a desert, then the daytime temperature rise is large, but, if water is readily available for transpiration, the daytime rise of temperature is greatly reduced, because most of the net radiation goes into the latent heat of vaporization. The surface sensible- and latent-heat fluxes have a large diurnal cycle, with a peak near local noon, as they are driven primarily by the incoming solar radiation. The surface temperature peaks a little later in the afternoon, and the surface sensible-heat flux goes negative once the surface cools sufficiently. The challenge in a numerical model is to get the surface net radiation correct (which depends on the cloud field, as well as the atmospheric structure and aerosols), and to predict the correct evaporation, which is constrained over land by vegetative and soil thermal and hydrological processes (which in turn depend on the model precipitation).

Illustrative Examples

Dependence of diurnal cycle on evaporation

Figure 6.1 illustrates this diurnal variation using data from sunny days in midsummer during a 1987 field experiment (the First International Satellite Land-surface Climatology Project Field Experiment (FIFE)) conducted over grassland near Manhattan, Kansas. The panels on the left show from top to bottom, net radiation, R_n, sensible heat flux, H, and latent heat flux, LE (data from Betts and Ball, 1995). The surface energy balance can be written as:

$$R_n = H + LE + G \tag{1}$$

where G is the storage in the ground and vegetation, which we do not show. In addition a small amount of energy goes into photosynthesis, which again we do not show. The time axis is local solar time (LST), which is UTC − 6 h.

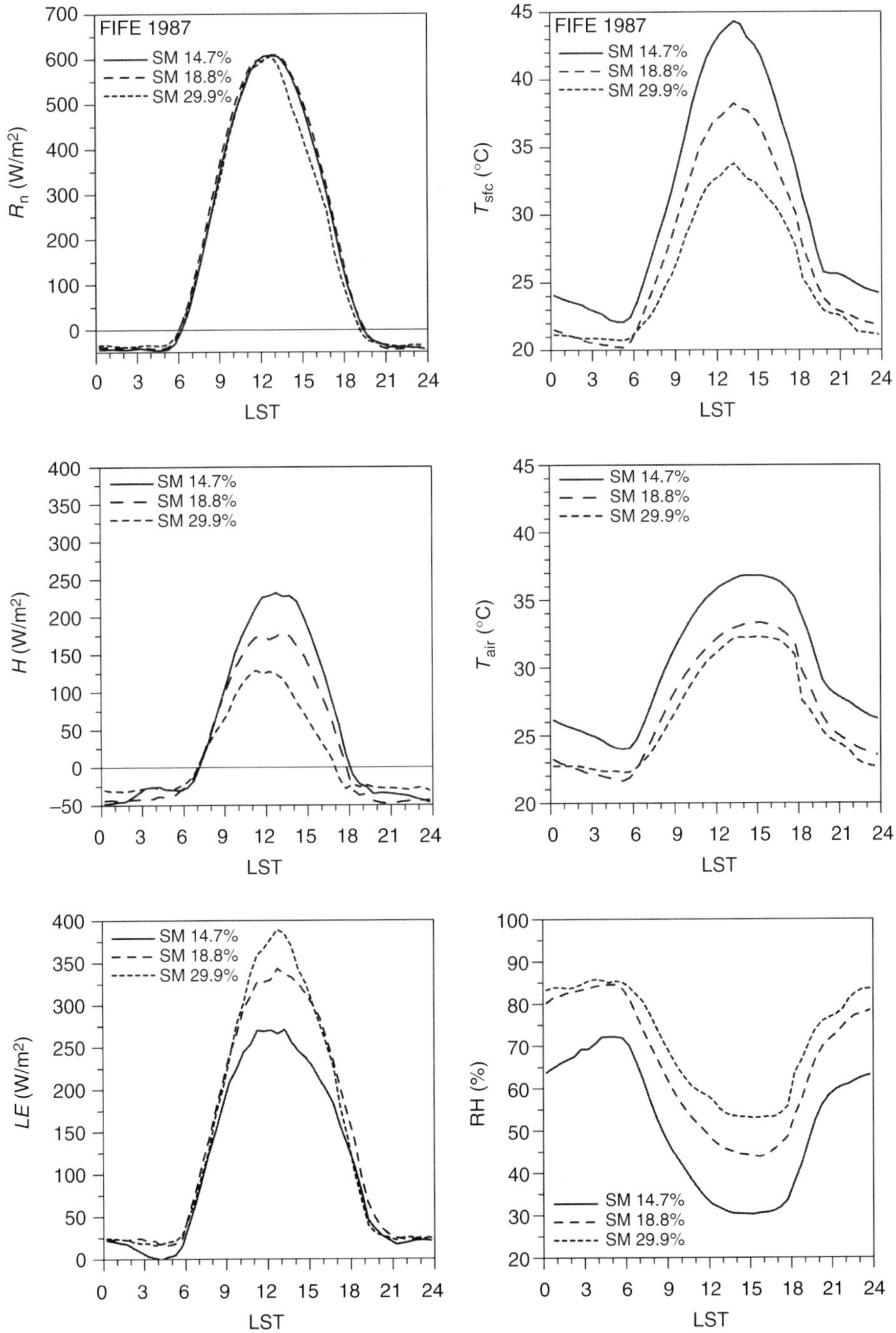

Fig. 6.1. Diurnal cycle of net radiation (R_n), sensible-heat flux (H) and latent-heat flux (LE) (left panels); surface temperature (T_{sfc}), air temperature (T_{air}) and relative humidity (RH) (right panels), stratified by percentage volumetric soil moisture (SM) in first 10 cm of the soil (data from Betts and Ball, 1995; figure from Betts, 2002).

The data have been grouped and averaged based on the percentage soil moisture (SM) in the first 10 cm of soil, so that there are three curves (each an average of about 10 days) representing dry, medium and wet soils. The upper left panel shows that the mean net radiation on these sunny days is very similar. However, because soil moisture is a major control on evaporation, the partition of the net radiation into sensible and latent heat is very different. When the soil is wet, the latent heat flux (or 'evaporative energy' flux) is about three times the sensible-heat flux, whereas, when the soil is dry, these two fluxes are nearly equal. The panels on the right side show the response to the different surface forcing. The upper right panel shows the surface temperature (measured by an infrared radiation thermometer, mounted on a tower and pointed downward at the grass). Although R_n is almost the same in all composites, on days when the soil is dry and water is not readily available for evaporation, the surface gets very hot, as warm as 44°C near noon. This warm surface temperature drives the large sensible-heat flux H and heats the air above the surface. The diurnal range of the surface temperature is more than 20°C on these days, while, for the air at 2 m above the surface in the middle panel, the diurnal range is only 12°C. As soil moisture increases, the daily maximum surface and air temperatures decrease. The upper two panels on the right are similar, except that the amplitude of the surface temperature is larger than that of the air temperature. The difference is related to the sensible-heat flux, H. Note that the air temperature has a broad afternoon maximum, because H is upward as long as the surface is warmer than the air. The surface temperature falls below the air temperature only in late afternoon; H then changes sign and at night the surface is cooler than the air. The lower right panel shows the diurnal cycle of relative humidity (RH) as a percentage. Over the wetter soils, the RH of the air at 2 m reaches 85% before sunrise and falls in the daytime as the surface and air warm. The fall of RH is smallest on the days with the greatest evaporation, LE. When evaporation is reduced because the soil is dry, daytime RH falls as low as 30% and even at night only reaches 72% at sunrise.

Coupling between the surface diurnal cycle and the atmospheric mixed layer

As the land surface is heated during the daytime, a dry convective BL grows in depth. This is called the 'mixed layer', because the turbulent dry convection rapidly stirs the layer to one of near-neutral buoyancy and near-constant water-vapour mixing ratio. The diurnal cycle of the surface and the mixed layer is tightly coupled. As a result, the pre-existing atmospheric structure above the surface at sunrise has a considerable impact on the daytime diurnal cycle, as illustrated in the following figures, using surface and sounding data collected over the boreal forest in Saskatchewan, Canada, during the Boreal Ecosystem–Atmosphere Study (BOREAS) in 1994. Figure 6.2 shows the surface diurnal cycle for 2 days in spring. The upper panel shows for each day the temperature at two levels, an upper level (T_U), which is at 21 m, about 5 m above the canopy of a jack-pine forest, and a lower level (T_L)

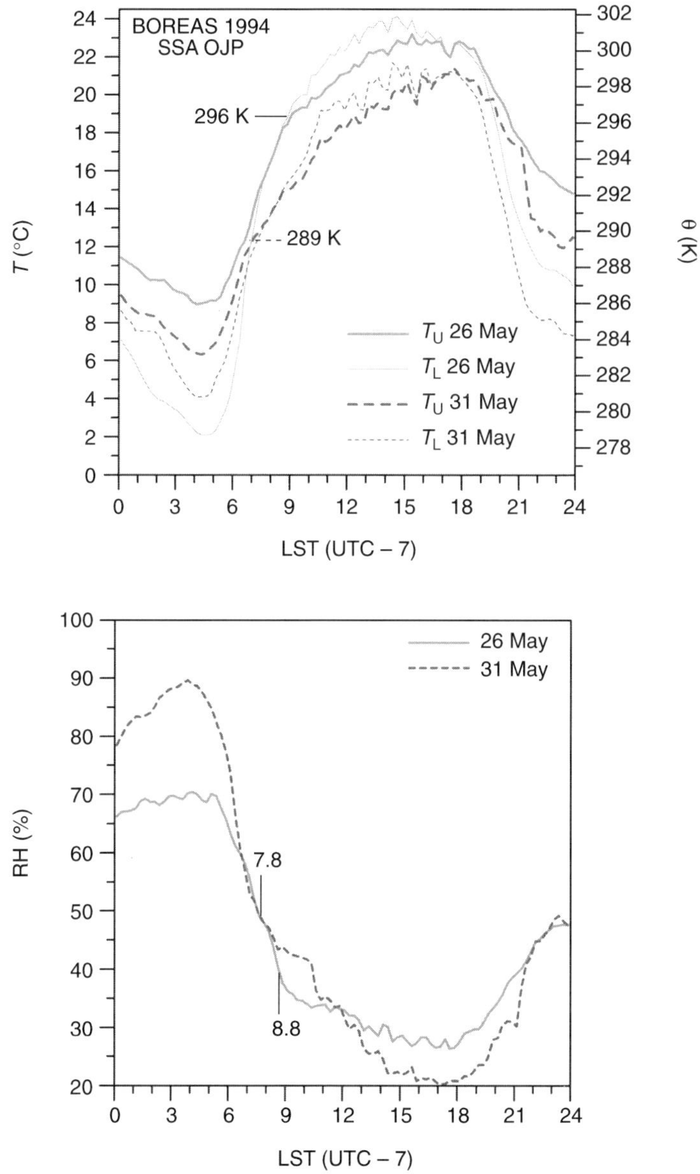

Fig. 6.2. Diurnal cycle of temperature above (T_U) and below (T_L) a boreal forest canopy (upper panel) and of relative humidity (RH) above the canopy (lower panel) for 2 days in May 1994 at the BOREAS southern study area (old jack-pine site (SSA OJP) (Betts, 2002).

about 5 m above the forest floor. On both days the surface cools strongly at night and rises steeply after sunrise, with a greater diurnal range than in Fig. 6.1. The diurnal range under the canopy is larger than above it. At night on 26 May, the winds are lighter, and the atmosphere above is more stable (see Fig. 6.3). The air under the canopy becomes effectively decoupled from the atmosphere above and the stable temperature gradient across the canopy at night reaches 7 K. There is very little evaporation from either the forest or the cold lakes at this time in spring. The lower panel shows RH measurements above the canopy. In the late afternoon, RH falls as low as 20% on 31 May. Before sunrise on this day, RH above the canopy reaches 90% as T_U falls to a minimum of 4°C. RH was not measured below the canopy, but the temperatures there are cold enough to saturate the air in the hours before sunrise. The dew-point is often used to estimate minimum night-time temperatures at the surface.

The right-hand scale of the upper panel shows the corresponding dry potential temperature, which is defined as:

$$\theta = (T + 273.15)(1000/p)\ 0.286 \qquad (2)$$

where p is the surface pressure (here about 950 hPa, since the observation site is about 500 m above sea level). The potential temperature, θ, is useful as a variable because it allows us to compare the surface and the atmosphere above. During the daytime the BL above the surface is mixed to almost constant potential temperature (see Fig. 6.3). The strong radiative cooling of the surface at night generates a stable layer close to the ground, typically only a few hundred metres deep. About 3–4 h after sunrise, the surface has warmed enough to remove this stable surface layer and reconnect to a deeper layer. When this happens, the rate of rise of temperature and fall of RH decreases sharply. In Fig. 6.2, this occurs on 26 May at a local time of 8.8 h, when θ reaches 296 K, while on 31 May, it occurs at 7.8 h, when $\theta = 289$ K, and on this day the change is smaller. Figure 6.3 shows sequences of seven profiles of potential temperature in the lower troposphere, measured by radiosonde ascents, nominally every 2 h from sunrise to late afternoon on the 2 days. The upper panel shows at sunrise (04.17 LST, light solid) a cold (stable) surface layer only about 25 hPa deep (200 m), with a deep layer above of constant θ, which is the residual or 'fossil' mixed layer from the previous day. At the surface the temperature warms rapidly, as the surface sensible-heat flux is trapped in this shallow surface layer. The profile at 08.24 LST shows a mixed layer with $\theta = 294.5$ K to 890 hPa. Shortly afterwards, when the surface potential temperature reaches $\theta = 296$ K, the new growing BL merges with the deep residual mixed layer. From then on, the surface and mixed layer warm much more slowly, as seen in Fig. 6.2. Even though H exceeds 300 W/m^2 at all the forest sites for several hours around local noon (not shown), this large heat flux is distributed through a deep layer.

The lower panel shows the time sequence on 31 May. Note that, at sunrise, the profile is quite different from that on 26 May. Instead of a deep layer of constant θ, produced by dry convection the previous day (a so-called dry adiabatic structure), there is a layer from 920 to 650 hPa in which θ increases steadily with height. In fact, this layer was produced by showers the previous evening (and it has a wet adiabatic structure). The change in slope of the early-morning profile at 920 hPa is at $\theta = 289$ K, and hence we see in Fig. 6.2 a change in the rate of warming once the surface reaches this potential temperature. This change of slope is more dramatic on 26 May, because the

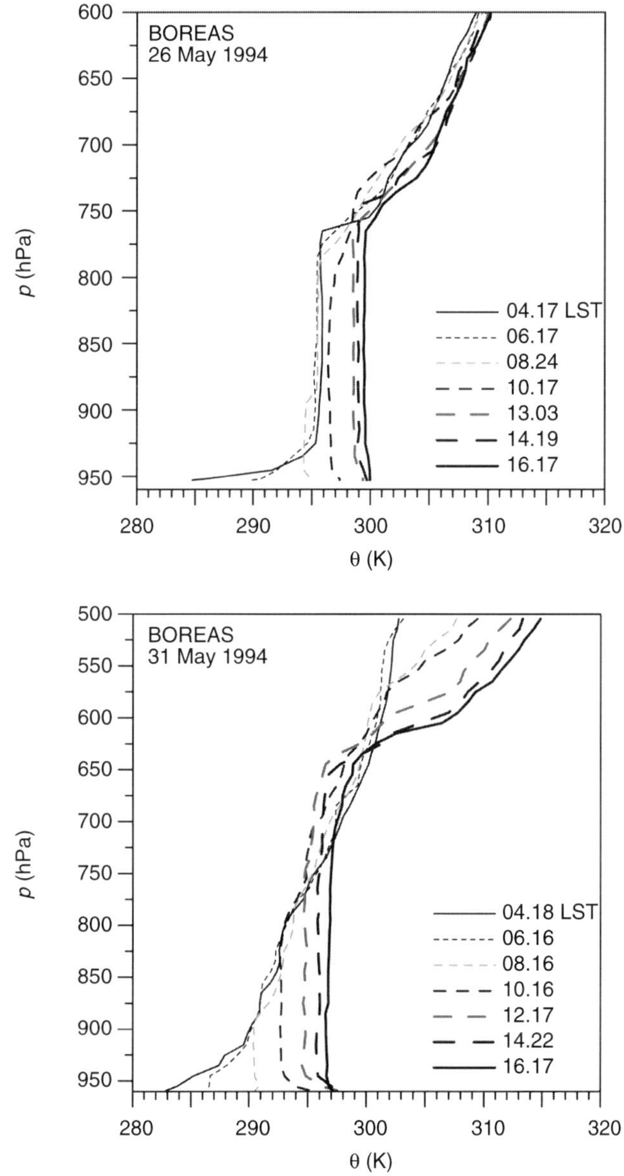

Fig. 6.3. Profiles of potential temperature in the lower troposphere on 26 and 31 May (Betts, 2002).

change in the vertical profile is also greater. On 31 May, the mixed layer grows steadily all day until it is 300 hPa deep (about 3000 m) in the late afternoon. On both these days, there is some broken cumulus cover in the afternoon at the top of the mixed layer. The rapid warming on 31 May that is seen between 500 and 600 hPa is related to the lowering and change in structure of a powerful jet stream above, not to surface processes.

Fundamental relationship between RH, saturation pressure and cloud base

Many introductory courses in meteorology imply that RH is not a fundamental variable. Rather, mixing ratio, a conserved quantity, is presented as a more basic variable. However, there are many parameters conserved in different processes in the atmosphere, and each tells its own story. RH does have a fundamental significance, because of its tight relationship to saturation pressure (which is conserved in dry and wet adiabatic processes (Betts, 1982)) and hence to lifting condensation level (LCL) and cloud base (the critical level of the liquid-phase transition, which affects radiative and microphysical processes). However, this link is generally poorly appreciated, and consequently the value of saturation pressure in modelling has not been fully explored. (Another unfortunate by-product, perhaps, is that RH has not traditionally been measured at climate stations.) This link is critical over land because the availability of water for evaporation is a major control (together, of course, with advection) on mixed-layer RH, and hence cloud base, and this relationship is largely independent of temperature. Figure 6.4a shows the relation between height of the LCL and RH as surface temperature varies, and Fig. 6.4b the corresponding relationship with $P_{LCL} = p_0 -$ the pressure height to the saturation level, p^*, or LCL. It is useful to keep in mind some characteristic values.

Over the ocean, as is well known, a typical cloud-base height of 500 m, or $P_{LCL} \approx 50$ hPa, corresponds to RH $\approx$ 80%. Over Amazonia in the rainy season, the afternoon cloud base may reach (in some wind regimes (Betts *et al.*, 2002)) around 800 m, or $P_{LCL} \approx 80$ hPa with RH $\approx$ 70%. Over the boreal forest in spring, as in the previous section, the cloud base may reach around 2500 m, or $P_{LCL} \approx 200$ hPa with RH $\approx$ 30%, while, over a desert, where water is largely unavailable for evaporation, the cloud base may be 3500 m, $P_{LCL} \approx 300$ hPa with RH $\approx$ 20% (or less).

Formally, P_{LCL} is directly related to (1 − RH) by the formula (Betts, 1997):

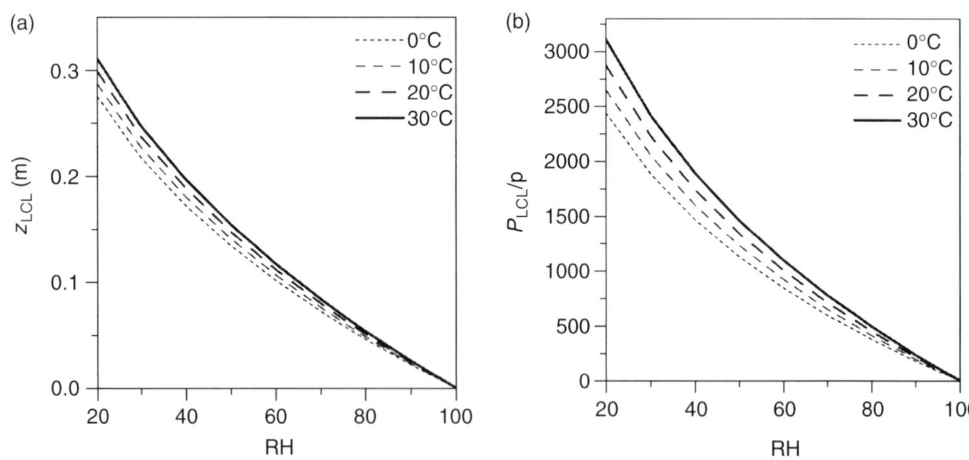

Fig. 6.4. (a) Relation between height of LCL (z_{LCL}) and RH as surface temperature varies. Note independence of surface pressure. (b) As (a) for ratio of P_{LCL} to surface pressure p. Note dependence on T is weak.

$$P_{LCL} = p(1 - RH)/(A + (A - 1)RH) \quad (3)$$

where $A \approx (0.622\ L/2\ C_p\ T)$ and increases with decreasing temperature from 2.6 at 25°C to 3.4 at −40°C, where L is the latent heat of vaporization and C_p is the specific heat of air at constant pressure.

Link between soil water (and resistance to evaporation) and P_{LCL}

Figure 6.5 shows how soil water, which is a primary control on 'resistance to evaporation' over land, controls the diurnal cycle of LCL (and RH). Figure 6.5a is the mean diurnal cycle of P_{LCL} from the European Centre 15-year reanalysis (ERA-15), averaged for nine Julys over the Missouri River basin (from Betts, 2000), and binned by soil water in the first model layer below ground (0–7 cm). There is a monotonic shift of the diurnal cycle of P_{LCL} and an increase in its amplitude for drier soils. RH goes down and LCL (related to cloud base) goes up as the resistance to evaporation at the surface, controlled by soil water, increases. (The model resistance actually depends on the whole root-zone soil water with bounds at the permanent wilting point of 0.131 and the field capacity of 0.323.) Figure 6.5b, for composites for the two summers of 1987 and 1988 from FIFE (1987 was shown in Fig. 6.1), shows data with a similar behaviour, although rather less pronounced than the model.

Diurnal cycle of CO_2

As we move towards fully coupled earth-system models capable of simulating the changing climate of the earth, the coupling of CO_2 with the meteorological

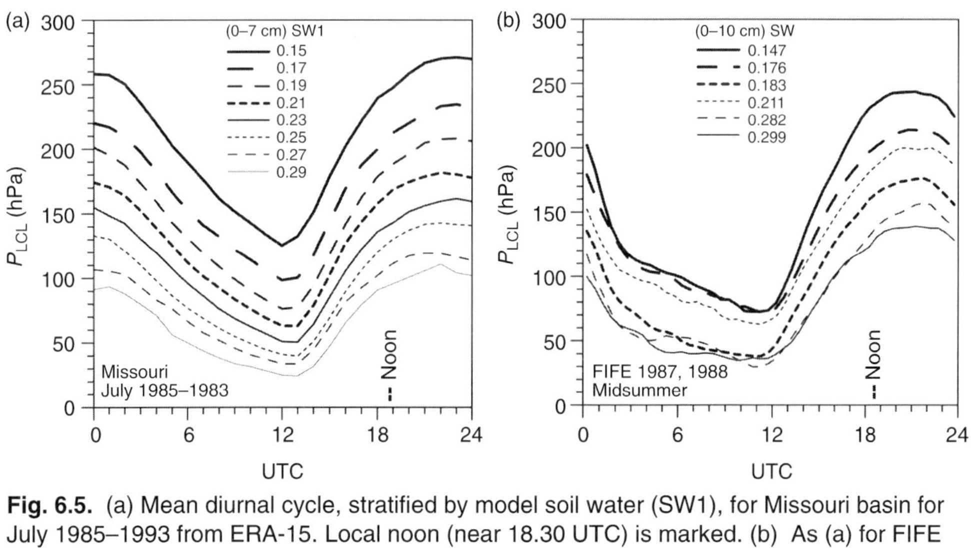

Fig. 6.5. (a) Mean diurnal cycle, stratified by model soil water (SW1), for Missouri basin for July 1985–1993 from ERA-15. Local noon (near 18.30 UTC) is marked. (b) As (a) for FIFE 1987 and 1988 midsummer composites, stratified by observed soil water (SW).

and climate fields is a critical issue. The diurnal cycle of the solar radiation drives a diurnal cycle in CO_2 through photosynthesis and respiration in plants. Respiration depends strongly on temperature, and at high latitudes the seasonal cycle of the diurnal cycle is large. Figure 6.6 shows the mean diurnal cycle over a young jack-pine canopy (about 5–6 m tall) near Thompson, Manitoba, from the 1996 BOREAS experiment for 3 months, June, August and October (from Betts *et al.*, 2001). During the summer months, CO_2 decreases during the daylight hours as it is taken up in photosynthesis, and it increases at night as it is released by respiration from both plants and soil. The amplitude of the diurnal cycle increases from June to August, as both photosynthesis and respiration increase, but the monthly mean decreases as there is a net CO_2 uptake by the entire northern hemisphere. By October of this year, however, the diurnal cycle is very small, as temperatures have dropped low enough for both photosynthesis and respiration to have almost ceased. The next generation of forecast models that assimilate CO_2 data need to get this diurnal cycle correct at the surface and in the BL. Evaporation and photosynthesis are tightly coupled and, consequently, there is a tight coupling between CO_2 and water vapour in the BL.

Coupling of water vapour and CO_2 within the BL

During the growing season, water-vapour and CO_2 gradients are strongly coupled through the daytime convective BL. Figure 6.7a shows the high correlation between BL fluctuations of mixing ratio, q, and CO_2, as air is mixed between the surface, where q is high and CO_2 low, and the free troposphere, where q is low and CO_2 is relatively high. The data are from the Canadian Twin Otter research aircraft (MacPherson, 1994). Figure 6.7b shows that the gradient in Fig. 6.7a within the BL is similar on different days near local noon (18.30 UTC), and is presumably controlled by stomatal transpiration and uptake of CO_2.

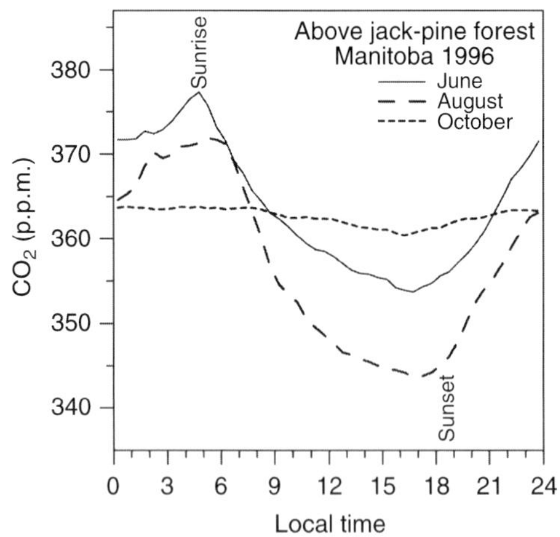

Fig. 6.6. Monthly mean diurnal cycle of CO_2 for June, August and October for a boreal jack-pine site.

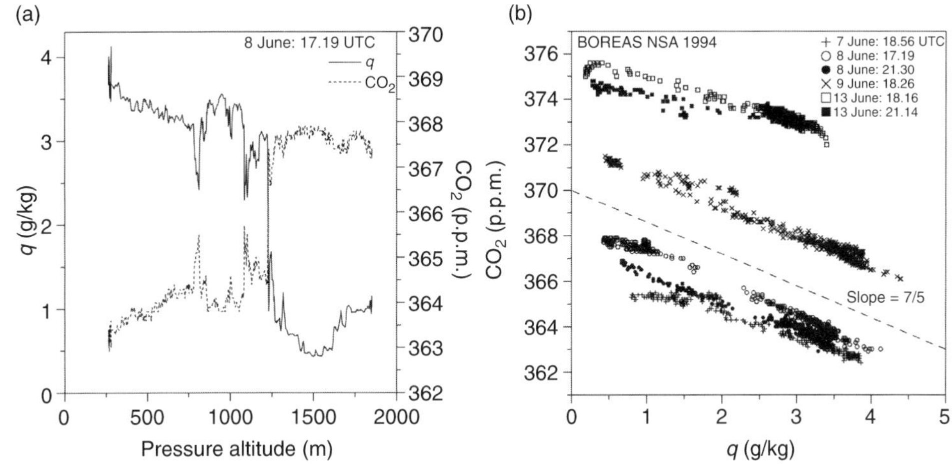

Fig. 6.7. (a) Profiles of mixing ratio, *q*, and CO_2 from surface to 1800 m. (b) Six aircraft profiles through the boundary layer, showing strong coupling between *q* and CO_2 fluctuations. The dotted line has a slope of 7 p.p.m. for a 5 g/kg change in *q*.

The Boreal Forest

An illustration from the boreal forest was used in Fig. 6.2 to show the coupling of the diurnal cycle to the residual BL. Here two other aspects are shown: the seasonal cycle of the diurnal cycle, and the dependence of the diurnal cycle of P_{LCL} on surface water.

Seasonal cycle of diurnal cycle at high latitudes

The freezing and thawing of the soil plays an important role in the climate at high latitudes. Winter temperatures are moderated by two processes. The first is the role of the snow pack as an insulator of the soil. The second is the freeze process, which increases the effective heat capacity of the soil by a factor of 20 (Viterbo *et al.*, 1999). This freeze–thaw introduces a significant lag into the climate system. In spring, a significant part of the net radiation goes into melting the snow pack, thawing the ground (Rouse, 2000) and melting the frozen lakes (and warming them). This energy becomes available in autumn and early winter, when the surface refreezes. In spring also, the ground thaw, which occurs when daytime temperatures rise well above freezing, is the key control on surface evaporation. Water is not available for transpiration in spring until the snow melts and the ground thaws, or for evaporation until the ground, wetlands and lakes warm in respect of the atmosphere. Jarvis and Linder (2000) suggest that it is the snow melt that first introduces liquid water into the soil profile. This unavailability of liquid water leads to very low evaporative fractions in spring, with large sensible-heat fluxes off the forest canopy, which in turn produce deep dry BLs in spring (as in Fig. 6.3). In autumn, when the lakes and ground are warm relative to the cooling atmosphere, the situation reverses.

The evaporative fraction is high for the conifers and lakes (but not for the deciduous species after leaf-fall). Net radiation is much lower by the time the surface freezes, so sensible-heat fluxes are very low, and BLs in autumn become very shallow, often capped by stratocumulus.

These seasonal changes in BL depth are very clear in graphs of P_{LCL}. Figure 6.8a (from Betts *et al.*, 2001) shows the mean diurnal cycle of temperature (averaged for the 3 years 1994–1996) from April (month 4) to October (month 10) for a site near Thompson, Manitoba (97.92°W, 55.80°N, elevation 221m). The peak daytime temperatures are barely above freezing in April and rise rapidly to a maximum in July and August, before falling again. In October, when the sun angle is low, the mean diurnal cycle is small and near freezing again. Figure 6.8b shows the seasonal trend of P_{LCL} for Thompson. P_{LCL} falls (and RH rises) almost monotonically from April to October, both during the day and at night. Conditions are so dry in spring, because the ground is still frozen and water is not available for evaporation. Because the trees have strong stomatal control, they can survive with high canopy temperatures (20–30°C) while their roots are still frozen. The particularly sharp fall from September to October is probably not just a local effect, but associated in part with the systematic advection of shallow BLs from the north.

Dependence of P_{LCL} on surface water availability

This site near Thompson has a stand of mixed spruce and poplar with a thick surface cover of moss. This acts as a reservoir for surface water, which has a large impact on evaporation, which falls on sequential days following significant rainfall events (> 5 mm). This impact can be seen in Fig. 6.9 on the diurnal cycle of P_{LCL} stratified by a wet-surface index (WS) (see Betts *et al.*, 1999, 2001). In this composite of all days from May to September, 1994–1996, WS = 0 repre-

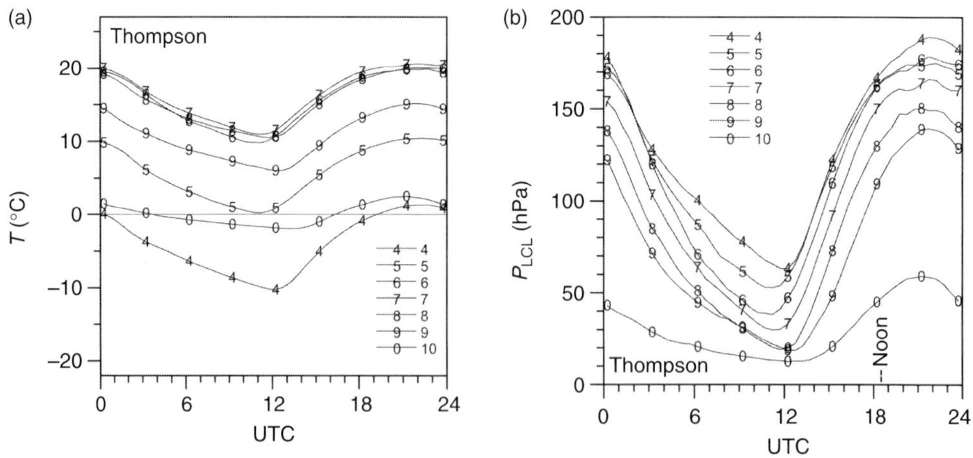

Fig. 6.8. (a) Diurnal cycle of above-canopy temperature for Thompson from April (month 4) to October (month 10). (b) As (a) for P_{LCL}.

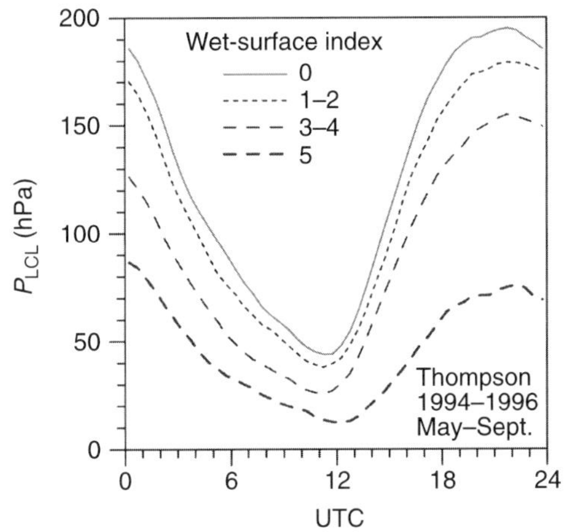

Fig. 6.9. Diurnal cycle of P_{LCL} for Thompson, Manitoba, stratified by wet-surface index.

sents days when the moss has dried out (negligible rain for 5 days) and WS = 5 represents days when ≥ 5 mm of rain fell the preceding day. The diurnal cycle depends strongly on the availability of water for evaporation at the surface (in part because evaporation from the conifers is subject to tight stomatal control). Within a few days following large rain events, mean afternoon cloud-base height rises dramatically as the surface dries out, until the characteristic deep dry BLs over the summer boreal forest are again established.

Diurnal Cycle in Rondônia in the Rainy Season

In this section model output from the European Centre for Medium-range Weather Forecasts (ECMWF) is compared with data collected at a pasture site located near Ouro Preto d'Oeste, Rondônia, Brazil (10.75°S, 62.37°W; about 30 km north-west of Ji-Parana), during the wet-season months of January and February 1999 as part of the large-scale biosphere–atmosphere experiment in Amazonia (LBA) wet-season campaign. The site is part of a large deforested area (> 250 km^2) dominated by short grass with isolated palm and hardwood trees scattered throughout the landscape. At this site, a micro-meteorological tower, eddy-correlation instrumentation and a gas analyser measured the surface meteorology and energy-balance components. For the comparison here, we use hourly averaged data from the analysis (Betts and Jakob, 2002). The forecast model outputs (for the nearest model grid point) used for comparison were from 12 to 36 h short-range forecasts, run at a triangular truncation of T319 and a vertical resolution of 60 levels, from each operational 12.00 UTC analysis. The forecast model was the operational ECMWF model in autumn 2000, which includes the tiled land-surface scheme (with the acronym TESSEL (Van den Hurk *et al.*, 2000)) and recent revisions to the convection, radiation and cloud schemes described in Gregory *et al.* (2000). The purpose of this section is to illustrate how time-series data from a field experiment can be used to check model-output time series and identify where and why a model fails to represent properly the observed atmospheric processes. This is critical research, as there remains a significant gap between real observable atmospheric processes and our ability to represent them parametrically in numerical models, and this limits our confidence in, for example, climate simulations.

Composites by lower-tropospheric wind regime

Composites are useful for the comparison between model and surface observations as they average over several days and the individual small-scale convective events within them, and give a picture of the mean diurnal cycle, more representative perhaps of the 60 km × 60 km grid square used in this global forecast model. We group the 40 days (from day of year 20 to 59) for which we have data at the pasture site into the five groups shown in Table 6.1. The first four correspond to the surface easterly and westerly lower tropospheric wind regimes in the analysis of Halverson *et al.* (2002). These two distinctly different lower tropospheric wind regimes in Rondônia were associated with significant differences in convection. Typically, the westerly regime was associated with a moister troposphere and convection with weak vertical development and weak electrification, while the easterly regime had a slightly drier troposphere and more strongly electrified convection with greater vertical development. The fifth group in Table 6.1 is a composite of 8 selected days when a strong rainband passed directly over the measurement site in the mid-afternoon. This is shown as a separate group, since strong convective down draughts in the afternoon produce such a distinct modification to the diurnal cycle, in order to see whether this feature is reproduced in the model. The last two columns compare the mean daily observed and model precipitation. Daily 'observed' precipitation was defined by first taking the mean of the rain-gauges in four networks (established to validate the Tropical Rainfall Measuring Mission satellite (TRMM)) and then averaging these four means (see Betts *et al.*, 2002). This basic grouping by the sequence of lower-tropospheric easterly and westerly components also represents a time progression of the rainy season.

Table 6.1. Surface diurnal composites.

		Mean daily precipitation (mm)	
Composite name	Days included (UTC)	Observed	ECMWF model
1. East 20–28	20–28	5.7	8.5
2. West 29–38	29–38	6.9	4.9
3. East 39–52	39–52	7.5	7.3
4. West 53–59	53–59	10.1	10.7
5. WET-8 days	20, 24, 30, 31, 38, 42, 44, 48 (afternoon rainband)	8.9	8.4
All days	20–59	7.4	7.6

For the entire 40-day period (bottom row), the mean daily precipitation for the TRMM rain-gauge networks is 7.4 mm, in close agreement with the model's 7.6 mm. However, the model does not reproduce well the increasing temporal trend of precipitation of the first four composites, and on a daily basis the correlation between model and rain-gauge observations is poor (not shown).

Diurnal cycle of precipitation

Despite reasonable agreement in mean precipitation during the rainy season, the current ECMWF model has a clear error in its diurnal cycle of precipitation over Rondônia, as shown in Fig. 6.10. The composites reflect the five groups of days in Table 6.1. On the left is the current ECMWF model using TESSEL as its land-surface model and a convective available potential energy (CAPE) convective closure for deep convection (Gregory *et al.*, 2000). Every model composite has a rainfall peak just after 12.00 UTC, about 2 h after sunrise, which is not observed in any of the composites on the right, derived from an average of four TRMM rain-gauge networks in Rondônia. Most days have afternoon rainfall maxima, some also have rain at night, while all have a rainfall minimum in the morning for the period 12.00–14.00 UTC (08.00–10.00 LST). This is also the time when the TRMM radars show a minimum in both fractional rain area and conditional rain intensity (Halverson *et al.*, 2002). We shall discuss below (Discussion of this model diurnal-cycle error) the reasons why the model convective parameterizations produce a precipitation maximum in the morning at a time when in reality rainfall is a minimum. The model produces a secondary rainfall maximum in late afternoon. For the WET-8 rainband composite, this peak is slightly higher than for the other composites, consistent with the obser-

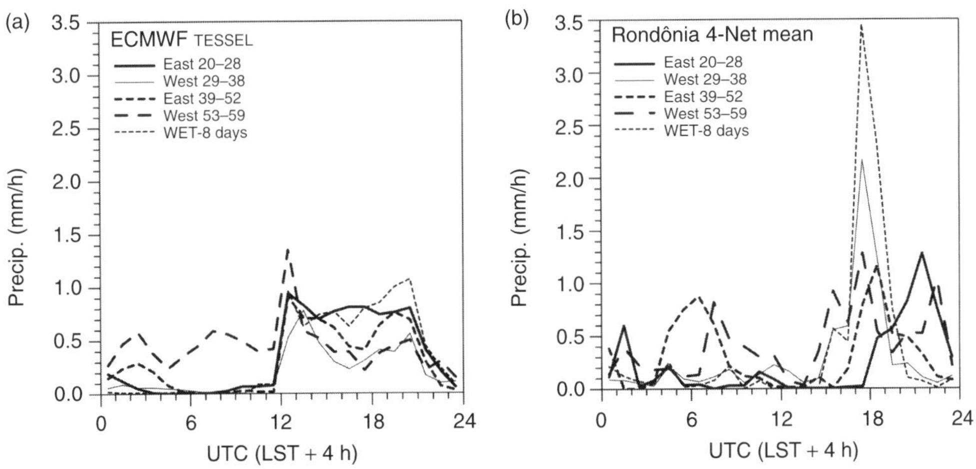

Fig. 6.10. (a) Mean diurnal cycle of precipitation over Rondônia for five convective classifications for the current ECMWF model. (b) As (a) for observed mean diurnal cycle of precipitation over Rondônia: an average of four rain-gauge networks.

vations (although the model peak is later in time and broader). For the last westerly composite (53–59), the model is consistent with the data in having a weaker diurnal cycle of precipitation, including more precipitation at night.

Comparison of surface thermodynamic cycle

Figure 6.11 compares the mean surface thermodynamic cycles of potential temperature, θ, mixing ratio, q, equivalent potential temperature, θ_E, and pressure height to the lifting condensation level, P_{LCL}, between the ECMWF model on the left and the pasture-site observations on the right. Overall, the model biases are rather small, typically cooler and wetter in the daytime. As a result, the model has a lower mean P_{LCL}, corresponding to a lower mean cloud base, and rather little bias in mean θ_E. In the model, the earlier onset of precipitation in the diurnal cycle is producing a cooler and moister BL in the daytime by the evaporation of falling precipitation. A distinct break in the model diurnal cycle can be seen, particularly in q and θ_E profiles, at 12.00 UTC with the onset of rain. Unlike the data, the model shows little variability in maximum temperature. The model does show some differences in q structure between easterly and westerly wind regimes (although the convective parameterizations are not directly aware of wind shear). The easterly regimes are wetter, with a higher afternoon θ_E, than the westerly regimes, and do not show a morning fall of q. However, this is not in agreement with the observations, which show the morning fall of q for the easterly regime and mean q increasing from the beginning to the end of the period (consistent with the mean precipitation increase). The θ_E comparison of the WET-8 rainband composite shows that the model does not represent the unsaturated down-draught process, which brings low-θ_E air down to the surface. Unlike the downward spike in the data at the time of the rainfall maximum, the model has a θ_E maximum at the time of its rainfall maximum, which is later at 20.00 UTC. The P_{LCL} comparison shows a much larger variation in the data, with the model resembling most closely the low-cloud-base case of the last westerly regime, which has more frequent weaker showers and the weakest diurnal cycle of rainfall. It is clear that the mechanisms by which the model convective parameterizations for shallow and deep convection are producing precipitation and modifying the BL, while they produce a plausible diurnal cycle of cloud base and θ_E, do not reproduce in detail the variability observed over Rondônia in these composites. The morning onset of precipitation is too early, which truncates the growth of the shallow cumulus BL (CBL), and the unsaturated down-draught process seems largely missing. It seems likely that improving the diurnal evolution of the CBL, so that the morning transition to deep convection is delayed, may significantly improve the model over land in the tropics.

Discussion of this model diurnal-cycle error

The purpose of these comparisons was to understand in more detail (and then correct) the error in the diurnal cycle of precipitation seen in Fig. 6.10. Some

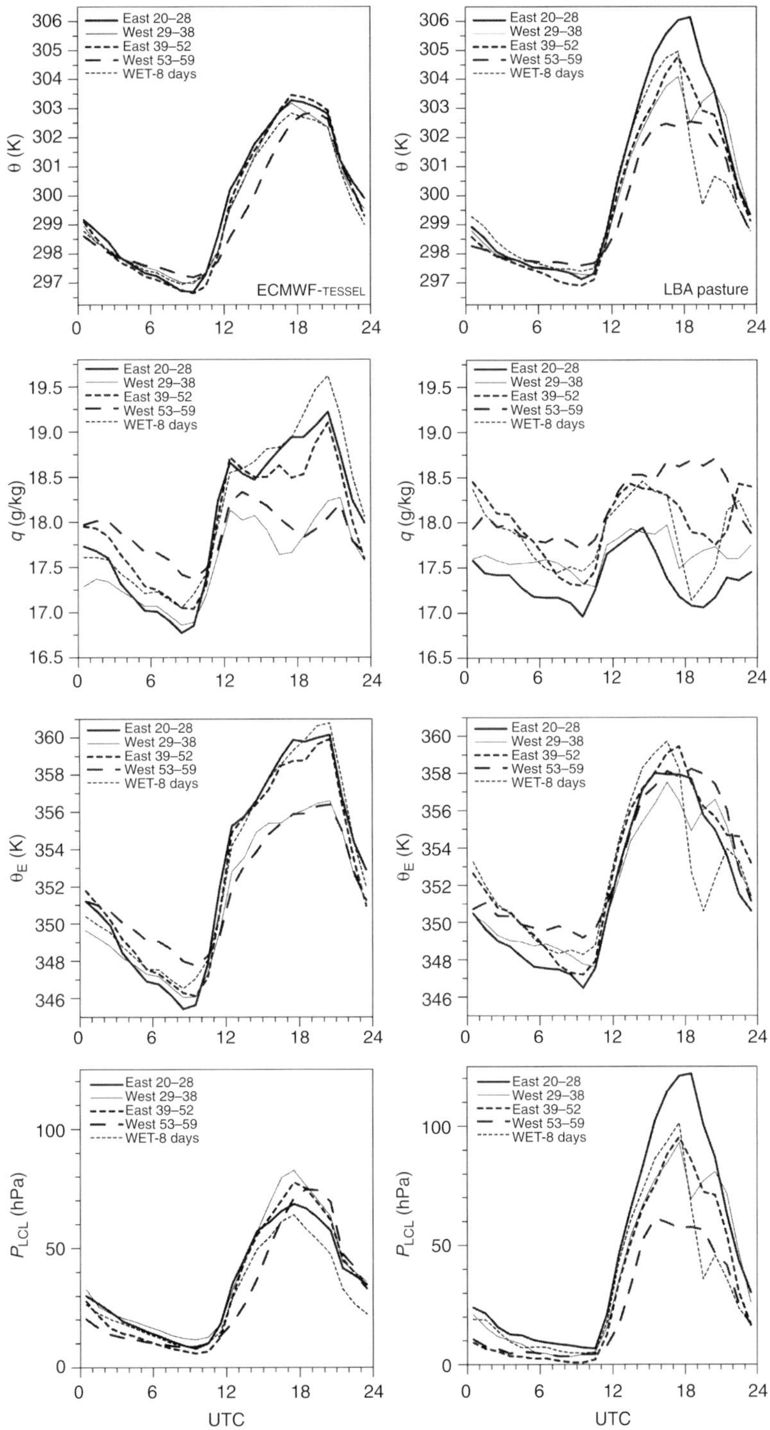

Fig. 6.11. Comparison of surface thermodynamic cycle in ECMWF model (left) with LBA pasture site (right).

progress has been made in understanding the early onset of precipitation in the model. The diurnal evolution of the tropical BL involves the tight interaction of many processes. At dawn, the layer near the surface is generally saturated and stabilized by precipitation late in the previous day and by radiative cooling at night (see Betts *et al.*, 2002). The surface sensible- and latent-heat fluxes are trapped initially in a shallow stable layer less than 400 m deep, and so initially θ and q rise rapidly in a shallow, growing, mixed layer. This process is reasonably represented in the model, but, only 2 h after sunrise, model representation and reality separate. At the Abracos pasture site, a shallow cumulus layer deepens rapidly once the nocturnal BL is penetrated, since the atmosphere above 950 hPa is conditionally very unstable and transports the large surface evaporation up and out of the subcloud layer. In the easterly wind regimes, mixed-layer q even starts to fall around 13.00–14.00 UTC (09.00–10.00 LST) as a result of this upward transport of moisture into non-precipitating clouds (Fig. 6.11). The developing cumulus grows deep enough to produce the first radar echoes about 15.00 UTC (11.00 LST), and the first showers often form near local noon. Organized convective bands typically take until around 18.00 UTC (14.00 LST) to develop in this Rondônia region (Silva Dias *et al.*, 2002). The model convective parameterizations do not describe this growing CBL stage at all well. In fact, as soon as the surface heating in the model breaks though the nocturnal BL about 2 h after sunrise, the deep convective parameterization 'sees' the deep conditionally unstable atmosphere, calculates a convective cloud top in the upper troposphere and produces convective rain (see Fig. 6.10a), computing a time-scale for the process from a CAPE closure. Although the shallow cloud parameterization (which has its own equilibrium closure, based on moist static-energy balance in the subcloud layer in a single time step) is activated intermittently, most of the morning growing shallow CBL phase is bypassed in the model. Calculating cloud top, or the depth of the CBL, during this morning growth phase is a challenging problem in a numerical model, as the tropical atmosphere over Rondônia in the rainy season is so conditionally unstable from 900 to 600 hPa. The reason is that, unless there is a strong inversion, as in the trade winds, the depth of the CBL, as it evolves, is determined by mixing or entrainment processes between the clouds and their surroundings. As yet, no suitable general formulation of this entrainment process has been found that will give the depth of the growing CBL in the very unstable atmosphere of the tropics over land. It seems clear also that most large-scale models, including the ECMWF model, are deficient in the way they separate convection and clouds into separate parameterizations. Rather than the continuum of convection seen in nature, in the model, the growing convective BL is broken into a dry convective process and a shallow cloud process; furthermore, shallow and deep convection are computed with two separate parameterizations with distinct closures.

The near-surface diurnal cycle of the thermodynamic variables and the surface fluxes (not shown) are influenced by the evaporation of falling precipitation, which occurs too early in the model. Cloud-base and near-surface temperatures are lower than observed in the model in the early afternoon by about 20 hPa and 1 K, respectively. These biases, however, are small by global-model standards. Interestingly enough, the model comes within ±2 K of replicating the maximum afternoon θ_E observed of around 358 K, although there is no evidence that the model represents well the convective down-draught process, which can be seen to bring low-θ_E air into the subcloud layer. The daytime diurnal cycle of the mixing ratio in the model is within ±0.5 g/kg of observations, which is also good for the moist tropical BL, although the detailed differences seen between the composites are not reproduced. The model has a small high bias in the evaporative fraction (not shown). One cause is that the early precipitation in the morning fills the model surface-water reservoir, which evaporates unimpeded by stomatal control.

Conclusions

This chapter has reviewed the nature of the diurnal thermodynamic cycle over land, using illustrations from high latitudes to the tropics. The coupling between different processes at the land surface is of fundamental importance, because in global models many processes are parameterized and are poorly constrained by routine observational inputs. Yet the diurnal cycle is observed synoptically, so it is an excellent indicator of whether the surface processes are correctly modelled, and of course it is a key forecast product.

The first section showed how, given a similar net radiation budget at the surface at a grassland site, soil moisture controls evaporation (through its impact on stomatal resistance), the diurnal cycles of temperature and RH. The pre-existing or residual BL above the nocturnal stable BL has a large impact on the details of the diurnal cycle as the morning mixed layer (ML) grows. This was illustrated comparing 2 days over the boreal forest in spring, when a deep BL develops because very little water is available for evaporation. The fundamental link between resistance to evaporation and the diurnal cycles of RH and P_{LCL} distinguishes the land surface from the ocean surface, where the surface is saturated. A section on the diurnal cycle of CO_2 and its daytime vertical gradients was included to emphasize the importance of CO_2 as a climate variable, not only for its radiative significance but also because it is tightly coupled in the BL to water vapour.

A section was included discussing further the diurnal cycle over the boreal forest: the seasonal cycle of the diurnal cycle and the dependence of the diurnal cycle of P_{LCL} on surface-water reservoirs. At high latitudes, the seasonal cycle of temperature is large and soil freezing plays a major climatic role. In spring, water is not available for evaporation until the ground melts. A monotonic increase of the diurnal cycle of RH (decrease of LCL) was seen in northern Manitoba from April to October. Surface-water stores, particularly in the moss layer, play an important role in surface evaporation in summer.

The comparisons between the ECMWF model and the Rondônia data illustrate the process of identifying model diurnal-cycle errors and their causes. The model diurnal cycle of precipitation is clearly in error in the wet season over the southern Amazon, and the cause is the poor representation of the growth of the shallow CBL. Although the model diurnal cycle of the near-surface thermodynamics is quite close to that observed, it is being produced in the model by a different mix of BL and surface processes, primarily more rainfall evaporation and less shallow cumulus convection than are observed. Consequently, the model diurnal cycle most closely resembles the westerly wind regime in late February, when showers were more frequent. Data comparisons of this type can be used both to identify errors and to systematically improve and unify the model convective parameterizations until they simulate reality more closely.

Acknowledgements

Alan Betts acknowledges support from the National Science Foundation under grant ATM-9988618 and from the National Aeronautics and Space Administration (NASA) under grants NAG5-7377 and NAG5-8364.

References

Betts, A.K. (1982) Saturation point analysis of moist convective overturning. *Journal of the Atmospheric Sciences* 39, 1484–1505.

Betts, A.K. (1997) The parameterization of deep convection. In: Smith, R.K. (ed.) *The Physics and Parameterization of Moist Atmospheric Convection.* NATO ASI Series C, Vol. 505, Kluwer Academic Publishers, Dordrecht, pp. 255–279.

Betts, A.K. (2000) Idealized model for equilibrium boundary layer over land. *Journal of Hydrometeorology* 1, 507–523.

Betts, A.K. (2002) Diurnal cycle. In: Holton, J.R., Pyle, J. and Curry, J.A. (eds) *Encyclopedia of Atmospheric Sciences.* Academic Press, London, pp. 640–643.

Betts, A.K. and Ball, J.H. (1995) The FIFE surface diurnal cycle climate. *Journal of Geophysical Research* 100, 25679–25693.

Betts, A.K. and Jakob, C. (2002) Evaluation of the diurnal cycle of precipitation, surface thermodynamics and surface fluxes in the ECMWF model using LBA data. *Journal of Geophysical Research* 107, 8045, doi: 10.1029/2001JD000427.

Betts, A.K., Goulden, M.L. and Wofsy, S.C. (1999) Controls on evaporation in a boreal spruce forest. *Journal of Climate* 12, 1601–1618.

Betts, A.K., Ball, J.H. and McCaughey, J.H. (2001) Near-surface climate in the boreal forest. *Journal of Geophysical Research* 106, 33529–33542.

Betts, A.K., Fuentes, J., Garstang, M. and Ball, J.H. (2002) Surface diurnal cycle and boundary layer structure over Rondônia during the rainy season. *Journal of Geophysical Research* 107, 8065, doi: 10.1029/2001JD000356.

Gregory, D., Morcrette, J.-J., Jakob, C., Beljaars, A.C.M. and Stockdale, T. (2000) Revision of the convection, radiation and cloud schemes in the ECMWF model. *Quarterly Journal of the Royal Meteorological Society* 126, 1685–1710.

Halverson, J.B., Rickenbach, T., Roy, B., Pierce, H. and Williams, E. (2002) Environmental characteristics of corrective systems during TRMM–LBA. *Monthly Weather Reviews* 130, 1493–1509.

Jarvis, P.G. and Linder, S. (2000) Constraints to growth of boreal forests. *Nature* 405, 904–905.

MacPherson, J.A. (1994) *NRC Twin Otter Operations in BOREAS 1994*. Report LTR-FR-129, Institute for Aerospace Research, National Research Council, Ottawa, Canada.

Rouse, W.R. (2000) The energy and water balance of high latitude wetlands. *Global Change Biology* 6, 59–68.

Silva Dias, M.A.F., Rutledge, S., Kabat, P., Silva Dias, P.L., Nobre, C., Fisch, G., Dolman, A.J., Zipser, E., Garstang, M., Manzi, A., Fuentes, J.D., Rocha, H., Marengo, J., Plana-Fattori, A., Sá, L., Alvalá, R., Andreae, M.O., Artaxo, P., Gielow, R. and Gatti, L. (2002) Clouds and rain processes in a biosphere atmosphere interaction context in the Amazon region. *Journal of Geophysical Research* 107, 8072, doi: 10.1029/2001JD000335.

Van den Hurk, B.J.J.M., Viterbo, P., Beljaars, A.C.M. and Betts, A.K. (2000) *Offline Validation of the ERA40 Surface Scheme*. ECMWF Technical Memo No. 295, ECMWF, Reading, UK, 43 pp.

Viterbo, P., Beljaars, A.C.M., Mahfouf, J.F. and Teixeira, J. (1999) The representation of soil moisture freezing and its impact on the stable boundary layer. *Quarterly Journal of the Royal Meteorological Society* 125, 2401–2426.

7

Medium- and Long-term Ecosystem Processes: Implications at the Forest–Atmosphere Interface

Frank Berninger
Department of Forest Ecology, PO Box 27, 00014 University of Helsinki, Finland

Introduction

Over 90% of the tree dry mass is derived from carbon dioxide and water. These substances are primarily exchanged between the ecosystem and the atmosphere through the plant canopy. The amount, properties and spatial distribution of foliage control this exchange. When analysing long-term changes in these processes we have to keep in mind that trees are large long-lived organisms. The size and age of trees result in a considerable inertia in the exchange processes and make it possible to have long delays and slow dynamic feedback loops in the reaction of ecosystems (and their relations to the atmosphere). These are due to internal, intrinsic properties of the forest stands and the trees they are composed of, as well as the interaction of these properties with each other and external factors, such as climate.

The present volume focuses on a discussion of the exchange process and its description. In this chapter, however, I shall describe how the properties of forests change in the long term. I shall point out emergent properties at the long-term level and describe modelling approaches to adequately describe these processes and to integrate the exchange processes into long-term dynamics. My focus will be at the scale of a few years to a few decades. I shall exclude the role of succession and stand-replacing disturbances from my discussion.

Research at the forest–atmosphere interface is at a critical stage. Measurement of forest carbon and water balances has moved from being a short-term, high-tech exercise towards the collection of consistent long-term monitoring data sets (as done in the EUROFLUX and AMERIFLUX networks). Medium- and long-term processes are likely to affect the carbon and water balances measured in these data sets and an understanding of the processes is required to analyse inter-annual variation of the stand fluxes.

Time Responses from Years to Decades

Forest–atmosphere interactions depend on the characteristics of the vegetation. These characteristics will not be constant over time, but the leaf-area index and the biological properties of foliage, of the trees and even of the soil are bound to change over time. Some variations will be due to the slow development of the stand and some due to variation in climate or other short-term influences.

Tree-ring research, confronted with similar problems in slow-growing trees, has made the distinction between long-term age-dependent growth trends, low-frequency variations in growth and high-frequency variations in growth. Age-dependent growth trends are regular slow changes in tree growth, due to ageing of the trees or increases in tree size. Low-frequency variations are observable at the scale of decades and tree-ring researchers attribute them to major disturbances of stand structure (such as thinnings or other partial cuttings). High-frequency variations have a time-scale of 1 year to a few years and are attributed to variations in climate (e.g. Fritts, 1976).

The clearest process in forest development is cumulative growth of trees over several years. From the point of view of forest–atmosphere interactions, this comes along with changes in biomass, increases in canopy height and changes in the spatial arrangement of foliage. These changes, however, are different for the different biomass compartments of the stand, i.e. while stand foliage biomass and fine-root biomass do not change very much after canopy closure, coarse-root, stem and branch biomass will continue to increase (Fig. 7.1). Another change in the stand structure consists of changes in plant canopy height and tree number (due to thinning or natural mortality). These changes bring along changes at the forest–atmosphere interface (as changes in the light interception due to changed clumping of the foliage and changes in atmospheric turbulence).

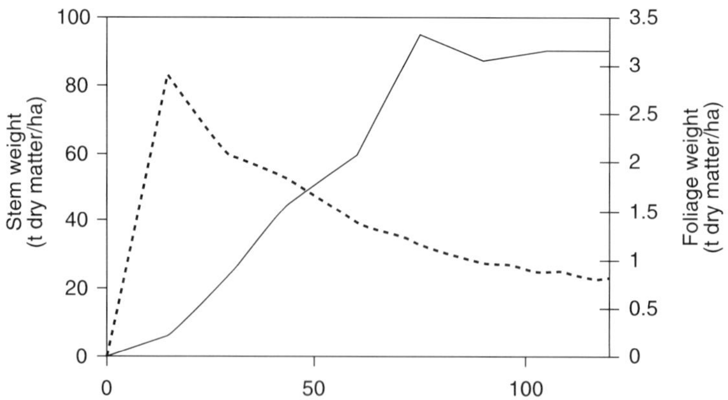

Fig. 7.1. Development of foliage (broken line) and stem (solid line) in an unthinned Scots-pine forest in southern Finland over time. Data were generated by applying the biomass equations of Marklund (1988) to the growth and the yield table of Ilvessalo (1920) for natural stands.

Short-term variation in forest growth has received much less emphasis from a biological point of view. We tried to investigate the inter-annual variation in net ecosystem exchange (NEE) for our Hyytiälä forest, but we did not find any good correlations of yearly NEE with climatic variables, such as temperature, irradiance, etc. However, year-to-year variations in climate will result in changes in foliage biomass, growth, etc. Year-to-year variations in carbon input to the soil through litter-fall may be an important source of variations in the soil carbon balance. For example, in the data of Cousens (1988) for a Scots-pine forest in Scotland, the litter input had a standard deviation of 819 kg/ha across 12 years of measurements. This would result in a variation of carbon input to the ecosystem of about 400 kg (Fig. 7.2). As foliage input is readily decomposed, its variation will probably cause corresponding variability in soil respiration fluxes. Variability in root-litter production is less understood but might result in similar variation.

Therefore, at least stand growth has a strong autoregressive component, i.e. growth of one year depends strongly on the growth of the previous year (Fig. 7.3). Besides leaf-level photosynthesis, stem growth of forest stands depends on other factors. Diameter growth, as such, could be lagged in respect of current environmental conditions. For example, Berninger *et al.* (2003) found that the diameter growth of subarctic Scots pine in northern Fennoscandia reacts to the photosynthetic production of the previous year (Fig. 7.4).

Where the Feedback is Taking Place

There are basically three locations where feedback between the atmosphere and the trees could take place: first, leaf properties, such as stomatal conductance or photosynthetic capacity, may change; secondly, the way a tree distributes its biomass among different plant organs (or the allocation) may change (with possible impacts on light interception and the distribution of foliage); and, thirdly, the stand structure (i.e. the number of trees and their sizes) may change. I shall discuss these three modes of feedback in more detail.

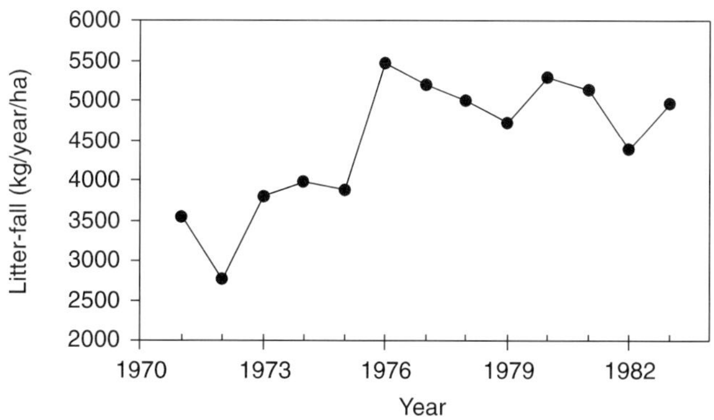

Fig. 7.2. Yearly variation in litter-fall for a Scots-pine stand in western Scotland (from Cousens, 1988).

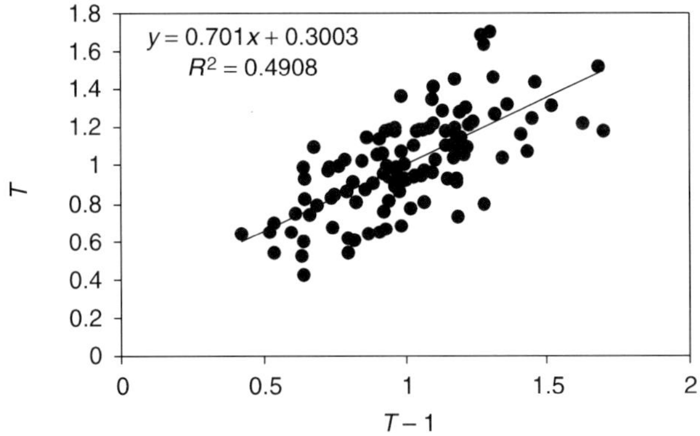

Fig. 7.3. Correlation of ring width with ring width from the previous year. The example is a detrended tree-ring chronology (so-called standard chronology) from Karasjok. (Data are from Lindholm, 1996. Data archived at the World Data Center for Paleoclimatology, Boulder, Colorado, USA.)

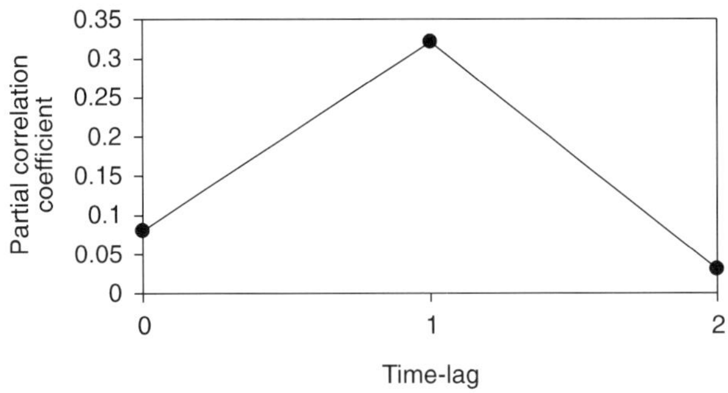

Fig. 7.4. Partial correlation between the tree-ring width index and photosynthesis at the time-lag of 0, 1 and 2 years (modified from Berninger *et al.*, 2003).

Feedback cycles affecting the gas exchange at the leaf level

The properties of leaves that affect forest–atmosphere interactions change during stand development. Yoder *et al.* (1994) proposed, in the so-called 'hydraulic limitation hypothesis of height growth', that water transport from the roots to the top of large trees is an obstacle for productivity. This transport takes place at large water tensions (up to several MPa) and there is evidence that the trees are operating close to the point where the transport system in the xylem fails due to excessive water stress (Tyree and Sperry, 1989). Mencuccini and Grace (1995) showed that leaf specific conductivity does indeed decrease with tree size in *Pinus sylvestris*. Indeed, foliar carbon-isotope and stomatal-conductance data indicate that old trees could be more limited by water than younger trees. However, Monserud and Marshall (1999) claim that there is interspecific variation. They investigated water-use efficiency of three Rocky Mountain conifers using the isotopic composition of tree rings. While there were indications of increasing water limitation for two of their species (*Pinus monticula* Dougl. and *Pinus ponderosa* Laws.), the third species (*Pseudotsuga menziesii* (Mirb.) Franco) showed no responses of water-use efficiency to an increase in size. The trees in this study were relatively young but they were growing up to 30 m during the analysis period. Bond (2000), however, showed, from pair-wise comparisons of old growth and young stands, decreases in stomatal conductance from middle age to old stands. However, it seems that some changes in leaf morphology, such as decreases in leaf mass per area, are evident with tree size (Thomas and Winner, 2002).

Short-term disturbances of a tree's structure may also result in altered leaf characteristics. Partial defoliation of trees may result in increased photosynthesis of the remaining parts. The responses of trees to such partial defoliation seem to be relatively rapid (i.e. take place in less than 2 weeks) and may persist for months or longer (Ovaska *et al.*, 1992). They form part of a larger response system of the plant involving changes in allocation and leaf biochemistry (Ovaska *et al.*, 1993).

Also Berninger *et al.* (2000) observed a prolonged reaction of water-use efficiency to a severe climatic event in the early 20th century. The climate events of 1902, a year with exceptional cold summer temperatures, which were associated with large foliage losses and widespread tree mortality, resulted in a decrease in water-use efficiency of Scots pine that persisted for several years. Water-use efficiency was estimated from the isotopic composition of tree-ring cellulose (Fig. 7.5).

Climatic variation will result in changes in decomposition and nutrient availability over time. I do not know of any publications where these are reported, and the effects of changes in the temporal rates of decomposition on foliage nutrient status and photosynthesis are not known. Fertilization results in rapid increases in leaf nitrogen content, photosynthesis and transpiration, with subsequent longer-term increases in foliage biomass.

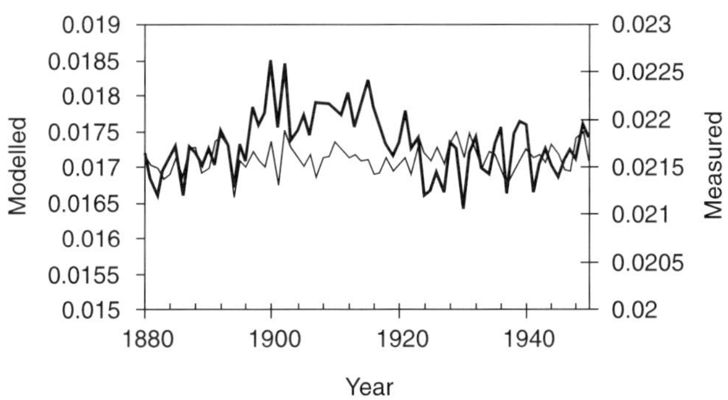

Fig. 7.5. Predicted and measured discrimination against ^{13}C for the period 1880–1950 for Karasjok, northern Norway. Measured data are from α-cellulose in pine rings. Simulated data were generated from a simple canopy gas exchange model. Thick lines denote measured istopic abundances, thin lines simulated isotopic abundances. Note that the offset between predicted and measured data is due to the post-photosynthetic discriminatory processes. (Data from Berninger *et al.*, 2000.)

Feedbacks via biomass distributions

Most of the living biomass in forests is in sapwood and only 'marginally' interacts with the atmosphere. Also the allocation, i.e. the distribution of new growth to different biomass compartments, changes with tree size and other factors. However, trees seem to maintain a certain allometry, i.e. the amounts of biomasses of different compartments are closely related to each other. This allometry is also maintained at the stand level. In particular, foliage biomass on a certain site type seems to be relatively independent of stand age (or stand density). Changes in allocation will result in differences between carbon that is returned 'immediately to the carbon cycle as litter' and carbon stored for some period in the form of wood. In other words, changes in allocation could affect the relationship between net primary production (NPP) and NEE.

At the tree level, a classical rule has been the pipe-model theory, i.e. the existence of a direct proportionality between sapwood area and foliage area. There has been a recent debate, involving mainly my friends in Edinburgh and ourselves, over the extent to which this ratio depends on climate and the extent to which it can be altered by climatic variation. Whitehead *et al.* (1984) proposed a theoretical relationship between foliage area, climate and sapwood area, based on the hydraulic properties of trees. Mencuccini and Grace (1995) showed that Scots pine from the same seed source planted at a cold and a warm location in Britain behaved, at least qualitatively, according to the 'Whitehead model', and similar trends were observed at a larger scale (Mencuccini and Bonosi, 2001). However, my co-workers and I found consistently different trends (Berninger and Nikinmaa, 1994) and we did not observe any influence of climate on the pipe-model relationships in Scots pine (Palmroth *et al.,* 1999). However, for another species, *Eucalyptus microtheca*, we found strong effects of drought on pipe-model ratios (Li *et al.*, 2000).

Overall, the pipe theory implies that large trees will allocate more growth to woody organs and less to foliage. It is not totally clear how quickly trees will restore their foliage–sapwood ratio after defoliation.

Feedbacks via stand structure and crown geometry

Stand structure and crown geometry will change over the development of stands as a response to the development of tree size or disturbances of stand structure. High-frequency changes induced by climate variability are not important. However, old forests will have fewer trees with larger crowns. Crown cover in old stands will generally be lower than in sapling stands (e.g. Easter and Spies, 1994). This implies that foliage is more clumped in these stands and light interception will be less efficient in them (Kellomäki and Oker-Blom, 1983). In other words, gross primary production in these stands will be less efficient at equal leaf-area index.

Modelling Medium-term Variation in Stand Growth

Experimental analysis of the causes underlying inter-annual variation of carbon fluxes is difficult. Also, most biomass measurements happen to have relatively large sampling errors and accurate determination of inter-annual variations in, for example, leaf-area index may be difficult.

In addition, we are dealing with several parallel changes going on in the system simultaneously. Process-based modelling as an analysis tool may have a better chance of succeeding, but, due to the fact that most process-based models have a large number of parameters and ecological hypotheses built into them, results might not be unambiguous.

Process-based models present an integrated framework to test the effects of short-term climatic variation or long-term trends during stand development on the forest–atmosphere interface. This section describes how to apply process-based models to analyse the linkages between stand-level processes and single physiological acclimatization mechanisms.

Process-based models differ from empirical models by being based on a causal description of the growth process, rather than on empirical relationships of growth and the environment. Process-based models are usually elaborated as dynamic non-linear models using differential or difference equations. Biomasses and their properties (such as nitrogen content) are usually part of the equations of state. This means that the value of such a variable at the time $t + 1$ depends on its value at time t.

In most process-based models we find the following six parts:

- A set of rules describing the uptake and loss of substances (usually nutrients and carbon).
- A set of rules describing the distribution (allocation) and redistribution of these substances (translocation) among different plant organs.
- A set of rules describing the death of plant organs or whole plants (senescence).
- A set of rules about how different plant organs interact with each other by modifying the environment (e.g. through shading in the canopy).
- A set of feedback rules on how internal states and properties of the plant affect its functioning.
- A set of rules about how litter is transformed in the soil (especially the carbon balance of litter and the rates of nutrient mineralization from litter).

This kind of system is shown in Fig. 7.6. A number of different ways have evolved as to how these rules can be formulated (e.g. Dixon *et al.*, 1978; Hari *et al.*, 1982; Valentine, 1999; Mäkelä and Hari, 1986; Mohren, 1987; Nikinmaa, 1992).

Conclusions

Continuous eddy-covariance measurements in forests for several years give a new long-term perspective on the forest–atmosphere exchange. Long-term flux

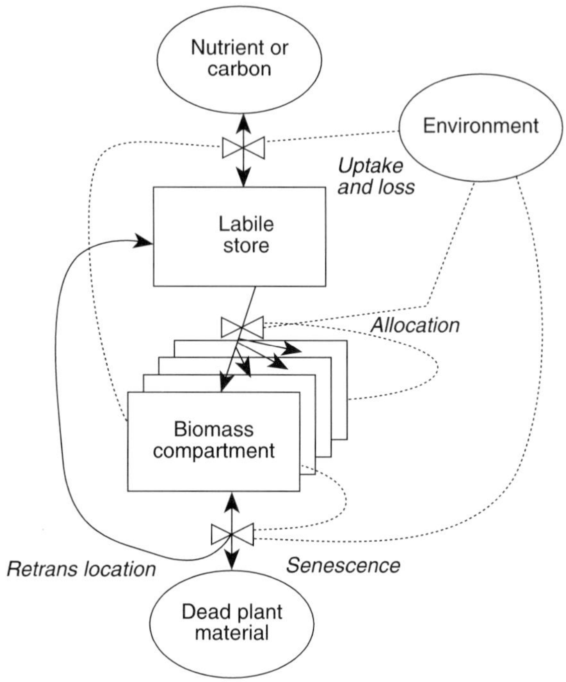

Fig. 7.6. Typical structure of a process-based model.

measurements spanning several years might, however, be affected by both variation of climate and long-term feedback processes, especially variations in soil respiration due to variations in litter input and feedback processes inside the forest. We have the experimental and modelling techniques to tackle both the long-term and the short-term feedback processes.

References

Berninger, F. and Nikinmaa, E. (1994) Within tree and between site variation in the foliage area/sapwood area relationship in Scots pine stands in different climatic conditions. *Canadian Journal of Forest Research* 24, 2263–2268.

Berninger, F., Sonninen, E., Aalto, T. and Lloyd, J. (2000) Modelling ^{13}C discrimination in tree rings. *Global Biogeochemical Cycles* 14, 213–223.

Berninger, F., Hari, P., Nikinmaa, E., Lindholm, M. and Meriläinen, J. (2003) Growth at Fennoscandian treelines is increasingly carbon saturated. *Tree Physiology* (in press).

Bond, B.J. (2000) Age-related changes in photosynthesis of woody plants. *Trends in Plant Science* 5, 349–353.

Cousens, I.E. (1988) Report of a twelve-year study of litter fall and productivity in a stand of mature Scots pine. *Forestry* 61, 255–266.

Dixon, K.R., Luxmoore, R. and Begovich, C.L. (1978) CERES – a model of forest stand biomass dynamics for predicting trace contaminant, nutrient and water effects. *Ecological Modelling* 5, 17–38.

Easter, M.J. and Spies, T.A. (1994) Using hemispherical photography for estimating photosynthetic flux density under canopies and in gaps in Douglas fir forests of the Pacific-Northwest. *Canadian Journal of Forest Research* 24, 2050–2058.
Fritts, H. (1976) *Tree Rings and Climate*. Academic Press, London.
Hari, P., Kellomaki, S., Mäkelä, A., Ilonen, P., Kanninen, M., Korpilahti, E. and Nygren, M. (1982) Metsikön varhaiskehityksen dynamiikka. *Acta Forestalia Fennica* 177, 1–82.
Ilvessalo Y. (1920) *Kasvu ja Tuottotaulut, Suomen eteläpuoliskon Mänty-, Kuusi J. Koivumetsille*. J. Simeliuksen Perilisten Kirjapaino, Helsinki, 96 pp.
Kellomäki, S. and Oker-Blom, P. (1983) Canopy structure and light climate in a young Scots pine stand. *Silva Fennica* 17, 1–21.
Li, C., Berninger, F., Koskela, J. and Sonninen, E. (2000) Different origins of *Eucalyptus microtheca* differ in their ability to acclimate to drought. *Australian Journal of Plant Physiology* 27, 231–238.
Lindholm, M. (1996) Reconstructions of past climate from ring-width chronologies of Scots pine (*Pinus sylvestris* L.) at the northern forest limit in Fennoscandia. PhD thesis, University of Joensuu, 169 pp.
Mäkelä, A. and Hari, P. (1986) A stand growth model based based on carbon uptake and allocation in individual trees. *Ecological Modelling* 205–229.
Marklund, L.G. (1988) *Biomassafunktioner for tall, gran och bjork i Sverige*. Rapport Institutionen for Skogstaxering, Sveriges Lantbruksuniversitet 45, Umeå, 73 pp.
Mencuccini, M. and Bonosi, L. (2001) Leaf/sapwood area ratios in Scots pine show acclimatization across Europe. *Canadian Journal of Forest Research* 31, 442–456.
Mencuccini, M. and Grace, J. (1995) Climate influences leaf area/sapwood area ratio in Scots pine. *Tree Physiology* 15, 1–10.
Mohren, G. (1987) *Simulation of Forest Growth, Applied to Douglas Fir Stands in The Netherlands*. Agricultural University, Wageningen, The Netherlands, 184 pp.
Monserud, R.A. and Marshall, J.D. (1999) Allometric crown relations in three Northern Idaho conifer species. *Canadian Journal of Forest Research* 29, 521–535.
Nikinmaa, E. (1992) Analysis of the growth of Scots pine: matching patterns with function. *Acta Forestalia Fennica* 235, 1–68.
Ovaska, J., Walls, M. and Mutikainen, P. (1992) Changes in leaf gas exchange properties of cloned *Betula pendula* saplings after partial defoliation. *Journal of Experimental Botany* 43, 1301–1307.
Ovaska, J., Walls, M. and Vapaavuori, E. (1993) Combined effects of partial defoliation and nutrient availability on cloned *Betula pendula* seedlings I. Changes in growth, partitioning and nitrogen uptake. *Journal of Experimental Botany* 44, 1385–1393.
Palmroth, S., Berninger, F., Nikinmaa, E., Lloyd, J., Pulkkinen, P. and Hari, P. (1999) Structural adaptation rather than water conservation was observed in Scots pine over a range of wet to dry climates. *Oecologia* 121, 302–309.
Thomas, S. and Winner, W. (2002) Photosynthetic differences between saplings and adult trees: an integration of field results by meta-analysis. *Tree Physiology* 22, 117–127.
Tyree, M.T. and Sperry, J.S. (1989) Vulnerability of xylem to cavitation. *Annual Review of Plant Physiology and Plant Molecular Biology* 40, 19–38.
Valentine, H.T. (1999) Estimation of the net primary productivity of even aged stands with a carbon allocation model. *Ecological Modelling* 122, 139–149.
Whitehead, D., Edwards, W.R.N. and Jarvis, P.G. (1984) Conducting sapwood area, foliage area and permeability in mature trees of *Picea sitchensis* and *Pinus contorta*. *Canadian Journal of Forest Research* 14, 940–947.
Yoder, B.J., Ryan, M.G., Waring, R.H., Schoettle, A.W. and Kaufmann, M.R. (1994) Evidence of reduced photosynthetic rates in old trees. *Forest Science* 40, 513–527.

8

A MAESTRO Retrospective

Belinda Medlyn
School of Biological Earth and Environmental Sciences, University of New South Wales, Sydney 2052, Australia

Introduction

It is sometimes said that the measure of a good scientist is not how much work they themselves do, but how much they stimulate in others. Accordingly, the MAESTRO model may be cited as one piece of evidence for the importance of Paul Jarvis's contribution to plant ecophysiology. The model was fostered by Paul throughout its development and its subsequent use in an extraordinary variety of applications. One may easily argue that it has strongly influenced the way in which we think about forest-canopy processes.

The occasion of Paul's retirement provides an opportunity for reflection on the development of forest-tree ecophysiology over the last few decades, and the history of the MAESTRO model embodies many aspects of this development. In this chapter I review the history of MAESTRO. I revisit the ideas leading to the development of the model and survey the wide range of applications for which it has been used. The history of the model takes us on a fascinating tour through forest-tree ecophysiology during the last three decades.

Development of MAESTRO

Although the name MAESTRO first appeared in print in Ying-Ping Wang's thesis in 1988, the model had a very long gestation period, stretching back to the early 1970s and involving work in several different countries. Inextricably associated nowadays with Edinburgh, the model's development also owes much to researchers in the USA and New Zealand.

The 'twinkle in the eye' that eventually led to the birth of MAESTRO, however, may be said to have taken place in Aberdeen, where John Norman came to do a post-doctoral associateship with Paul Jarvis from March 1971 to May 1972. They were studying light interception by shoots of Sitka spruce with the aim of modelling photosynthesis and transpiration in conifers. John Norman writes:

 Forests at the Land–Atmosphere Interface (eds M. Mencuccini, J. Grace, J. Moncrieff and K.G. McNaughton)

> From a model of shoot light interception that we never published, we knew that the distribution of light surrounding a spruce shoot was very important to predicting the photosynthesis and stomatal conductance of that shoot. Therefore we needed to model the light distribution in the spruce canopy in order to get a good estimate of what shoots were doing.
>
> (John Norman, 2001, personal communication)

The work they did led to two of the publications in the well-known 'Photosynthesis in Sitka spruce' series (Norman and Jarvis, 1974, 1975). Contemporary models of canopy radiation transmission generally represented canopies either as arrays of solid geometric objects or as horizontally homogeneous layers of randomly distributed elements (Lemeur and Blad, 1974). The major advance contributed by Norman and Jarvis (1974, 1975) was to measure and characterize non-randomness in forest-canopy structure and to incorporate this in a model. Reviews of radiation models of the time were strongly critical of 'armchair' models developed at the desk while field observations were almost non-existent (Lemeur and Blad, 1974; Norman, 1975). The work of Norman and Jarvis (1974, 1975) met this criticism with detailed measurements of shoot, whorl and crown structure (despite the 'arduous and self-defeating' nature of the task (Norman, 1975)). This work laid the foundation for a continued emphasis on empirical validation of theoretical results throughout the development of MAESTRO (Grace *et al.*, 1987a; Wang and Jarvis, 1990a).

Returning from Scotland, John Norman took up a position at Penn State in the USA, where, together with an MSc student, Jon Welles, he developed the model feature that was to distinguish MAESTRO from other models of its time, namely, the treatment of the canopy as a three-dimensional array of ellipsoidal tree crowns (Fig. 8.1). Jon Welles describes how this came about:

> I was a masters student at Penn State, looking for a thesis topic and an advisor, and somehow got connected with John Norman, one of the two faculty members (of about 15) there who did actual field work in micrometeorology. John did a good sales job on me, and we set out to do a radiative model of a heated orchard. This

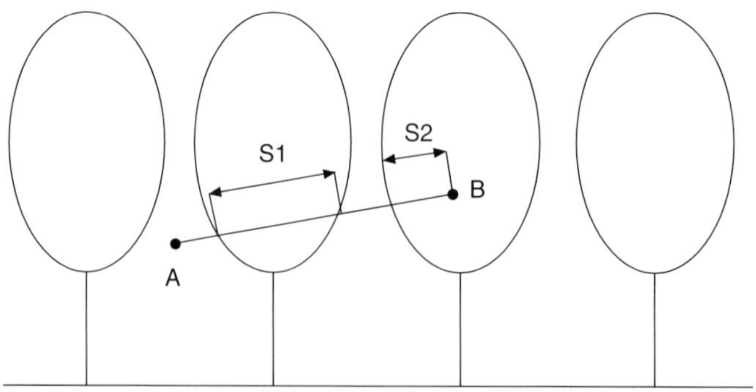

Fig. 8.1. Representation of the canopy in the Norman and Welles (1983) model. The canopy consists of an evenly spaced array of ellipsoidal tree crowns. The path length of radiation through crowns (S1 + S2) is calculated from geometrical considerations. Radiation transmission along the path length is calculated using Beer's law.

> was in the days when it was not uncommon to protect orchards from frost damage by interspersing fuel oil powered heaters among the trees. There were controversies about the most efficient heater arrangements and protocols (the first oil embargo having recently happened), and we thought we could provide some answers. The model was an exercise in geometry; one had an array of isolated tree canopies, each containing foliage at some density and orientation distribution. Interspersed among the trees was an array of heaters. The model computed foliage temperature distributions based on the radiation balance of foliage elements, which was determined by the element's relative view of the cold sky, hot heaters, ground, and other foliage. There were three publications that came out of that work, in addition to the thesis: 'An orchard foliage temperature model' (Welles *et al.*, 1979); 'Modeling the radiant output of orchard heaters' (Welles *et al.*, 1981); and 'Radiative transfer in an array of canopies' (Norman and Welles, 1983). I believe it was the material in this third publication that went into the model that came to be known as Maestro.
>
> (Jon Welles, 2001, personal communication)

The model was known as the General Array Model (GAR). Welles went on to develop the model further for his PhD, on bidirectional reflectance and transmittance, and renamed it BIGAR (Welles and Norman, 1991). Later he joined Li-Cor and notes that his experience in radiation modelling served him well when developing the LAI2000 Plant Canopy Analyzer.

Paul Jarvis remained in touch with Norman and was aware of the work that he and Welles were doing. Around 1982, Paul went to visit Forest Research in New Zealand for a period of several months. At that time the tree physiology section at Forest Research had as a goal:

> To model the impact of silviculture on crown and stem growth in radiata pine stands with the objective of providing the silviculturist with a predictive tool for managing his crop through pruning/thinning/fertilisation/disease control to optimise the desired timber product.

Paul suggested that the model of Norman and Welles (1983) would be a useful way to examine the effects of thinning and pruning on forest-tree growth, and a copy of the code was duly obtained from Welles. Jenny Grace was a post-doctoral researcher at Forest Research at the time and was given the job of modifying the model for *Pinus radiata*. She made many improvements that came to be fundamental to the model, including (Fig. 8.2): the specification of individual tree-crown positions and dimensions; the introduction of non-random foliage distributions within the crown; modelling of crowns as truncated ellipsoids, useful for simulating pruning; and the use of grid points evenly spaced through the crown for calculating photosynthesis (Grace *et al.*, 1987a,b). The model at this time was known as RADIATE.

Although Grace's work modifying the model was successful, she notes that the overall aim of the project, to develop a process-based growth model, was not achieved. She writes:

> [The aims of the model] were really in direct competition with those of mensurational models, and these could be developed far quicker and included data from a wide range of sites. The geology and climate within New Zealand is far too variable for data from one site to hold for all other sites.
>
> (Jenny Grace, 2001, persononal communication)

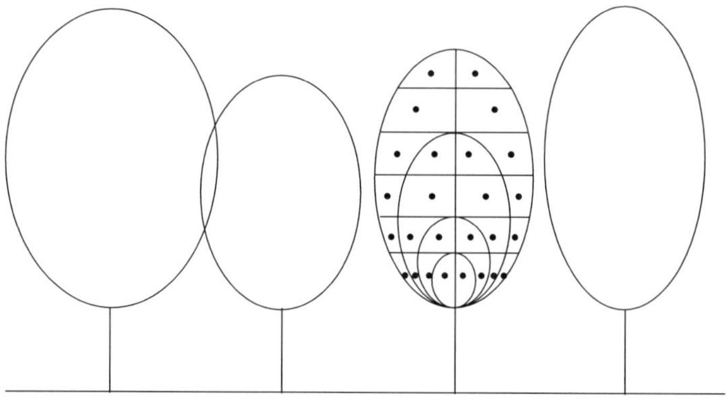

Fig. 8.2. Representation of the canopy in the Grace *et al.* (1987a,b) model. Positions and dimensions of each crown are now specified. Grid volumes within the target crown are used for crown photosynthesis calculations. Inner ellipsoids within crowns are used to specify leaf-area distribution.

The aim of developing a good predictive process-based growth model was very ambitious and remains something of a Holy Grail today.

Grace's background was in empirical forest modelling and she was enthusiastic to work on a process-based model in order to bring more physiology into her work. This emphasis continues to influence her work: since 1991 she has been studying branch development in radiata pine and has made a deliberate attempt to include elements of both empirical and process models.

Meanwhile, Paul Jarvis had returned to Edinburgh and was on the lookout for someone to continue building on Grace's work in his laboratory. Russ Sinclair, of Adelaide University, was on study leave in Edinburgh during the second half of 1983. He hoped to use the model as part of his study on whole-tree transpiration rates, but notes that the program was very involved and he made little progress. The model evidently needed some dedicated hard work. The man for the job arrived in Edinburgh in early 1985. Fresh from China, Ying-Ping Wang was planning to do a thesis on water relations with Paul. Instead, Paul managed to persuade him to work on modelling radiation use efficiency in Sitka spruce. Armed with a copy of Grace's model on tape and a Fortran textbook translated into Chinese, Wang set about reprogramming the model to make it work on the university mainframes. Andrew Sandford, a postdoctoral associate in the laboratory at the time, played an important role as day-to-day supervisor, computer adviser and interpreter for Wang. John Norman also returned to visit at this time and was influential in helping Wang to test the model, suggesting such experiments as observing model behaviour with the leaf reflectance set to 1.

Wang developed a new method to describe the distribution of leaf-area density within canopies, using two-dimensional beta functions (Wang *et al.*, 1990), and implemented leaf incidence-angle distributions (Wang and Jarvis, 1988; Fig. 8.3). However, Wang's main contribution was in developing the model to the point where it could be carefully validated against measurements

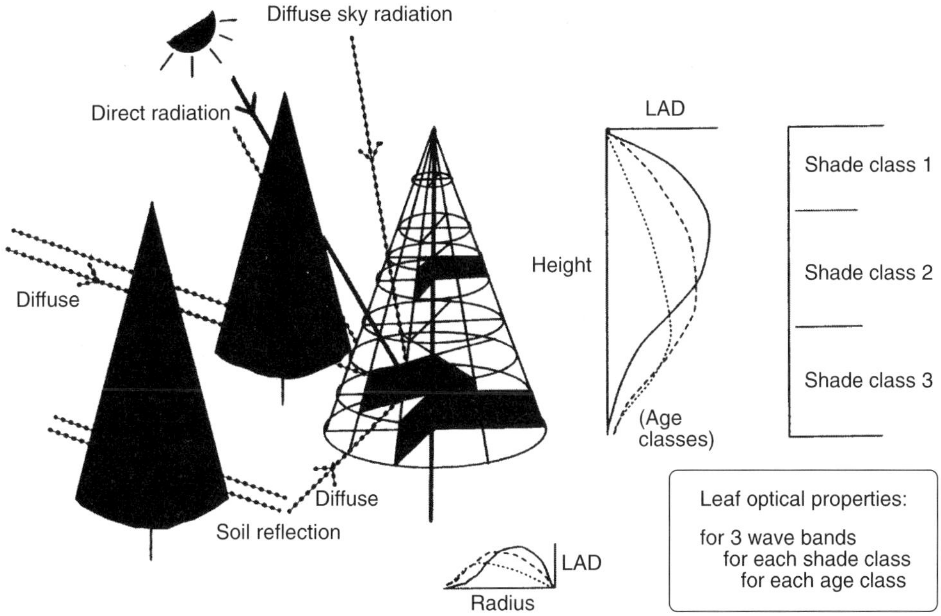

Fig. 8.3. Representation of the canopy in MAESTRO (diagram courtesy of Bart Kruijt). Tree crowns can now be cones. The leaf-area distribution (LAD), for several different age classes, is specified both horizontally and vertically. Leaf physiological properties can be specified by age class and foliage height.

of photosynthetically active radiation (PAR) transmittance in the field. Wang spent several weeks camped in Tummel Forest making these measurements. He also empirically validated the leaf-area estimates. Andrew Sandford recalls:

> I remember Paul breaking the news that YP [Wang] had to validate the leaf area estimates manually which involved a lot of work in harvesting several trees and feeding them through a Li-Cor leaf area meter. Sitka spruce has rather sharp needles and that was not a fun thing to have to do! This was one of the few periods I remember YP lost that enthusiastic smile he had (and still has to this day). I seem to recall Paul disappeared on a sabbatical or something similar during this process, so missed most of the mess of having several trees spread across the lab.
>
> (Andrew Sandford, 2001, personal communication)

Paul soon realized that the model Wang was developing had great potential. When he went to Australia for several months in 1986, he took Wang with him to apply and validate the model against the highly detailed data set from the biology of forest growth experiment in Canberra (Wang *et al.*, 1990). During this visit, Wang worked with another major influence, Ross McMurtrie, who was developing the BIOMASS model at the time.

Wang's thesis appeared in 1988. Several papers from this thesis were published soon after (Wang and Jarvis, 1990b,c), including the key paper Wang and Jarvis (1990a), in which MAESTRO was fully described and validated. The exact origin of the name MAESTRO is now lost in time, but it seems likely that it emerged from the Jarvis household. The name is, in fact, an acronym, although the acronym is generally conveniently forgotten, since it is unlikely to go down well with funding bodies. However, I can here reveal that it stands for: multi-array evaporation stand tree radiation orgy.

Having developed this useful tool, the Jarvis laboratory was not slow to put it to use, and published applications of the model followed very soon after. In 1989 Jarvis *et al.* used the model to estimate effects of water stress on *Eucalyptus globulus* plantations in Portugal, while Dick *et al.* (1990) used the model to look at the effect of cone-bearing on photosynthesis in *Pinus contorta*. These two early papers are indicative of the diversity of applications in which the model has proved useful. Wang notes that the most curious application he heard of was that of a landscape architect in Sheffield who wanted to predict shading of buildings by street trees.

Although the structure of the model remained essentially the same as that described by Wang and Jarvis (1990a), it continued to be developed during the years that followed. Craig Barton and Jon Massheder, working in collaboration with Bob Teskey of the University of Georgia, added responses to ozone and a water balance to the model. Bart Kruijt added responses to CO_2, foliar nitrogen and acclimatization to PAR. As part of a European Union (EU)-funded collaborative project, ECOCRAFT, I rewrote the model to make it easier to use and set up a website to disseminate the code. As I was working with an Italian student, Sabina Dore, at the time, we jokingly renamed the rewritten model as MAESTRA.

The continuing success of the model is somewhat surprising to those who worked on it in the early days. Wang reflects that it may be attributed to several factors – perhaps not least of which being Jarvis's ability as a salesman! But also, Wang notes, the core of the model has stood the test of time because it is based on sound physical principles describing radiation transmission through canopies. The model was unique when developed because of the three-dimensional description of the canopy, making it a versatile tool to study canopy processes in detail. Its use as a research tool was strongly encouraged, and it was flexible enough for a very wide variety of applications, all of which made it attractive to many researchers. In what follows, I survey some of the major fields of application of the model.

Applications of MAESTRO

Canopy structure

Perhaps the most important feature of the MAESTRO model is the level of detail it uses to represent the canopy. This level of detail makes it possible to explore in a concrete way the interactions between canopy structure and canopy processes. The obvious application of this detailed model is to examine the direct influence of canopy structure on radiation interception and photosynthesis. Thus, one of the first exercises with the model that came to be known as MAESTRO was to examine the sensitivity to stocking, foliar density and crown shape (Rook *et al.*, 1985). Similarly, an inaugural application of the newly named MAESTRO (Wang and Jarvis, 1990b) was a detailed investigation of the importance of crown shape, leaf area, leaf-area distribution and leaf inclination angles for crown radiation interception and photosynthesis. Wang and Jarvis (1990b) were able to show that total leaf area and leaf-area distribution were the most important properties for canopy processes. Using a variant of Grace's model, Whitehead *et al.* (1990) came to a similar conclusion, that within-crown leaf-area distribution had an important effect on radiation interception.

An important, related practical application was the investigation of the importance of changes in canopy structure due to silvicultural treatments. Hence, Grace's main applications of her model were to examine the effects of thinning, pruning, tree arrangement and defoliation by insects. Grace *et al.* (1987a) demonstrated that pruned *P. radiata* stands intercepted more radiation than thinned stands with the same leaf-area index. Grace (1988b) showed that green-crown pruning led to an exponential reduction in intercepted PAR and photosynthesis, and estimated the maximum effect on canopy photosynthesis of defoliation by a major pathogen. Grace (1990b) investigated the effects of tree spacing and arrangement on radiation interception, a study with important implications for agroforestry.

Silvicultural applications have been less common since Grace's work, although the model was recently applied by Ibrom (1999) to examine thinning strategies for Sitka spruce. As Grace and colleagues state many times, the model only calculates radiation interception and photosynthesis, not growth, owing to a lack of understanding of allocation patterns. Hence long-term responses to silvicultural treatments cannot be predicted, a major limitation to model application. I return to this problem of linking photosynthesis and growth later in this chapter.

Agroforestry

The three-dimensional representation of the canopy in MAESTRO also had obvious relevance to agroforestry, and several attempts have been made over the years to use the model to predict or interpret radiation interception, photosynthesis and water use in agroforestry systems. An early example was given by Grace (1988a), who showed that, while row orientation had little effect, alternative arrangements of trees could vary annual canopy photosynthesis by up to 11% in a *P. radiata* system with 100 stems/ha.

In the early 1990s, as part of the Agroforestry Modelling Project, MAESTRO was coupled to a crop model, PARCH, to provide a representation of the entire agroforestry system (Lawson *et al.*, 1995). The coupled model was used to predict crop yield of a maize/eucalyptus system in different climates. The coupled-model predictions differed from those of a second model assuming a horizontally homogeneous canopy, in that at the driest site the spatial distribution of PAR interception and evapotranspiration allowed small areas distant from the tree sufficient water to produce a modest yield, while the second model predicted complete crop failure. However, these model predictions were not followed up with experimental tests. Levy (1994) noted several problems with the approach taken in this project, including the difficulty of evaluating tree–crop competition for water because of lack of information on root spatial distribution. As Grace (1988a) commented, 'the full potential of process-based model will not be realised until there is a sound theory for allocation of carbon … At present the model is most suited to research studies.'

The use of the model as a research tool is exemplified in the study by Broadhead (2000) on agroforestry systems involving two contrasting indigenous tree species in Kenya. MAESTRA (the model had undergone its sex change by this time) was first validated against quantum-sensor measurements made beneath the tree canopies. The model was then used to scale leaf-level measurements to estimate canopy-scale photosynthesis and transpiration, which could be compared for the two different systems. The model thus played an important role in the synthesis and interpretation of field data. J.S. Broadhead (2001, personal communication) wrote: 'MAESTRO was a Godsend because I had collected a lot of data not really thinking that the models available would be so constraining in respect of the high level of aggregation of data they required.'

Physiology

Many researchers have found MAESTRO to be a highly useful tool to integrate measurements of leaf physiology over the canopy, thus obtaining canopy-scale responses. Applications of MAESTRO along these lines include the effects of water, nutrient and low-temperature stress; elevated ozone levels; elevated atmospheric [CO_2]; and climate change. The first such study was relatively simplistic: Jarvis *et al.* (1989) simulated water, nutrient and low-temperature stress by halving the quantum efficiency, the stomatal conductance and the carboxylation efficiency, singly and in combination, depending on the type of stress. This exercise gave a basic understanding of the sensitivity of canopy photosynthesis to each of these leaf photosynthetic parameters.

The studies that followed generally adopted a more detailed approach. Modelling of canopy photosynthetic responses to ozone, for example, was based on long-term field studies of branch-scale ozone responses in loblolly pine (Teskey *et al.*, 1991). These studies were used to characterize responses of leaf phenology and physiology to light, temperature, vapour-pressure deficit (VPD), predawn water potentials and ozone levels. MAESTRO was then used to estimate the effect of ozone on annual canopy photosynthesis and to investigate interactions with incident PAR and water stress (Jarvis *et al.*, 1990; Dougherty *et al.*, 1992).

There have also been numerous studies in which MAESTRO has been used to examine the effects of elevated atmospheric [CO_2] and elevated temperature on canopy photosynthesis. McMurtrie *et al.* (1992) and McMurtrie and Wang (1993) first illustrated how leaf-level responses to [CO_2] and temperature could be incorporated into models of canopy photosynthesis. They estimated the responses of canopy photosynthesis to changes in [CO_2] and temperature and showed how these could be interpreted in terms of the proportions of light-saturated and non-saturated foliage. These papers were extremely important in showing how information from small-scale experimental studies on elevated [CO_2] could be extrapolated to the canopy. Two important limitations to the use of MAESTRO were noted, however. First, the absence of nutrient cycling in the model means that long-term nutrient feedbacks could not be estimated. McMurtrie *et al.* (1992) overcame this limitation by using a hierarchy of models; the detailed model MAESTRO was used to develop a simple relationship that could be fed into a simpler model of plant carbon and nutrient cycling. This strategy is an example of hierarchical modelling, a common use of MAESTRO, to which I return below.

A second shortcoming noted by McMurtrie and Wang (1993) was that the model did not incorporate phenomena of acclimatization of photosynthesis to [CO_2]. This shortcoming was addressed in papers following soon after. Medlyn (1996) used MAESTRO to investigate how the response of canopy photosynthesis to elevated [CO_2] was modified by down-regulation due to reduced leaf nitrogen content or leaf ribulose 1,5-bisphosphate carboxylase-oxygenase (Rubisco) activity, both acclimatization responses that had been observed in high-[CO_2] experiments. Kruijt *et al.* (1999) included feedbacks from changes in leaf phenology, leaf area and nutrition in their study of the sensitivity of canopy photosynthesis and transpiration to changes in atmospheric [CO_2] and climate. In this study, two tree species with contrasting canopy structure were considered, *Picea sitchensis* and *Betula pendula*. It was found that canopy photosynthesis in *B. pendula* was more sensitive to the hypothesized feedbacks, owing to the deciduous nature of the canopy.

The above studies on elevated [CO_2] were theoretical studies designed to investigate implications of responses measured in experimental studies to mature forest canopies. The model has also proved extremely useful in directly interpreting results of experiments. Wang *et al.* (1998) reported an analysis of data collected on a study of birch trees grown for 4 years in ambient and elevated [CO_2]. Using an extensive data set on plant structure, leaf-area development, photosynthesis, stomatal conductance and respiration measured during the final year of the experiment, Wang *et al.* constructed a carbon balance for both ambient- and elevated-[CO_2]-grown trees. They showed that net canopy photosynthesis was increased by 110% in elevated [CO_2], as a result of the direct effect of increased [CO_2] on photosynthesis plus the indirect effect of enhanced leaf area, but that biomass increment only increased by 59% over the year, implying substantial losses of carbon to fine-root turnover and mycorrhizae in the elevated-[CO_2] treatment.

The approach used by Wang *et al.* (1998) was applied by Laitat *et al.* (1999) to analyse a series of four other experiments. Responses of above-ground biomass to elevated [CO_2] in these experiments varied from zero to a 140% increase. The MAESTRA model was used to analyse the reasons for the different responses in different experiments; it was found that canopy structure (single trees vs. plant canopies) and allocation patterns played key roles in determining the observed responses.

Finally, a similar study is under way on the Duke Forest free-air CO_2 exchange (FACE) experiment. Luo *et al.* (2001) used MAESTRA to scale up from leaf-scale physiology measurements to estimate canopy carbon uptake in both ambient- and elevated-[CO_2] treatments. Eddy-flux measurements were available to validate the estimates in the ambient treatment. It was estimated that treatment with elevated [CO_2] increased gross primary productivity by 35–40%; it remains to compare these estimates with biomass measurements to determine elevated-[CO_2] effects on partitioning.

Acclimatization to PAR

In addition to studies that integrate leaf physiology over the canopy, MAESTRO has also been used to study the distribution of leaf physiological properties within the canopy. A question of particular interest is whether or not the distribution of leaf nitrogen content and photosynthetic capacity are related to the distribution of intercepted PAR. Many so-called 'big-leaf' models of forest canopies rest on this assumption (e.g. Sellers *et al.*, 1992). Kruijt *et al.* (1999) modified MAESTRO such that photosynthetic capacity followed intercepted PAR averaged over a given time frame, and showed that, under this assumption, the 'big-leaf' simplification held even for a complex canopy.

However, other studies that compared measured distributions of leaf nitrogen content with the distribution of intercepted PAR modelled using MAESTRO generally concluded that the two distributions were not correlated (Leuning *et al.*, 1991; Livingston *et al.*, 1998). Livingston *et al.* (1998), for example, showed that, while leaf nitrogen content declined consistently with height, there was a strong azimuthal variation in intercepted radiation, implying that the distribution of leaf nitrogen was influenced by other factors, potentially including the sink strength of the leader. Bart Kruijt used MAESTRO to investigate the potential error involved in making the 'big-leaf' assumption of perfect correlation between leaf nitrogen and intercepted PAR; however, these results were never published.

Analysis of flux data

An area of forest ecophysiology that has gained dramatically in momentum over the last 10 years is the use of eddy covariance to measure carbon uptake rates by forests. Naturally MAESTRO has been brought into this area as well. In several studies the eddy-covariance data have simply been used to validate baseline model estimates of canopy carbon exchange (or the other way round, as some modellers might claim!) (Jarvis, 1995; Rayment, 1998; Ibrom, 1999; Luo *et al.*, 2001). However, the model also has great potential for interpreting eddy-covariance measurements in terms of underlying physiological processes, particularly for sparse canopies (Levy *et al.*, 1997). For example, work is under way to use MAESTRA to analyse reasons for measured differences in carbon sequestration between contrasting forest stands (Medlyn *et al.*, unpublished observations).

Model hierarchies

MAESTRO is a highly complex and detailed model and is thus inappropriate for many studies involving large spatial scales or long time-scales. Simpler models with fewer parameters must be used in such cases. However, MAESTRO has still proved to be highly useful in studies on such scales, being used either to develop or to test the simplifications used. One example already mentioned above is the work of McMurtrie *et al.* (1992), who used MAESTRO to develop relationships between light-use efficiency, leaf nitrogen concentration and atmospheric [CO_2]. These relationships were then fed into the nutrient-cycling model G'DAY to predict long-term forest responses to elevated [CO_2]. The light-use-efficiency model is a good example of a simple model, useful at large spatial and time-scales, which can readily be calibrated using MAESTRO (e.g. Wang *et al.*, 1991, 1992; Kirschbaum *et al.*, 1994). The validity of the light-use-efficiency approach was also thoroughly analysed using MAESTRO (Medlyn, 1998).

Other examples of the use of MAESTRO to test or calibrate the canopy component of larger-scale models include the work of Wang and Polglase (1995), who developed a long-term model of forest carbon balance under climate change, and Luxmoore *et al.* (2000), who attempted to assess forest response to environmental change at the regional scale. Finally, there have also been several studies where output from MAESTRO has been used to develop models of tree growth (e.g. Ludlow *et al.*, 1990; Baldwin *et al.*, 2001). The issue of modelling tree growth has been one that has troubled MAESTRO modellers throughout the history of MAESTRO, so I examine it in more detail in the following section.

Modelling Forest Growth

As described above, much of the early development of the MAESTRO model was carried out by the New Zealand Forest Research Institute, who were ultimately aiming to produce a process-based model of forest growth. MAESTRO was to be one component of this model (Rook *et al.*, 1985). However, although Grace *et al.*'s (1987a,b) work with MAESTRO was highly successful, the aim of producing a predictive process-based forest-growth model was never achieved, largely because of the difficulty of modelling carbon partitioning. The realization of the magnitude of this obstacle may be traced in the series of published articles. Grace *et al.* (1987b) wrote blithely: 'The addition of routines for respiration, allocation of carbon from photosynthesis, and tree dimensional growth will improve our understanding of the factors controlling tree growth.' Compare this with Grace (1990a): 'to make full use of the potential of process models will require a long commitment to ecophysiological research'. Although predictions of canopy photosynthesis were correlated with measurements of above-ground net primary productivity (Grace *et al.*, 1987b), these correlations were specific to forest types, and hence the goal of being able to apply the model to predict growth in any given forest stand was unattainable (Goulding, 1994).

The group in Edinburgh faced the same problems. Andrew Sandford writes:

> At the time I thought it probably a little premature trying to implement more detailed photosynthesis models when the program (or model) was weak in other areas. I guess other groups with expertise in partitioning and growth have now improved those sections, so by now it should be pretty good?
>
> (Andrew Sandford, 2001, personal communication)

However, despite Jarvis's continuing efforts to have allocation routines added to MAESTRO, the model remains without them. I suspect that most modellers would find that it would be unbalanced to incorporate our sketchy, empirical understanding of allocation processes into the highly detailed, complex (and, yes, still time-consuming) radiation-transmission scheme of MAESTRO. The best alternative seems to be the hierarchical approach, whereby MAESTRO is nested within a growth model. Baldwin and colleagues (Baldwin *et al.*, 1993, 1998, 2001) have made the most progress using this approach. They linked MAESTRO with a distance-dependent, individual-tree model, PTAEDA2, and obtained a coupled model system able to predict changes in forest-stand growth and yield in response to environmental changes. The linkage was achieved by passing tree and stand descriptive information from PTAEDA2 to MAESTRO and passing changes in canopy photosynthesis and thus stand index back the other way. The coupled system gives reasonable results for the system for which it was developed, namely loblolly-pine stands (Baldwin *et al.*, 1998), but considerable effort would be required to adapt it for other forest stands.

Conclusion: MAESTRO – Past and Future

From this overview it is clear that, despite the failure to create the hoped-for process-level growth model, MAESTRO has been a highly successful model. Its most important role has not been as a predictive model, but rather as a research tool, allowing a generation of workers to explore ideas and refine thinking in many different areas of research.

And what of the model's future? MAESTRA is currently provided free of charge on the World-wide Web (see Bibliography). Despite concerted efforts to make the model more user-friendly, it remains highly complex, a fact likely to deter the casual user. However, good documentation is provided, and, with its well-established core of radiation transmission, it seems likely that the model will continue to see application in the future. The year 2000 bug was recently fixed; one wonders if the year 2050 bug that was thereby introduced will ever need to be corrected!

Acknowledgements

Many, many thanks to the following for interviews, written comments, photographs, articles and assorted memorabilia: Paul Jarvis, University of Edinburgh, UK; John Norman, University of Wisconsin, USA; Jon Welles, Li-Cor, USA; Jenny Grace, Forest Research Institute, New Zealand; Ying-Ping

Wang, Department of Atmospheric Research, CSIRO, Australia; Bart Kruijt, Alterra, The Netherlands; Clark Baldwin, US Forest Service, USA; Andrew Sandford, Campbell Scientific, UK; Andreas Ibrom, Institut für Bioklimatologie, Göttingen, Germany; Russ Sinclair, University of Adelaide, Australia; Jeremy Broadhead, Food and Agriculture Organization, Rome, Italy; Gerry Lawson, Centre for Ecology and Hydrology, UK; Jan Dick, Centre for Ecology and Hydrology, UK; Jon Massheder, University of Edinburgh, UK; Mark Rayment, University of Edinburgh, UK; Craig Barton, State Forests New South Wales, Australia; Fraser Gemmell, Centre for Ecology and Hydrology, UK; Ian Gilbert, University of Edinburgh, UK.

MAESTRO Bibliography

Website

http://www.maestra.unsw.edu.au

Model development

Grace, J.C., Jarvis, P.G. and Norman, J.M. (1987a) Modelling the interception of solar radiant energy in intensively managed stands. *New Zealand Journal of Forest Science* 17, 193–209.

Grace, J.C., Rook, D.A. and Lane, P.M. (1987b) Modelling canopy photosynthesis in *Pinus radiata* stands. *New Zealand Journal of Forest Science* 17, 210–228.

Norman, J.M. and Jarvis, P.G. (1974) Photosynthesis in Sitka spruce (*Picea sitchensis* (Bong.) Carr.) III. Measurements of canopy structure and interception of radiation. *Journal of Applied Ecology* 11, 375–398.

Norman, J.M. and Jarvis, P.G. (1975) Photosynthesis in Sitka spruce (*Picea sitchensis* (Bong.) Carr.) V. Radiation penetration theory and a test case. *Journal of Applied Ecology* 12, 839–878.

Norman, J.M. and Welles, J.M. (1983) Radiative transfer in an array of canopies. *Agronomy Journal* 75, 481–488.

Wang, Y.-P. (1988) Crown structure, radiation absorption, photosynthesis and transpiration. PhD thesis, University of Edinburgh, UK.

Wang, Y.-P. and Jarvis, P.G. (1990a) Description and validation of an array model – MAESTRO. *Agricultural and Forest Meteorology* 51, 257–280.

Welles, J.M. and Norman, J.M. (1991) Photon transport in discontinuous canopies: a weighted random approach. In: Myneni, R.B. and Ross, J. (eds) *Photon–Vegetation Interactions*. Springer-Verlag, Berlin, pp. 389–414.

Model components

Day-length calculation

Barkstrom, B. (1981) What time does the sun rise and set? *BYTE* July, 94–114.

Partitioning of direct and diffuse radiation

Spitters, C.J.T., Toussaint, H.A.J.M. and Goudriaan, J. (1986) Separating the diffuse and direct component of global radiation and its implications for modelling canopy photosynthesis. I. Components of incoming radiation. *Agricultural and Forest Meteorology* 38, 217–229.

Weiss, A. and Norman, J.M. (1985) Partitioning solar radiation into direct and diffuse, visible and near-infrared components. *Agricultural and Forest Meteorology* 34, 205–213.

Irradiance on slopes

Steven, M.D. and Unsworth, M.H. (1979) The diffuse solar irradiance of slopes under cloudless skies. *Quarterly Journal of the Royal Meteorological Society* 105, 593–602.

Leaf-angle distribution

Campbell, G.S. (1986) Extinction coefficients for radiation in plant canopies calculated using an ellipsoidal inclination angle distribution. *Agricultural and Forest Meteorology* 36, 317–321.

Campbell, G.S. (1990) Derivation of an angle density function for canopies with ellipsoidal leaf angle distributions. *Agricultural and Forest Meteorology* 49, 173–176.

Wang, Y.-P. and Jarvis, P.G. (1988) Mean leaf angles for the ellipsoidal inclination angle distribution. *Agricultural and Forest Meteorology* 43, 319–321.

Crown shape

Baldwin, V.C., Jr and Peterson, K.D. (1997) Predicting the crown shape of loblolly pine trees. *Canadian Journal of Forest Research* 27, 102–107.

Leaf-area density distribution (beta functions)

Baldwin, V.C., Jr, Peterson, K.D., Burkhart, H.E., Amateis, R.L. and Dougherty, P.M. (1997) Equations for estimating loblolly pine branch and foliage weight and surface area distributions. *Canadian Journal of Forest Research* 27, 918–927.

Porté, A., Bosc, A., Champion, I. and Loustau, D. (2000) Estimating the foliage area of Maritime pine (*Pinus pinaster* Ait.) branches and crowns with application to modelling the foliage area distribution in the crown. *Annales des Sciences Forestières* 57, 73–86.

Wang, Y.-P., Jarvis, P.G. and Benson, M.L. (1990) Two-dimensional needle-area density distribution within the crowns of *Pinus radiata*. *Forest Ecology and Management* 32, 217–237.

Model applications

Baldwin, V.C., Jr, Burkhart, H.E., Dougherty, P.M. and Teskey, R.O. (1993) *Using a Growth and Yield Model (PTAEDA2) as a Driver for a Biological Process Model (MAESTRO)*. Research Paper No. 276, Southern Forest Experiment Station, Forest Service, USDA, New York.

Baldwin, V.C., Jr, Dougherty, P.M. and Burkhart, H.E. (1998) A linked model for simulating growth processes and stand development of loblolly pine. In: Mickler, R.A. and Fox, S. (eds) *The Productivity and Sustainability of Southern Forest Ecosystems in a Changing Environment*. Springer-Verlag, New York, pp. 305–325.

Baldwin, V.C., Jr, Burkhart, H.E., Westfall, J.A. and Peterson, K.D. (2001) Linking growth and yield and process models to estimate impact of environmental changes on growth of loblolly pine. *Forest Science* 47, 77–82.

Broadhead, J.S. (2000) The ecophysiology of indigenous trees in agroforestry systems in the semi-arid tropics. PhD thesis, Nottingham University, UK.

Dick, J.M., Jarvis, P.G. and Barton C.V.M. (1990) Influence of male and female cones on assimilate production of *Pinus contorta* trees within a forest stand. *Tree Physiology* 7, 49–63.

Dougherty, P.M., Teskey, R.O. and Jarvis, P.G. (1992) Development of MAESTRO, a process based model for assessing the impact of ozone on net carbon exchange of loblolly pine trees. In: Flagler, R.B. (ed.) *The response of Southern Commercial Forests to Air Pollution: Transactions 21.* Air and Waste Management Association, Pittsburgh, Pennsylvania, pp. 37–48.

Grace, J.C. (1988a) Effect of tree arrangement on intercepted radiant energy and photosynthesis. In: Ek, A.R., Shifley, S.R. and Burk, T.E. (eds) *Forest Growth Modelling and Prediction*. General Technical Report NC-120, Forest Service, US Department of Agriculture, St Paul, Minnesota, pp. 285–292.

Grace, J.C. (1988b) Effect of foliage distribution within crowns on intercepted radiant energy and photosynthesis. In: Werger, M.J.A., van der Aart, P.J.M., During, H.J. and Verhoeven, J.T.A. (eds) *Plant Form and Vegetation Structure: Adaption, Plasticity and Relation to Herbivory.* SPB Academic Publishing, The Hague, The Netherlands, pp. 153–159.

Grace, J.C. (1990a) Modeling the interception of solar radiant energy and net photosynthesis. In: Dixon, R.K., Meldahl, R.S., Ruark, G.A. and Warren, W.G. (eds) *Process Modeling of Forest Growth Responses to Environmental Stress.* Timber Press, Portland, Oregon, pp. 142–158.

Grace, J.C. (1990b) Process-level models for investigating alternative spacing patterns. In: James, R.N. and Tarlton, G.L. (eds) *New Approaches to Spacing and Thinning in Plantation Forestry*. FRI Bulletin No. 151, Ministry of Forestry, Rotorua, New Zealand, pp. 229–232.

Ibrom, A. (1999) *Simulation of CO_2 and Water Vapour Fluxes in a Young Sitka Spruce (*Picea sitchensis *(Bong.) Carr.) Plantation*. Internal Report, Institute of Ecology and Resource Management, University of Edinburgh, Edinburgh.

Jarvis, P.G. (1995) The role of temperate trees and forests in CO_2 fixation. *Vegetatio* 121, 157–174.

Jarvis, P.G., Wang, Y.P., Borralho, N.M.G. and Pereira, J.S. (1989) Simulation of the role of stress on radiation absorption, assimilation, transpiration and water use efficiency of stands of *Eucalyptus globulus*. In: Pereira, J.S. and Landsberg, J.J. (eds) *Biomass Production by Fast-growing Trees.* Kluwer Academic Publishers, Dordrecht, pp. 169–179.

Jarvis, P.G., Barton, C.V.M., Dougherty, P.M., Teskey, R.O. and Massheder, J.M. (1990) Development and use of tree and forest response models: MAESTRO. In: Kiester, A.R. (ed.) *Acidic Deposition: SOS/T,* Vol. III. The US National Acid Precipitation Assessment Program, Washington, DC, pp. 167–179.

Kirschbaum, M.U.F., King, D.A., Comins, H.N., McMurtrie, R.E., Medlyn, B.E., Pongracic, S., Murty, D., Keith, H., Raison, R.J. and Khanna, P.K. (1994) Modelling forest response to increasing CO_2 concentration under nutrient-limited conditions. *Plant, Cell and Environment* 17, 1081–1099.

Kruijt, B., Barton, C.V.M., Rey, A. and Jarvis, P.G. (1999) The sensitivity of stand-scale photosynthesis and transpiration to changes in atmospheric CO_2 concentration and climate. *Hydrology and Earth System Sciences* 3, 55–69.

Laitat, E., Ceulemans, R., Medlyn, B.E., LeFevre, F., Forstreuter, M., Jach, M.E., Rey, A., Strassemeyer, J. and Jarvis, P.G. (1999) Carbon uptake and allocation. In: Jarvis, P.G. (coordinator) *ECOCRAFT: Predicted Impacts of Rising Carbon Dioxide and Temperature on Forests in Europe at Stand Scale. Final Report to the EU.* Edinburgh University, Edinburgh, pp. 87–96.

Lawson, G.J., Crout, N.M.J., Levy, P.E., Mobbs, D.C., Wallace, J.S., Cannell, M.G.R. and Bradley, R.G. (1995) The tree–crop interface – representation by coupling of forest and crop process-models. *Agroforestry Systems* 30, 199–221.

Leuning, R., Wang, Y.-P. and Cromer, R.N. (1991) Model simulations of spatial distributions and daily totals of photosynthesis in *Eucalyptus grandis* canopies. *Oecologia* 88, 494–503.

Levy, P.E. (1994) *Application of MAESTRO to Tropical Agroforestry Systems.* Internal Report, Institute of Ecology and Resource Management, University of Edinburgh, Edinburgh.

Levy, P.E., Moncrieff, J.B., Massheder, J.M., Jarvis, P.G., Scott, S.L. and Brouwer, J. (1997) CO_2 fluxes at leaf and canopy scale in millet, fallow and tiger bush vegetation at the HAPEX-Sahel southern super-site. *Journal of Hydrology* 189, 612–632.

Livingston, N.J., Whitehead, D., Kelliher, F.M., Wang, Y.-P., Grace, J.C., Walcroft, A.S., Byers, J.N., McSeveny, T.M. and Millard, P. (1998) Nitrogen allocation and carbon isotope fractionation in relation to intercepted radiation and position in a young *Pinus radiata* D. Don tree. *Plant, Cell and Environment* 21, 795–803.

Ludlow, A.R., Randle, T.J. and Grace, J.C. (1990) Developing a process-based growth model for Sitka spruce. In: Dixon, R.K., Meldahl, R.S., Ruark, G.A and Warren, W.G. (eds) *Process Modeling of Forest Growth Responses to Environmental Stress.* Timber Press, Portland, Oregon, pp. 249–262.

Luo, Y., Medlyn, B.E., Hui, D., Ellsworth, D., Reynolds, J. and Katul, G. (2001) Gross primary productivity in the Duke forest: modelling synthesis of CO_2 experiment and eddy-flux data. *Ecological Applications* 11, 239–252.

Luxmoore, R.J., Hargrove, W.W., Tharp, M.L., Post, W.M., Berry, M.W., Minser, K.S., Cropper, W.P., Jr, Johnson, D.W., Zeide, B., Amateis, R.L., Burkhart, H.E., Baldwin, V.C., Jr and Peterson, K.D. (2000) Signal-transfer modelling for regional assessment of forest responses to environmental changes in the south-eastern United States. *Environmental Modeling and Assessment* 5, 125–137.

McMurtrie, R.E. and Wang, Y.-P. (1993) Mathematical models of the photosynthetic response of tree stands to rising CO_2 concentrations and temperatures. *Plant, Cell and Environment* 16, 1–13.

McMurtrie, R.E., Comins, H.N., Kirschbaum, M.U.F. and Wang, Y.-P. (1992) Modifying existing forest growth-models to take account of effects of elevated CO_2. *Australian Journal of Botany* 40, 657–677.

Medlyn, B.E. (1996) Interactive effects of atmospheric carbon dioxide and leaf nitrogen content on canopy light use efficiency – a modelling analysis. *Tree Physiology* 16, 201–209.

Medlyn, B.E. (1998) Physiological basis of the light use efficiency model. *Tree Physiology* 18, 167–176.

Rayment, M.B. (1998) Carbon and water fluxes in a boreal forest ecosystem. PhD thesis, University of Edinburgh, 145 pp.

Rook, D.A., Grace, J.C., Beets, P.N., Whitehead, D., Santantonio, D. and Madgwick, H.A.I. (1985) Forest canopy design: biological models and management implications. In: Cannell, M.G.R. and Jackson, J.E. (eds) *Attributes of Trees as Crop Plants*. Institute of Terrestrial Ecology, Natural Environment Research Council, Abbots Ripton, Huntingdon, pp. 507–524.

Wang, Y.-P. and Jarvis, P.G. (1990b) Influence of crown structural properties on PAR absorption, photosynthesis, and transpiration in Sitka spruce – application of a model (MAESTRO). *Tree Physiology* 7, 297–316.

Wang, Y.-P. and Jarvis, P.G. (1990c) Effect of incident beam and diffuse radiation on PAR absorption, photosynthesis and transpiration of Sitka spruce – a simulation study. *Silva Carelica* 15, 167–180.

Wang, Y.-P. and Polglase, P.J. (1995) The carbon balance in the tundra, boreal and humid tropical forests during climate change – scaling up from leaf physiology and soil carbon dynamics. *Plant, Cell and Environment* 18, 1226–1244.

Wang, Y.-P., Jarvis, P.G. and Taylor, C.M.A. (1991) PAR absorption and its relation to above-ground dry-matter production of Sitka spruce. *Journal of Applied Ecology* 28, 547–560.

Wang, Y.-P., McMurtrie, R.E. and Landsberg, J.J. (1992). Modelling canopy photosynthetic productivity. In: Baker, N.R. and Thomas, H. (eds) *Crop Photosynthesis: Spatial and Temporal Determinants*. Elsevier, Amsterdam, pp. 43–67.

Wang, Y.-P., Rey, A. and Jarvis, P.G. (1998) Carbon balance of young birch trees grown in ambient and elevated atmospheric CO_2 concentrations. *Global Change Biology* 4, 797-807.

Whitehead, D., Grace, J.C. and Godfrey, M.J.S. (1990) Architectural distribution of foliage in individual *Pinus radiata* D. Don crowns and the effects of clumping on radiation interception. *Tree Physiology* 7, 135–155.

Additional references

Goulding, C.J. (1994) Development of growth models for *Pinus radiata* in New Zealand – experience with management and process models. *Forest Ecology and Management* 69, 331–343.

Lemeur, R. and Blad, B.L. (1974) A critical review of light models for estimating the shortwave radiation regime of plant canopies. *Agricultural Meteorology* 14, 255–286.

Norman, J.M. (1975) Radiative transfer in vegetation. In: deVries, D.A. and Afgan, N.H. (eds) *Heat and Mass Transfer in the Biosphere.* Part I: *Transfer Processes in the Plant Environment.* Scripta Book Company, Washington, DC, pp. 187–205.

Sellers, P.J., Berry, J.A., Collatz, G.J., Field, C.B. and Hall, F.G. (1992) Canopy reflectance, photosynthesis, and transpiration. 3. A reanalysis using improved leaf models and a new canopy integration scheme. *Remote Sensing of Environment* 42, 187–216.

Teskey, R.O., Dougherty, P.M. and Wiselogel, A.E. (1991) Design and performance of branch chambers suitable for long-term ozone fumigation of foliage in large trees. *Journal of Environmental Quality* 20, 591–595.

Welles, J.M., Norman, J.M. and Martsolf, J.D. (1979) An orchard foliage temperature model. *Journal of the American Society for Horticultural Science* 104, 602–610.

Welles, J.M., Norman, J.M. and Martsolf, J.D. (1981) Modeling the radiant output of orchard heaters. *Agricultural Meteorology* 23, 275–286.

9

Thermal Radiation, Canopy Temperature and Evaporation from Forest Canopies

Hamlyn G. Jones, Nicole Archer and Eyal Rotenberg*
Division of Environmental and Applied Biology, Biological Sciences Institute, School of Life Sciences, University of Dundee, Dundee DD1 4HN, UK

Abstract

We review the remote sensing of canopy temperature using thermal radiation and outline the main sources of error in the measurement of emitted thermal radiation and its use in canopy temperature estimation. The theory underlying different approaches to the use of these remote estimates of canopy temperature for the estimation of evaporation from forest and other canopies is discussed. The data requirements for the different approaches for remote estimation are compared. We illustrate the application of different approaches to local (branch-level) thermal remote sensing for the estimation of evaporation rate from *Quercus rotundifolia* trees in Portugal.

Introduction

Thermal, or long-wave, radiation and evaporation are intimately linked through their separate contributions to the overall canopy energy balance and hence effects on canopy temperature. Although the amount of thermal radiation emitted by a canopy is related to the canopy surface temperature by means of the Stefan–Bolzmann law, so that the emitted radiation can be used to estimate the canopy temperature, the precise relationship between temperature and radiation depends on several factors, as discussed below. The canopy temperature itself can be used in a number of ways for estimation of evapotranspiration rates (E).

*Present address: ESER, Weizmann Institute of Science, Rehovot 76100, Israel.

The vast range of techniques available for the measurement or estimation of evapotranspiration from plant and crop communities have been well reviewed over the past few years (see, for example, Jones, 1992; Allen *et al.*, 1998; Campbell and Norman, 1998). In addition to methods based on direct measurement of soil water-balance components or the integration of sap-flow measurements on individual trees, there are a wide range of possible micrometeorological and remote-sensing techniques available for the estimation of *E* from plant canopies. These range from conventional energy-balance and combination approaches (e.g. Penman–Monteith (P-M) equation; see Monteith and Unsworth, 1990; Allen *et al.*, 1998) through methods based on remote sensing of canopy temperature to more sophisticated micrometeorological approaches, such as eddy correlation and the 'surface-renewal' approach. The latter is based on the rate of temperature change in ramp-like coherent features exhibited in the temperature time series close to the canopy–atmosphere interface. Such structures have been observed over agricultural and forest canopies and successfully used for estimation of heat flux (e.g. Paw U *et al.*, 1992; Snyder *et al.*, 1996; Spano *et al.*, 2001) using 'surface-renewal' analysis. The approach can be used either directly (from measurements of humidity changes) or indirectly via a residual-energy approach to estimate *E* (Spano *et al.*, 2001). In spite of the range of approaches available, conventional approaches all have significant limitations for application to small plots of trees, as in free-air CO_2 exchange (FACE) experiments, or in heterogeneous vegetation stands.

In this chapter we shall concentrate on a discussion of the use of thermal remote sensing of leaf or canopy temperature for estimation of evaporation rates from forests and other vegetation. Nevertheless, it is worth comparing these methods with the other methods available, so as to identify the situations where remote sensing of temperature may have particular advantages. Although eddy-covariance technology is probably the measurement method of choice when extensive areas of relatively homogeneous canopy are available, its fetch requirement restricts its value for small plots. All other methods have some limitations. For example, the widely used P-M combination equation (see Jones, 1992; Allen *et al.*, 1998) requires detailed surface meteorological information for estimation of boundary-layer transfer and an estimate of 'physiological conductance' (g_s) of the vegetation. Similarly, porometer measurement of leaf conductance, which varies with depth in the canopy, is tedious and extremely difficult to integrate to give the corresponding canopy conductance (e.g. Shuttleworth and Wallace, 1985). In contrast, estimates of surface temperature by thermal imagery (e.g. Jones, 1999a) or other remote-sensing techniques can potentially provide a powerful approach to the estimation of evapotranspiration rates or for estimating canopy conductance for substitution in the P-M or other energy-balance equations.

Remote Sensing of Surface Temperature using Thermal Radiation

As a first step we review the basics of remote sensing of surface temperature using emitted thermal radiation. The upwelling long-wave radiation (L_u) at the surface is related to the surface skin temperature (T_s) by the Stefan–Boltzmann equation

$$L_u = \varepsilon\sigma \cdot (T_s)^4 \quad (1)$$

where ε is the surface emissivity, often taken to be close to 1, and σ is the Stefan–Boltzmann constant ($\sigma = 5.67 \times 10^{-8}$ W/m^2/K^4). This equation can be inverted, with measurements of L_u being used to estimate the corresponding surface temperature. In practice most thermal-radiation sensors and infra-red (IR) thermometers or imagers do not respond equally to all long-wave radiation, so specific conversions are required to estimate T_s from the thermal radiation detected. Furthermore, as discussed below, there are a range of other potential errors in estimation of effective surface temperatures from satellite or other remote observations. These include the following.

1. Effects of the intervening atmosphere. It is extremely difficult to get satellite estimates of surface temperatures to an accuracy of even 1 K. For example, errors in estimated surface temperature from the use of single-channel GOES satellite IR data can be of the order of 5 K, though this can be decreased by using multiple-channel information (Gu *et al.*, 1997). The errors largely arise from the effect of the intervening atmosphere on estimates of both upward and downward fluxes. Aspects of the estimation of land-surface temperatures from satellites, including the use of split-window algorithms, are discussed in some detail by Qin and Karnieli (1999).

2. Anisotropy of the radiation field. It is widely recognized that the incoming atmospheric radiation and the outgoing vegetation radiation, both short-wave and long-wave, are anisotropic and depend on the view elevation and azimuth (see, for example, Unsworth and Monteith, 1975; Kimes *et al.*, 1980; McGuire *et al.*, 1989; Paw U *et al.*, 1989; Monteith & Unsworth, 1990; Lagouarde *et al.*, 1995; Snyder & Wan, 1998; Otterman *et al.*, 1999; Menenti *et al.*, 2001). These effects are dependent on the angular-radiation properties of soil surfaces, leaf-area index, canopy structure and the proportion of soil likely to be seen by the sensor (Norman *et al.*, 1995) and can lead to apparent temperature variation with view azimuth as great as 9.3 K in sunflower (Paw U *et al.*, 1989) or even 13 K (Kimes *et al.*, 1980), while an apparent temperature variation with view elevation of 4–6 K (for coniferous forest (McGuire *et al.*, 1989)) and 4 K (for various agricultural crops (Lagouarde *et al.*, 1995)) is typical. It is therefore difficult to obtain the full hemispherical fluxes from the directional radiances usually measured by remote detectors, though Otterman *et al.* (1995) have suggested using a rather simple model where measurements made at an angle of 50° from the zenith should estimate the hemispherical emission within a few per cent.

3. Errors related to emissivity. The surface emissivity is frequently assumed to be constant and close to 1. Yet it is well known that vegetation comprises a mixture of surfaces, with emissivities normally varying between 0.94 and 0.995 (Idso *et al.*, 1969; Kimes *et al.*, 1981), though lower values can occur where vegetation cover is thin, partially exposing soil beneath. The emissivity of bare soils varies widely, with typical values for dolomite of 0.958 and those for granite and sand, respectively, being reported as 0.815 and 0.6 (Bramson, 1968; Sabins, 1978). The barren emissivity value in the National Aeronautics and Space Administration (NASA) emissivities map, where the emissivity is a function of wavelength, is between 0.84 ($8 < \lambda < 9$ μm) and 0.92 ($\lambda > 16$ μm) (Gupta *et al.*, 1999). The emissivities of soil and leaf surfaces are also functions of water content, among other factors (Bramson, 1968; Fuchs and Tanner, 1968; Salisbury and Milton, 1988; Chen *et al*., 1989; Lagouarde *et al*., 1995). In addition, for most surfaces emissivity is a weak function of the wavelength (Sutherland, 1986). Effects of surface structure on the emissivity are reviewed further elsewhere (Norman *et al*., 1995). Approximate corrections for emissivity can be derived using information on vegetation cover as obtained from remotely sensed vegetation indices such as the normalized-difference vegetation index (NDVI) (Van de Griend and Owe, 1993).

Although the emissivity of a given surface is also a function of the emission angle, this has been little studied over soil surfaces (Sobrino and Cuenca, 1999) and even less over vegetation (Lagouarde *et al*., 1995). As a consequence of these and other factors, the surface emissivity of an area of land surface may vary substantially, with consequent implications for estimates of the surface-temperature distribution. It is difficult to estimate independently the emissivity of any surface from satellite images (see Norman *et al*., 1995; Qin and Karnieli, 1999). As a result, the current surface-emissivity values used by NASA's Clouds and the Earth Radiation Energy System (CERES) and in many other radiative codes are taken from laboratory measurements. However, according to Van de Griend and Owe (1993), measurements of emissivity in the field can only be taken for small-scale areas (e.g. several square metres). For larger-scale measurements, they suggest that pixel-average emissivity, using information from NDVI, reduces the problems of the spatial variability of emissivity differences. At 300 K, an emissivity uncertainty of 1% may lead to a temperature error of 0.6 K (Sobrino and Cuenca, 1999). It is also worth noting that any errors in estimation of surface temperature also have implications for model estimates of other components of the surface energy budget and lower-atmosphere conditions, with feedback effects on the algorithms used for retrieving surface temperature itself (Taconet *et al*., 1986).

4. Distribution of sources/sinks for radiant and other forms of energy. A further complication that makes validation of remote-sensing temperature estimates difficult relates to differences in the sources and sinks for different forms of energy transfer. The temperature measured at ground stations is not necessarily equal either to the effective radiative surface temperature for substitution into the Stefan–Boltzmann equation or to the effective aerodynamic temperature for other forms of energy exchange (Garratt, 1995; Norman *et al*., 1995; Anderson *et al*., 1997). This discrepancy arises because the emitted thermal radiation comes largely from the upper surfaces of the canopy rather than from throughout the canopy profile, which may be more appropriate for other forms of energy exchange. Moreover, standard ground stations measure air temperature at a given height above the surface; in some cases a few degrees' deviation may exist between the measured air temperature and that of the effective skin temperature. The variation in radiative temperature detected using different view angles can usefully be used to separate temperatures of soil and vegetation components (e.g. François, 1997).

5. Heterogeneous surfaces. A less obvious error can arise when the surface in the field of view is composed of an ensemble of surfaces at different temperatures (e.g. soil and leaves), because, even if each has an emissivity close to unity, the wavelength distribution of the emitted thermal radiation from the ensemble will not correspond exactly to that of a black body. Temperature differences greater than 10°C between soil and canopy can consequently lead to errors in radiometric estimates of surface temperature of the order of 1 K (Norman *et al.*, 1995).

6. Limited temporal sampling. The limited and potentially biased (e.g. as a result of cloud cover) temporal sampling by satellites can also lead to difficulties in estimating true mean surface temperatures (e.g. Seguin *et al.*, 1999) and hence difficulties in estimating the resulting E over longer periods.

Estimation of Evapotranspiration from Canopy Temperature

The majority of methods available for the estimation of E are based on formulations of the canopy energy balance, which can be written: $\lambda E = R_n - G - H$, where E is the evaporation rate, λ is the latent heat of vaporization, R_n is the net radiation absorbed, G is the soil heat flux (often assumed to be negligible over 24 h periods) and H is the sensible-heat loss to the air. The most usual formulation for meteorological estimates of E is the P-M combination equation (see, for example, Monteith and Unsworth, 1990; Jones, 1992). Since the early use of canopy temperature to estimate E (e.g. Bartholic *et al.*, 1972), there have been a number of attempts using satellite thermal channels, aircraft or ground-based sensors (see reviews by Norman *et al.*, 1995; Boegh *et al.*, 1999; Jiang and Islam, 1999; Li and Lyons, 1999). We concentrate here on those methods dependent on remotely sensed canopy temperatures. Unfortunately, there is often a lack of good ground data against which these estimates can be validated. Furthermore, many of the approaches themselves require information on variables such as R_n, T_a, D (humidity deficit) and u (wind speed) whether from parallel ground observation or from sophisticated soil–vegetation–atmosphere transfer (SVAT) models based on remote-sensing data. To aid comparison between different possible approaches for the estimation of E from remote-sensing data, we summarize some of the main methods available in Table 9.1.

Basic Theory – Single Leaves

Using boundary-layer conductance

In principle, energy-balance considerations enable one to estimate evaporation rates from leaf or canopy temperature (as measured by IR thermometry or other remote sensing), though some other ancillary measurements are required. One approach that has been used by several workers for the estimation of leaf transpiration (e.g. Harrison-Murray, 1991; Jones, 1992; Qiu *et al.*, 1996) is to compare the temperature of a non-transpiring leaf or model (T_{dry}) with that of the transpiring leaf (T_{leaf}). Indeed, it has long been recognized that E is proportional to ($T_{dry} - T_{leaf}$), with the main difficulty being estimation of the constant of proportionality for any given conditions.

Table 9.1. A summary of some methods for estimation of E, outlining the variables that are required or not required for each approach.

Method	Variables eliminated	Variables needed	Comment/authors
Methods not requiring T_s explicitly			
Combination (Penman–Monteith) equation $\lambda E = \frac{(s(R_n - G) + \rho\, c_p\, g_a\, D)}{(s + (\gamma\, g_a/g_s))}$	T_s	G, R_n, g_a (or u), g_c, T_a, D	Many, see Monteith and Unsworth (1990); Allen *et al.* (1998) Several authors base work on this using SVAT models to estimate g_c (e.g. Choudhury *et al.*, 1986)
Equilibrium evaporation (Priestley–Taylor) (Eq. 8) $\lambda E_{eq} = \alpha\, R_n\, (s/(s + \gamma))$	T_s, $^*T_a$, g_a (or u), g_c, D	G, R_n, (T_a for estimation of physical constants), ground calibration for Priestley–Taylor constant, α	For large areas of 'smooth' canopy: Priestley and Taylor (1972); Brutsaert (1982)
Imposed evaporation (McNaughton and Jarvis, 1983) $\lambda E = (\rho c_p/g_s\gamma)\, g_s\, D$	G, R_n, $^*T_a$, T_s, g_a, g_c	D, g_s (T_a for estimation of physical constants)	For well-coupled canopies such as forests: Nemani and Running (1989)
Methods utilizing surface temperature			
Energy balance with dry reference and elimination of R_n (Eq. 5) $\lambda E = \rho c_p(g_{aH} + g_R)(T_{dry} - T_{leaf})$	(G not relevant for leaves), R_n, T_a, D	g_a, T_{dry}, T_{leaf}	Most appropriate at the leaf scale, though possible to scale up with good boundary-layer estimate: Jones (1992)
Energy balance with dry reference and elimination of boundary layer (Eq. 7) $\lambda E = R_{ni}\, (T_{dry} - T_{leaf})/(T_{dry} - T_a)$	(G not relevant for leaves), g_a, D	R_n, T_a, T_{dry}, T_{leaf}	Again appropriate at leaf scale: Brough *et al.* (1986); Qiu *et al.* (1996)
Combined CWSI and VI (Inoue *et al.*, 1994) (Eq. 13) $E_a = A\, R_s\, (1 - \text{CWSI})\, \text{VI}$	G, R_n, g_a, g_s, T_a, D	R_s, VI, A (a constant including λ), CWSI (either from full energy balance or from T_{dry}, T_{leaf}, T_p (often use a crop-specific empirical relation with D))	Inoue *et al.* (1994); Inoue and Moran (1997)

Residual-energy-balance methods			Many incorporate more or less complex SVAT models: see, for example, Brown and Rosenberg (1972); Soer (1980); Reginato *et al.* (1985); Choudhury (1994)
$\lambda E = R_{ni} - H$	D, g_s	G, R_n, T_s, T_a, g_a	
Jackson's semi-empirical method (Jackson *et al.*, 1977) (Eqs 9, 10) $\lambda E = R_n - B(T_s - T_a) + A$	No explicit ground information (e.g. D, g_a (or u))	R_n, T_s, T_a, ground calibration for constants B	Jackson *et al.* (1977, 1983); Seguin and Itier (1983); Lagouarde (1991); Courault *et al.* (1996); Caselles *et al.* (1998)
Use of ground reference with potential transpiration (Nieuwenhuis *et al.*, 1985) $\lambda E = \lambda E_p - B(T_c - T_p)$	G, R_n, T_a, g_a, g_c, D	E_p and T_p from a reference crop, T_s, ground calibration for constant B	Nieuwenhuis *et al.* (1985); Gash, 1987; Kustas *et al.* 1990; Kalluri *et al.* (1998)
Dual-source models	Vary	T_{soil}, T_{veg} and many other variables, depending on model	Kustas (1990); Anderson *et al.* (1997); Kustas and Norman (1997)
Specific techniques for important variables			
T_c from NDVI/T_s relationship	g_s, D	NDVI, T_s, R_n, constants for g_a, T_a	Goward *et al.* (1985); Kalluri *et al.* (1998); Boegh *et al.* (1999)
g_c from slope of NDVI/T_s		NDVI, T_s	Nemani and Running (1989), Kalluri *et al.* (1998)
CWSI from trapezoid		NDVI, T_s	Moran *et al.* (1994); Yang *et al.* (1997)
Fully empirical			
$E = A + B\,T_s + C\,\mathrm{NDVI}$	R_n, G, D, g_a, g_s, T_a	NDVI, T_s, calibration of constants	Di Bella *et al.* (2000)

*The physical 'constants' c_p, g_a, s, λ and γ have a more or less minor dependence on T_a.
A, B, C, n, empirical constants; α, Priestley–Taylor constant (= *c.* 1.26); ρ, mean air density at constant pressure; c_p, specific heat of air; s, slope of saturation vapour pressure–temperature curve; γ, psychometric constant; λ, latent heat of vaporization; CWSI, crop water-stress index; D, vapour-pressure deficit; E_p, potential evapotranspiration; g_a, aerodynamic conductance; g_c, bulk surface conductance; G, soil heat flux; H, sensible-heat flux; R_n, net radiation; R_{ni}, net isothermal radiation; R_s, solar radiation; SVAT, soil–vegetation–atmosphere transfer model; T_a, air temperature; T_{dry}, temperature of non-transpiring leaf; T_{leaf}, temperature of a transpiring leaf; T_p, temperature of canopy transpiring at a potential rate; T_s, surface temperature; T_{soil}, temperature of the soil; T_{veg}, temperature of the transpiring vegetation; u, wind speed; VI, vegetation index (usually soil-adjusted vegetation index (SAVI) or normalized-difference vegetation index (NDVI)).

The relevant relationship can be derived as follows. For a dry reference surface (denoted by subscript $_{dry}$), such as a non-transpiring leaf, one can write for the steady state (there is no soil heat flux (G) for single leaves):

$$R_{ndry} = H_{dry} = \rho c_p\, g_{aH}\, (T_{dry} - T_a) \tag{2}$$

where R_n is the net radiation absorbed, H is the sensible-heat loss, ρ is the density of air, c_p is the specific heat capacity of air, T_a is the air temperature, and g_{aH} is the boundary-layer conductance for heat transfer. For a corresponding evaporating surface (denoted by subscript $_{leaf}$):

$$R_{nleaf} = H_{leaf} + \lambda E = \rho c_p\, g_{aH}\, (T_{leaf} - T_a) + \lambda E \tag{3}$$

Making the substitution:

$$R_n = R_{ni} - \rho c_p\, g_R (T_{leaf} - T_a) \tag{4}$$

where R_{ni} is the net isothermal radiation (defined as the net radiation that would be received by an identical surface in an identical environment if it were at air temperature (see Jones, 1992; p. 108)) and g_R is a radiative conductance ($= 4\epsilon\sigma T_a^{\,3}/\rho c_p$), and subtracting the resulting version of Equation (2) from Equation (3) gives after rearrangement (Jones, 1992):

$$\lambda E = \rho c_p (g_{aH} + g_R)(T_{dry} - T_{leaf}) \tag{5}$$

It follows from this equation that E can be estimated from knowledge of T_{leaf} and of a dry reference as long as an estimate of the boundary-layer conductance is available and the temperature is known (for estimation of g_R). Inoue *et al.* (1990) estimated g_{aH} in this equation directly from measurements of wind speed and leaf size. Other possible approaches for estimation of g_{aH} at a leaf scale include the use of heated and unheated leaves (e.g. Stanghellini, 1987; Brenner and Jarvis, 1995) or dynamic cooling of preheated leaves (Jones, 1992: Appendix 8).

Using net radiation

An alternative formulation of the same energy-balance equations leads to an equation that does not require knowledge of the boundary-layer conductance (Brough *et al.*, 1986; Qiu *et al.*, 1996). Eliminating g_{aH} from Equations (2) and (3) and rearranging gives:

$$\lambda E = R_{nleaf} - R_{ndry}(T_{leaf} - T_a)/(T_{dry} - T_a) \tag{6}$$

Substituting the appropriate formulation of isothermal net radiation for R_{nleaf} and R_{ndry} and rearranging leads to:

$$\lambda E = R_{ni}\, (T_{dry} - T_{leaf})/(T_{dry} - T_a) \tag{7}$$

In this formulation it is clear that all that is required to estimate E is the measurement of air, leaf and reference surface temperatures, together with an estimate of net isothermal radiation appropriate for the surface being considered.

This formulation is particularly valuable for modelling, as the net isothermal radiation is an environmentally dependent variable, unaffected by transpiration rate (or leaf temperature).

An essential requirement for all the above equations is that the reference surfaces have the same short-wave (and long-wave) radiative properties as the real leaves.

Canopy-level Methods

Remotely sensed canopy temperature may be substituted into the full P-M equation or its derivatives, but there is a major requirement for subsidiary information, such as wind speed (or boundary-layer resistance), vapour-pressure deficit, R_n, soil heat flux and canopy conductance (see, for example, Norman *et al.*, 1995). Therefore there has been great interest in the derivation of simplified approaches, especially for the estimation of daily values of λE from instantaneous measurements of temperature. For these purposes, it is common to assume that the soil heat flux is negligible over 24 h periods. In this case, the available energy ($\approx R_n$) determines the rate of evaporation according to the equilibrium evaporation rate (McNaughton and Jarvis, 1983; Monteith and Unsworth, 1990):

$$\lambda E_{eq} = \alpha R_n (s/(s + \gamma)) \quad (8)$$

where s is the slope of the curve relating saturation vapour pressure to temperature, γ is the psychrometer constant and α, the so-called Priestley–Taylor constant, is an empirically determined constant that is used to describe evaporation from large areas of homogeneous vegetation where often $\alpha \cong 1.26$ (see, for example, Priestley and Taylor, 1972; Brutsaert, 1982).

Many remote-sensing estimates of E are based on the use of the energy-balance equation involving surface-temperature estimates, using what has been termed the residual energy-balance approach, and these are reviewed by Choudhury (1994). λE is estimated by difference after R_n and G (soil heat flux) have been measured and H is estimated from the canopy–air-temperature difference and the boundary-layer conductance, g_a. Of course, a key requirement for this approach remains the need to obtain an accurate estimate of g_a, while appropriate estimates of canopy temperature and air temperature are also somewhat problematic from satellite or airborne sensors.

Semi-empirical relationships

Using the residual-energy-balance concept, Jackson and colleagues proposed a simple semi-empirical approach (e.g. Jackson *et al.*, 1977, 1983). In this they assumed that λE is given by the difference between net radiation and a sensible-heat-loss term (proportional to the canopy–air-temperature difference), so that they wrote:

$$\lambda E = R_n - B\,(T_s - T_a)^n \tag{9}$$

where B and n are empirical coefficients. Seguin and Itier (1983) found that B was a function of surface roughness and wind speed, while Paloscia and Pampaloni (1986) and others have shown that n is often near unity and they added an extra constant to give:

$$\lambda E = R_n - B(T_s - T_a) + A \tag{10}$$

This general approach has been widely used for estimation of E by means of satellite data, particularly where meteorological data on T_a are available (see, for example, Lagouarde, 1991; Courault *et al.*, 1996; Caselles *et al.*, 1998). Others have eliminated the need for R_n and T_a in Equation (10) by replacing $(T_s - T_a)$ by $(T_s - T_{ref})$, where T_{ref} is the temperature of a reference crop transpiring at a potential rate (Niewenhuis *et al.*, 1985; Gash, 1987; Kustas, 1990).

Methods incorporating vegetation indexes

As the vegetation gets sparser, the remotely sensed average surface temperature in any pixel becomes more and more biased towards that of the soil rather than the transpiring vegetation. A second, but related effect is that a greater proportion of the incoming energy is absorbed by the soil rather than by the vegetation, so it does not contribute directly to evaporation. In order to correct for these effects, various remote-sensing approaches for the estimation of vegetation cover or leaf-area index (LAI) have been introduced. General use is made of the fact that vegetation indexes calculated from the ratio of red/near-infrared (NIR) reflectances, such as the NDVI or the soil-adjusted vegetation index (SAVI), vary in approximate proportion to the fractional vegetation cover.

Inoue and Moran

Using the available energy as a basis, Inoue *et al.* (1994) extended their earlier work on the energy balance of single leaves to the estimation of daily canopy evapotranspiration by assuming that E, for a well-watered canopy ($= E_p$), is proportional to the amount of radiation intercepted by the vegetation canopy, which in turn could be estimated as the incident short-wave radiation (R_s) multiplied by the SAVI. Therefore:

$$E_p = a\,R_s\,\mathrm{SAVI} \tag{11}$$

where a is a constant including the latent heat of vapourisation (λ) and a correction for canopy reflectance to obtain the net radiation absorbed. Some examples where SAVI or NDVI have been used in this way for estimation of absorbed incoming energy in the estimation of E have been outlined by Choudhury (1994).

In order to determine the actual evaporation in most situations it is necessary to correct Equation (11) for any reduction of actual evaporation (E_a) below the potential rate due to stomatal closure. As was pointed out by Idso and Jackson and co-workers (Idso *et al.*, 1981; Jackson, 1982), this stomatal closure should be reflected in a rise in the canopy temperature and hence should be detectable through changes in what they termed the crop water-stress index (CWSI). Jackson (1982) showed that:

$$E_a/E_p = (1 - \text{CWSI}) \quad (12)$$

where the CWSI is defined as $(T_{leaf} - T_p)/(T_{dry} - T_p)$, T_{dry} is the temperature of a non-transpiring canopy and T_p is the temperature of a canopy transpiring at the potential rate.

Combining Equations (11) and (12) gives (Inoue *et al.*, 1994):

$$E_a = a\,R_s\,(1 - \text{CWSI})\ \text{SAVI} \quad (13)$$

In the original formulation, CWSI was estimated from the explicit version of the energy-balance equation as outlined by Jackson (1982) and therefore requires independent measurements of atmospheric vapour-pressure deficit (*D*), R_n and g_a. Note that the correct canopy temperature to substitute in CWSI in Equation (12) is that of the leaves and not the average of soil and leaf temperatures within the view. Calculations using this model agreed well with independent estimates of *E* (Inoue and Moran, 1997).

This approach can be modified to use within-image references such as water bodies or bare soil (Jones, 1999a,b), or crops transpiring under optimal soil-moisture conditions (Nieuwenhuis *et al.*, 1985; Kustas *et al.*, 1990).

Regression of T_s *on NDVI*

Spatial variations in surface radiant temperature can be attributed to density of vegetation, due to the high latent-heat transfer associated with vegetation cover. A number of workers have used the NDVI–T_s relationship to estimate vegetation temperature by extrapolating the NDVI–T_s regression line to the NDVI representative for dense vegetation (Kalluri *et al.*, 1998; Boegh *et al.*, 1999). Using such an underlying principle, Nemani and Running (1989) used the slope of the relationship between T_s and NDVI to infer canopy conductance, while Yang *et al.* (1997) used similar data to determine the vegetation-index temperature trapezoid for estimation of the CWSI.

More than one of the methods shown in Table 9.1 can be used together to estimate *E*. For example, Kalluri *et al.* (1998) and Boegh *et al.* (1999) incorporated the NDVI–T_s relationship into the energy-balance approach, where *E* is estimated as a residual. Other variables, such as emissivity and the radiation emitted by the surroundings, must also be accounted for where T_s is evaluated from thermal images.

Temporal and spatial extrapolation

For most applications, one is interested in obtaining estimates of evapotranspiration over periods of weeks to months; yet remotely sensed data are normally available only once a day, while, in cloudy conditions or when using higher-resolution satellites, there may be more than a week between observations. This shortage of diurnal data has resulted in the development of a range of methods to integrate the energy balance over a whole day. A common approach to predicting daily *E* is to use instantaneous midday measurements to represent daily values, using simple sinusoidal approximations, coupled with the assumption that *G* is negligible for 24 h periods (Jackson *et al.*, 1983; Seguin and Itier, 1983; Moran *et al.*, 1989; Inoue and Moran, 1997). In some cases intensive measurements of *E* for a reference area for a 24 h period have been used to calibrate the midday measurements, using thermal imagery (Kustas *et al.*, 1990). More sophisticated models, such as TEGRA, which uses transport equations for heat and moisture flow in the soil, plant and atmosphere, can also be used to transform instantaneous measurements from thermal images to daily measurements (Soer, 1980; Nieuwenhuis *et al.*, 1985), though effective use of these models tends to require much detailed ground information.

The use of remotely sensed data has the benefit of accessing spatial information over large areas. Unfortunately, many of the assumptions used for estimation of fluxes such as *E* from remotely sensed data are only really appropriate for homogeneous vegetation, as many subpixel properties, such as emissivity, do not necessarily scale linearly. Another problem is that transfer coefficients (e.g. g_a) depend on atmospheric stability, wind speed and surface roughness, but, over heterogeneous vegetation, at least one of these variables may change. To alleviate such problems, many workers have extrapolated ground-based reference evaporation measurements to larger areas by using horizontal changes in surface temperature measured by remote sensing (e.g. Reginato *et al.*, 1985; Gash, 1987; Moran *et al.*, 1989; Kustas *et al.*, 1994; Reicosky *et al.*, 1994; Courault *et al.*, 1996).

With the increasing use of remote-sensed variables, spatial information can be calculated over large areas; for example, Caselles *et al.* (1998) used a pixel-based method to determine maximum *E* for crops in an area approximately 100 km^2. This method depended on three main inputs: ground measurements, from a meteorological station, the difference between each pixel and the pixel with the maximum *E* using NOAA-AVHRR thermal images, and crop distribution, using Landsat thematic mapper (TM) images. With greater understanding of the relationships between the ground surface and remotely sensed measurements, there are algorithms, such as the surface energy-balance alogorithm for land (SEBAL) (Bastiaanssen *et al.*, 1998), which can be applied with little or no ground-based weather data. However, for accurate predictions of *E*, actual measurements for solar radiation and wind speed on the days of remote measurements are still needed (Morse *et al.*, 2000).

Preliminary Application to Montado Woodland

Methods

We illustrate some aspects of the estimation of canopy temperatures and their application to estimation of *E* using images obtained with a digital camera (Canon

Digital Ixus) and a thermal IR imager (SnapShot 225, Infrared Solutions Inc., Minneapolis, supplied by Alpine Components, Littlehampton, UK) of *Quercus rotundifolia* Lam. in a Montado woodland at Mitra near Evora in Portugal on a sunny day in October 2001. The necessary values of canopy temperature (= T_{leaf}) and the temperature of areas of canopy sprayed with water ($T_{wet} = T_p$) or covered with petroleum jelly (T_{dry}) were measured using the SnapView camera, viewing from an adjacent flux tower. Six images were taken into the sun (forming images of mainly shaded canopy) and six images were taken away from the sun (forming images of mainly sunlit canopy). These images were taken at approximately 1 min intervals. The standard digital images were cropped and warped within ENVI to co-register them to the thermal images. In each tree an aluminium mast carrying detached branches, which were used as reference areas, was placed in the canopy. The leaves on these detached branches were either sprayed with water to provide a wet reference (T_{wet}) or were covered with petroleum jelly (Vaseline) to prevent transpiration and provide a non-transpiring reference (T_{dry}) with similar spectral properties to the normal canopy leaves.

Figure 9.1 shows representative examples of the corresponding digital (A, D) and thermal (C, F) images of the mainly sunlit and mainly shaded canopies, respectively. The centre image in each triplet (B, E) shows the areas identified as 'sunlit leaves', 'shaded leaves', 'sunlit soil' and 'shaded soil' as classified using a

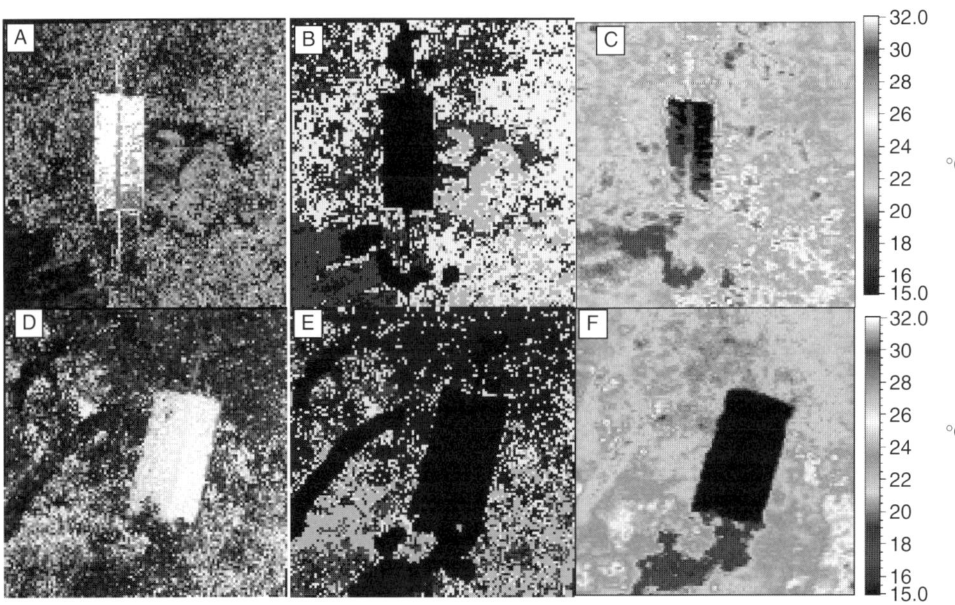

Fig. 9.1. A, B and C are images taken on a sunlit *Q. rotundifolia* canopy and D, E and F are taken on a shaded side of the *Q. rotundifolia* canopy. A and D are digital images cropped and warped to the corresponding thermal image (C or F). B and E are the corresponding images classified using supervised classification in ENVI. The black areas have been masked out of the analysis and represent areas of aluminium foil, reference areas and tree trunks; light grey areas represent sunlit canopy; dark grey represents shaded canopy; medium-dark grey represents shaded soil; medium-light grey represents sunlit soil.

supervised classification in ENVI (version 3.4, Research Systems). Surface temperatures of the different classes of surface within the image were extracted either using the ENVI classification or directly in SNAPVIEW software (version 3.1, Infrared Solutions Inc.), using representative areas selected by eye.

Ancillary environmental data were obtained from a nearby meteorological station, supplemented by R_n, obtained from a net radiometer positioned 15 cm above the sunlit canopy of an oak tree, T_a, obtained using fine-wire (40 gauge) thermocouples hanging freely in the shade within the canopy, and g_a, estimated from the temperatures (also measured using thermocouples) of paired artificial heated and unheated model leaves (Stanghellini, 1987) displayed within the canopy.

Results and discussion

For this work we estimated E for sunlit and shaded canopy using Equations (5) and (7). In addition we estimated the E_a as a fraction of the potential evaporation, E_p, using Equation (12), where T_p was assumed equal to the temperature of the wet leaves. Results are summarized in Table 9.2, which shows the average temperatures from the six images, along with corresponding averages for R_n, T_a and g_a. As R_n was only measured over the sunlit canopy, Equation (5) was only applicable to the sunlit areas. Equation (7), which eliminates R_n, was applicable to both sunlit and shaded areas. The different columns show the results of different methods for extracting the relevant temperatures. The first approach was to visually select representative areas in the SNAPVIEW images, while the second and third approaches used the ENVI classification to extract either the modal or the mean temperature for each classification.

The values of E estimated using Equation (5) and Equation (7) agreed surprisingly well, even though they used rather different input data. As one might expect, the calculated evaporation rates were significantly higher for sunlit than for shaded leaves. A comparison of different approaches to the estimation of the temperatures of canopy components showed little difference between the different methods tested in this work (compare the estimates for the single images) and the resulting estimates of E were closely comparable. It is apparent from the large standard deviation of calculated E and from the large difference between the sample image and the mean that the calculated E varied markedly over short periods. It is difficult to be certain whether this large variation in apparent E over short periods was real (and related to the quite large and rapid variation in local wind speed) or an artefact, possibly related to the rather large variation in canopy temperature across each image (see, for example, Fig. 9.2). Nevertheless, there was a high correlation between the results estimated using the two rather different approaches, which supports the hypothesis that the short-term variation in E was real.

The canopy-temperature data when combined with the reference temperatures can also be used to estimate E_a/E_p. It is important to emphasize that this is based on an estimate of the 'local' potential evaporation and will significantly overestimate the evaporation from an extensive area of vegetation, where the evaporation itself will modify the environment so that E_p will tend to approach the equilibrium value determined by the Priestley–Taylor equation.

Table 9.2. Summary of temperatures and other essential environmental variables observed for *Quercus rotundifolia* canopies and estimated *E*, together with standard deviations where appropriate. The first block of data refer to the means of six images each for sunlit and shaded canopies; the remaining columns refer to a single image for each illumination, using either ENVI or SNAPVIEW.

	Temperature averages of six images extracted by SNAPVIEW software				Mean and mode extracted from the ENVI classification of a digital image overlain on analogous thermal image				Corresponding values extracted by SNAPVIEW software	
	Sunlit leaves		Shaded leaves		Sunlit canopy		Shaded canopy		Sunlit leaves mean	Shaded leaves mean
	Mean	Standard deviation	Mean	Standard deviation	Mode	Mean	Mode	Mean		
T_{leaf} (°C)	23.8	0.8	19.5	0.41	22.7	23.0	20.4	20.7	23.1	20.2
T_{wet} (°C)	18.0	0.5	15.9	0.58	17.5	17.5	16.6	16.6	17.5	16.6
T_{dry} (°C)	24.9	0.6	20.4	0.84	25.6	25.6	21.6	21.6	25.6	21.6
T_{air} (°C)	20.9	0.3	20.8	0.19	21.4	21.4	20.9	20.9	21.4	20.9
R_n (W/m^2)	488	13	*	*	502	502	*	*	502	*
g_a (m/s)	0.10	0.01	0.09	0.00	0.10	0.10	0.09	0.09	0.10	0.09
R_{ni} (W/m^2)	503	8	*	*	509	510	*	*	511	*
Equation (7)										
λE (W/m^2)	147	98	*	*	356	326	*	*	303	*
Equation (5)										
λE (W/m^2)	147	98	99	56	383	350	143	111	325	170
E_a/E_p	0.16	0.09	0.18	0.09	0.36	0.33	0.24	0.19	0.29	0.31

* Indicates no relevant information.

Equation (7) is $\lambda E = R_{ni}(T_{dry} - T_{leaf})/(T_{dry} - T_a)\lambda$ and Equation (5) is $\lambda E = \rho c_p(g_a + g_r)(T_{dry} - T_{leaf})$, while E_a/E_p was calculated from (1 − CWSI) (Equation (12) with T_p assumed equal to T_{wet}).

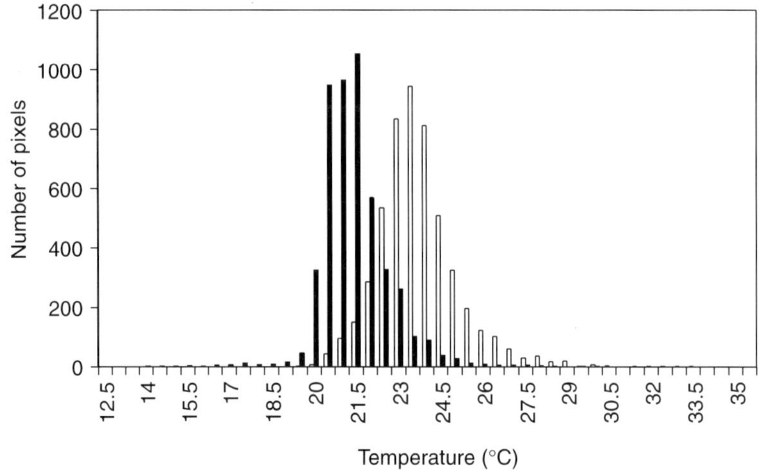

Fig. 9.2. Temperature distributions in relation to pixel number of a shaded canopy area (black bars) and a sunlit canopy area (clear bars) taken from overlaying the classified images (Fig. 9.1B and E) over the thermal images (Fig. 9.1C and F).

Conclusions

The choice of method to estimate E by means of remote sensing depends on three main factors.

1. Objectives. Is E estimated for hourly, daily or monthly intervals? What is the scale being studied: leaf, canopy or a whole region?
2. Type of environment. Is it a homogeneous crop, heterogeneous woodland?
3. Available variables to input into the energy balance. What is possible to measure on the ground or remotely?

These three factors have to be considered when choosing a method to calculate E (Fig. 9.3).

The preliminary analysis of data for a single period on 1 day in October presented here illustrates different ways of extracting appropriate temperature means from thermal images and also illustrates different ways of arranging the energy-balance equation to estimate E, eliminating a requirement for either R_n or g_a. Unfortunately, we do not have the necessary ground data on the physiological behaviour of *Q. rotundifolia* at the time of the measurements to fully compare the calculated E (Fig. 9.3), though the absolute values calculated appear reasonable and are at least consistent between different approaches.

In order to scale up the type of measurements made in this study from the branch scale used here to the whole canopy, it is likely to be necessary to use two-source models (Brenner and Incoll, 1997). It will therefore be necessary to obtain estimates of the areas of each canopy and soil class and their respective temperatures. We are in the process of analysing images from a series of flights made under the European Union (EU) Hysens programme using the DAIS 7915 and ROSIS hyperspectral imagers for our site. The conversion

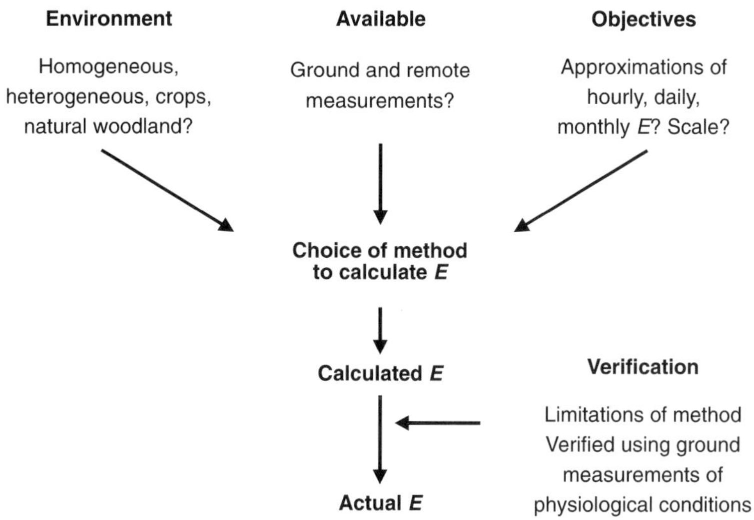

Fig. 9.3. Flow chart of the process involved in selecting an appropriate method for estimation of evaporation from remote thermal data.

from planar images to the estimation of total leaf area in the sunlit and shaded classes and the consequent development of an appropriate two-source model will require the inversion of simple canopy-structure models, using the multi-angular data we obtained in these flights (e.g. Anderson *et al.*, 1997; Kustas and Norman, 1997).

Acknowledgements

We are grateful to the European Commission for funding this work under the Environment Programme (Project WATERUSE EVK1-2000-22061) and to Professor J.S. Pereira and Dr J. Soares David for providing the field site.

References

Allen, R.G., Perreria, L.S., Raes, D. and Smith, M. (1998) *Crop Evapotranspiration: Guidelines for Computing Crop Water Requirements*. FAO 56, FAO, Rome.

Anderson, M.C., Norman, J.M., Diak, G.R., Kustas, W.P. and Mecikalski, J.R. (1997) A two-source time-integrated model for estimating surface fluxes using thermal infrared remote sensing. *Remote Sensing of Environment* 60, 195–216.

Bartholic, J.F., Namken, L.N. and Wiegand, C.L. (1972) Aerial thermal scanner to determine temperatures of soils and crop canopies differing in water stress. *Agronomy Journal* 64, 603–608.

Bastiaanssen, W.G.M., Menenti, M., Feddes, R.A. and Holtslag, A.A.M. (1998) A remote sensing surface energy balance algorithm for land (SEBAL) – 1. Formulation. *Journal of Hydrology* 213, 198–212.

Boegh, E., Soegaard, H., Hanan, N., Kabat, P. and Lesch, L. (1999) A remote sensing study of the NDVI–T_s relationship and the transpiration from sparse vegetation in the Sahel based on high-resolution satellite data. *Remote Sensing of Environment* 69, 224–240.

Bramson, M.A. (1968) *Infrared Radiation*. Plenum Press, New York.

Brenner, A.J. and Incoll, L.D. (1997) The effect of clumping and stomatal response on evaporation from sparsely vegetated shrublands. *Agricultural and Forest Meteorology* 84, 187–205.

Brenner, A.J and Jarvis, P.G. (1995) A heated leaf replica technique for determination of leaf boundary layer conductance in the field. *Agricultural and Forest Meteorology* 72, 261–275.

Brough, D.W., Jones, H.G. and Grace, J. (1986) Diurnal changes in water content of the stems of apple trees, as influenced by irrigation. *Plant, Cell and Environment* 9, 1–7.

Brown, K.W. and Rosenberg, N.J. (1972) A resistance model to predict evapotranspiration and its application to a sugar beet field. In: *Research in Evapotranspiration 1969–1972*. Horticulture Process Report no. 96, University of Nebraska, Lincoln, Nebraska.

Brutsaert, W.H. (1982) *Evaporation into the Atmosphere.* D. Riedel, Dordrecht, The Netherlands.

Campbell, G.S. and Norman, J.M. (1998) *An Introduction to Environmental Biophysics,* 2nd edn. Springer, New York.

Caselles, V., Artigao, M.M., Hurtado, E., Coll, C. and Brasa, A. (1998) Mapping actual evapotranspiration by combining Landsat TM and NOAA-AVHRR images: application to the Barrax area, Albacete, Spain. *Remote Sensing of Environment* 63, 1–10.

Chen, J.M., Yang, B.J. and Zhang, R.H. (1989) Soil thermal emissivity as affected by its water content and surface treatment. *Soil Science* 148, 433–435.

Choudhury, B.J., Idso, S.B. and Reginato, R.J. (1986) Analysis of a resistance energy-balance method for estimating daily evaporation from wheat plots using one-time-of-day infrared temperature observations. *Remote Sensing of Environment* 19, 253–268.

Choudhury, J.C. (1994) Synergism of multispectral satellite observations for estimating regional land surface evaporation. *Remote Sensing of the Environment* 49, 264–274.

Courault, D., Aloui, B., Lagouarde, J.-P., Clastre, P., Nicolas, H. and Walter, C. (1996) Airborne thermal data for evaluating the spatial distribution of actual evapotranspiration over a watershed in oceanic climatic conditions – application of semi-empirical models. *International Journal of Biometeorology* 17, 2281–2303.

Di Bella, C.M., Rebella, C.M. and Paruelo, J.M. (2000) Evapotranspiration estimates using NOAA AVHRR imagery in the Pampa region of Argentina. *International Journal of Remote Sensing* 21, 791–797.

François, C. (1997) The potential of directional radiometric temperatures for monitoring soil and leaf temperature and soil moisture status. *Remote Sensing of Environment* 80, 122–133.

Fuchs, M. and Tanner, C.B. (1968) Surface temperature measurements of bare soils. *Journal of Applied Meteorology* 7, 303–305.

Garratt, J.R. (1995) Observed screen (Air) and GCM surface screen temperatures – implications for outgoing long-wave fluxes at the surface. *Journal of Climate* 8, 1360–1368.

Gash, J.H.C. (1987) An analytical framework for extrapolating evapotranspiration measurements by remote sensing surface temperature. *International Journal of Remote Sensing* 8, 1245–1249.

Goward, S.N., Cruickshanks, G.D. and Hope, A.S. (1985) Observed relation between thermal emission and reflected spectral radiance of a complex vegetated landscape. *Remote Sensing of Environment* 18, 137–146.

Gu, J., Smith, E.A., Hodges, G. and Cooper, H.J. (1997) Retrieval of daytime surface net longwave flux over BOREAS from GOES estimates of surface solar flux and surface temperature. *Canadian Journal of Remote Sensing* 23, 176–187.

Gupta, S.K., Wilber, A.C. and Kratz, D.P. (1999) Surface emissivity maps for satellite retrieval of the longwave radiation budget. In: *10th Conference on Atmospheric Radiation*. American Meteorological Society, Madison, Wisconsin.

Harrison-Murray, R.S. (1991) An electrical sensor for potential transpiration: principle and prototype. *Journal of Horticultural Science* 66, 141–149.

Idso, S.B., Jackson, R.D., Ehrler, W.L. and Mitchell, S.T. (1969) A method for determination of infrared emittance of leaves. *Ecology* 50, 899.

Idso, S.B., Jackson, R.D., Pinter, P.J., Reginato, R.J. and Hatfield, J.L. (1981) Normalizing the stress-degree-day parameter for environmental variability. *Agricultural Meteorology* 24, 45–55.

Inoue, Y. and Moran, M.S. (1997) A simplified method for remote sensing of daily canopy transpiration – a case study with direct measurements of canopy transpiration in soybean canopies. *International Journal of Remote Sensing* 18, 139–152.

Inoue, Y., Kimball, B.A., Jackson, R.D., Pinter, P.J. and Reginato, R.J. (1990). Remote estimation of leaf transpiration rate and stomatal-resistance based on infrared thermometry. *Agricultural and Forest Meteorology* 51, 21–33.

Inoue, Y., Moran, M.S. and Pinter, P.J. (1994) Remote sensing of potential and actual daily transpiration of plant canopies based on spectral reflectance and infrared thermal measurements – concept with preliminary test. *Journal of Agricultural Meteorology* 49, 237–246.

Jackson, R.D. (1982) Canopy temperature and crop water stress. *Advances in Irrigation* 1, 43–85.

Jackson, R.D., Reginato, R.J. and Idso, S.B. (1977) Wheat canopy temperature: a practical tool for evaluating water requirements. *Water Resources Research* 13, 651–656.

Jackson, R.D., Hatfield, J.L., Reginato, R.J., Idso, S.B. and Pinter, P.J., Jr (1983) Estimation of daily evapotranspiration from one time-of-day measurement. *Agricultural Water Management* 7, 35.

Jiang, L. and Islam, S. (1999) A methodology for estimation of surface evapotranspiration over large areas using remote sensing observations. *Geophysical Research Letters* 26, 2773–2776.

Jones, H.G. (1992) *Plants and Microclimate*, 2nd edn. Cambridge University Press, Cambridge.

Jones, H.G. (1999a) Use of thermography for quantitative studies of spatial and temporal variation of stomatal conductance over leaf surfaces. *Plant, Cell and Environment* 22, 1043–1056.

Jones, H.G. (1999b) Use of infrared thermometry for estimation of stomatal conductance in irrigation scheduling. *Agricultural and Forest Meteorology* 95, 135–149.

Kalluri, S.N.V., Townshend, J.R.G. and Doraiswamy, P. (1998) A simple single layer model to estimate transpiration from vegetation using multi-spectral and meteorological data. *International Journal of Remote Sensing* 19, 1037–1053.

Kimes, D.S., Idso, S.B., Pinter, P.J., Reginato, R.J. and Jackson, R.D. (1980) View angle effects in the radiometric measurement of plant canopy temperatures. *Remote Sensing of Environment* 10, 273–284.

Kimes, D.S., Smith, J.A. and Link, L.E. (1981) Thermal IR excitance model of a plant canopy. *Applied Optics* 20, 623–632.

Kustas, W.P. (1990) Estimates of evapotranspiration with one- and two-layer model of heat transfer over partial land cover. *Journal of Applied Meteorology* 29, 704–715.

Kustas, W.P. and Norman, J.M. (1997) A two-source approach for estimating turbulent fluxes using multiple angle thermal infrared observations. *Water Resources Research* 33, 1495–1508.

Kustas, W.P., Moran, M.S., Jackson, R.D., Gay, L.W., Duell, L.F.W., Kunkell, K.E. and Matthias, A.D. (1990) Instantaneous and daily values of the surface energy balance over agricultural fields using remote sensing and a reference field in an arid environment. *Remote Sensing of Environment* 32, 125–141.

Kustas, W.P., Perry, E.M., Doraiswamy, P.C. and Moran, M.S. (1994) Using satellite remote sensing to extrapolate evapotranspiration estimates in time and space over a semiarid rangeland basin. *Remote Sensing of Environment* 49, 275–286.

Lagouarde, J.-P. (1991) Use of NOAA AVHRR data combined with an agrometeorological model for evaporation mapping. *International Journal of Remote Sensing* 12(9), 1853–1864.

Lagouarde, J.-P., Kerr, Y.H. and Brunet, Y. (1995) An experimental-study of angular effects on surface-temperature for various plant canopies and bare soils. *Agricultural and Forest Meteorology* 77, 167–190.

Li, F.Q. and Lyons, T.J. (1999) Estimation of regional evapotranspiration through remote sensing. *Journal of Applied Meteorology* 38, 1644–1654.

McGuire, M.J., Balick, L.K., Smith, J.A. and Hutchison, B.A. (1989) Modelling directional thermal radiance from a forest canopy. *Remote Sensing of Environment* 27, 169–186.

McNaughton, K.G. and Jarvis, P.G. (1983) Predicting the effects of vegetation changes on transpiration and evaporation. In: Kozlowski, T.T. (ed.) *Water Deficits and Plant Growth*, Vol. VII. New York, Academic Press, pp. 1–45.

Menenti, M., Jia, L., Li, Z.L., Djepa, V., Wang, J.M., Stoll, M.P., Su, Z.B. and Rast, M. (2001) Estimation of soil and vegetation temperatures with multiangular thermal infrared observations: IMGRASS, HEIFE, and SGP 1997 experiments. *Journal of Geophysical Research – Atmospheres* 106, 11997–12010.

Monteith, J.L. and Unsworth, M.H. (1990) *Principles of Environmental Physics*, 2nd edn. Edward Arnold, London.

Moran, M.S., Kustas, W.P., Viodal, A., Stannard, D.I., Blanford, J.H. and Nichols, W.D. (1994) Use of ground-base remotely sensed data for surface-energy balance evoluation of a semiarid rangeland. *Water Resources Research* 40, 1339–1349.

Moran, S.M., Jackson, R.D., Raymond, L.H., Gay, L.W. and Slater, P.N. (1989) Mapping surface energy balance components by combining Landsat Thematic Mapper and ground-based meteorological data. *Remote Sensing of Environment* 30, 77–87.

Morse, A., Tasumi, M., Allen, R.G. and Kramber, W.J. (2000). *Application of the SEBAL Methodology for Estimating Consumptive Use of Water and Streamflow Depletion in the Bear River Basin of Idaho through Remote Sensing*. Final Report, Idaho Department of Water resources and University of Idaho, Department of Biological and Agricultural Engineering, Kimberley submitted to the Raytheon Systems Company, Earth Observation System Data and Information System Project, 15 December.

Nemani, R.R. and Running, S.W. (1989) Estimation of regional surface resistance to evapotranspiration from NDVI and thermal-IR AVHRR data. *Journal of Applied Meteorology* 28, 276–284.

Nieuwenhuis, G.J.A., Smidt, E.H. and Thunnissen, H.A.M. (1985) Estimation of regional transpiration of arable crops from thermal infrared images. *International Journal of Remote Sensing* 6, 1319–1334.

Norman, J.M., Divakarla, M. and Goel, N.S. (1995) Algorithms for extracting information from remote thermal-IR observations of the Earth's surface. *Remote Sensing of Environment* 51, 157–168.

Otterman, J., Susskind, J., Brakke, T., Kimes, D., Pielke, R. and Lee, T.J. (1995) Inferring the thermal-infrared hemispheric emission from a sparsely-vegetated surface by directional measurements. *Boundary-Layer Meteorology* 74, 163–180.

Otterman, J., Brakke, T., Fuchs, M., Lakshmi, V. and Cadeddu, M. (1999) Long-wave emission from a plant/soil surface as a function of the view direction: dependence on the canopy architecture. *International Journal of Remote Sensing* 20, 2195–2201.

Paloscia, S. and Pampaloni, P. (1986) The measurement of evapotranspiration using infrared remote sensing methods. *Alta Frequenza* 55, 395–400.

Paw U, K.T., Ustin, S.L. and Zhang, C.A. (1989) Anisotropy of thermal infrared exitance in sunflower canopies. *Agricultural and Forest Meteorology* 48, 45–58.

Paw U, K.T., Brunet, Y., Collineau, S., Shaw, R.H., Maitani, T., Qiu, J. and Hipps, L. (1992) On coherent structures in turbulence above and within agricultural plant canopies. *Agricultural and Forest Meteorology* 61, 55–68.

Priestley, C.H.M. and Taylor, R.J. (1972) On the assessment of surface heat flux and evaporation using large-scale parameters. *Monthly Weather Review* 100, 81–92.

Qin, Z. and Karnieli, A. (1999) Progress in remote sensing of land surface temperature and ground emissivity using NOAA-AVHRR data. *International Journal of Remote Sensing* 20, 2367–2393.

Qiu, G.-Y., Momii, K. and Yano, T. (1996) Estimation of transpiration by imitation leaf temperature – theoretical consideration and field verification (I). *Transactions of Japanese Society of Irrigation, Drainage and Reclamation Engineering* 183, 47–56.

Reginato, R.J., Jackson, R.D. and Pinter, P.J. (1985) Evapotranspiration calculated from remote multispectral and ground station meteorological data. *Remote Sensing of Environment* 18, 75–89.

Reicosky, D.C., Brown, P.W. and Moran, M.S. (1994) Diurnal trends in wheat canopy temperature, photosynthesis and evapotranspiration. *Remote Sensing of Environment* 49, 235–245.

Sabins, F.F. (1978) *Remote Sensing*. Freemans, San Francisco.

Salisbury, J.W. and Milton, N.M. (1988) Thermal infrared (2.5 to 13.5 μm) directional hemispherical reflectance of leaves. *Photogrammetric Engineering and Remote Sensing* 54, 1301–1304.

Seguin, B. and Itier, B. (1983) Using midday surface temperature to estimate daily evaporation from satellite thermal infrared data. *International Journal of Remote Sensing* 4, 371–383.

Seguin, B., Becker, F., Phulpin, T., Gu, X.F., Guyot, G., Kerr, Y., King, C., Lagouarde, J.-P., Ottle, C., Stoll, M.P., Tabbagh, A. and Vidal, A. (1999) IRSUTE: a minisatellite project for land surface heat flux estimation from field to regional scale. *Remote Sensing of Environment* 68, 357–369.

Shuttleworth, W.J. and Wallace, J.S. (1985) Evaporation from sparse crops – an energy combination theory. *Quarterly Journal of the Royal Meteorological Society* 111, 839–855.

Snyder, W.C. and Wan, Z.M. (1998) BRDF models to predict spectral reflectance and emissivity in the thermal infrared. *IEEE Transactions on Geoscience and Remote Sensing* 36, 214–225.

Snyder, R.L., Spano, D. and Paw U, K.T. (1996). Surface renewal analysis for sensible and latent heat flux density. *Boundary-Layer Meteorology* 77, 249–266.

Sobrino, J.A. and Cuenca, J. (1999) Angular variation of thermal infrared emissivity for some natural surfaces from experimental measurements. *Applied Optics* 38, 3931–3936.

Soer, G.J.C. (1980) Estimation of regional evapotranspiration and soil moisture conditions using remotely sensed crop surface temperature. *Remote Sensing of Environment* 9, 27–45.

Spano, D., Snyder, R.L., Duce, P. and Paw U, K.T. (2001) Estimating sensible and latent heat flux densities from grapevine canopies using surface renewal. *Agricultural and Forest Meteorology* 104, 171–183.

Stanghellini, C. (1987) Transpiration of greenhouse crops: an aid to climate management. PhD dissertation, Landbouwuniversiteit, Wageningen.

Sutherland, R.A. (1986) Broadband and spectral emissivities (2–18 μm) of some natural soils and vegetation. *Journal of Atmospheric and Oceanic Technology* 3, 199–202.

Taconet, O., Carlson, T., Bernard, R. and Vidal-Madjar, D. (1986) Evaluation of a surface/vegetation parameterization using satellite measurements of surface temperature. *Journal of Climate and Applied Meteorology* 25, 1752–1767.

Unsworth, M.H. and Monteith, J.L. (1975) Geometry of long-wave radiation at the ground. I. Angular distribution of incoming radiation. *Quarterly Journal of the Royal Meteorological Society* 101, 13–24.

Van de Griend, A.A. and Owe, M. (1993) On the relationship between thermal emissivity and the normalized difference vegetation index for natural surfaces. *International Journal of Remote Sensing* 14, 1119–1131.

Yang, X., Zhou, Q. and Melville, M. (1997) Estimating local sugarcane evapotranspiration using Landsat TM image and a VITT concept. *International Journal of Remote Sensing* 18, 453–459.

10

Forest–Air Exchange in Non-ideal Conditions: the Role of Horizontal Flux and its Divergence

Xuhui Lee
School of Forestry and Environmental Studies, Yale University, 370 Prospect Street, New Haven, CT 06511, USA

Abstract

The objective of this chapter is to examine the role of the horizontal eddy flux and its horizontal divergence in the surface-layer mass conservation. It is shown that the horizontal flux is generated by gradient diffusion and by a correlation between the vertical and horizontal velocities, the latter of which is the dominant mechanism in the roughness sublayer over a forest, where the shear stress is large. Thus, a horizontal flux divergence of the scalar of interest can arise from a streamwise gradient of its vertical flux and/or the momentum flux. An example is given for the temperature flux divergence over a forest on rolling terrain to illustrate that it is possible to measure experimentally the divergence in non-ideal conditions. Numerical simulations using an advection–diffusion model show that a compensating mechanism exists between horizontal advection and the horizontal flux divergence and that experimental attempts to correct the vertical eddy flux for either horizontal advection or the flux divergence alone do not necessarily lead to a complete surface-layer mass/energy-balance closure.

In the appendix, an approximation equation is presented that allows the horizontal flux divergence to be calculated, with measurements made by a vertical array of sensors on a single tower.

Introduction

This chapter represents a continued effort to assess micrometeorological theory for observational studies of the forest–air exchange of energy, water and trace gases. One fundamental element of the theory is the mass-conservation equation, written here in two-dimensional form as:

(eds M. Mencuccini, J. Grace, J. Moncrieff and K.G. McNaughton)

$$\frac{\partial \bar{c}}{\partial t} + \frac{\partial \overline{u'c'}}{\partial x} + \bar{u}\frac{\partial \bar{c}}{\partial x} + \bar{w}\frac{\delta \bar{c}}{\delta z} + \frac{\partial \overline{w'c'}}{\partial z} = s \tag{1}$$

where c is concentration of a scalar quantity, t is time, u and w are the velocity components in the streamwise (x) and vertical (z) directions, respectively, s is source strength, and overbar and prime denote the time-averaged and fluctuating quantities, respectively. In most of the studies of this type reported in the literature, the atmosphere near the vegetation is assumed to be horizontally homogeneous and thus all atmospheric quantities vary only in the vertical direction. With this one-dimensional, idealized approximation, the net ecosystem exchange (NEE) is taken as the sum of two easily measured quantities, the vertical eddy flux and air storage. Violation of the assumption may occur in non-ideal conditions, such as free convection, rainy weather, strongly stratified air, horizontal heterogeneity of the source strength or uneven terrain. Even at sites that satisfy the usual micrometeorological fetch requirement, horizontal homogeneity is not always attained, leading to large errors in the surface exchange measurement. In recognition of this, several recent studies have been devoted to the mechanisms through which horizontal and vertical advection (terms 3 and 4 on the left of Equation 1) affects the measurements of vegetation–air exchange (Lee, 1998; Finnigan, 1999; Paw U *et al.*, 2000; Massman and Lee, 2002). So far little attention has been given to the horizontal eddy-flux divergence term (term 2 on the left of Equation 1).

The objective of this chapter is to examine the role of the horizontal eddy-flux divergence in the surface-layer mass budget. It begins in the next section-with a discussion of the mechanism of the horizontal eddy-flux generation. This is followed by an analysis (A Field Observation of the Flux Divergence) of a data set on the horizontal temperature-flux divergence obtained in the roughness sublayer over a forest on rolling terrain. Evidence is presented to suggest that direct field observation of the divergence term is possible. In the Modelling section, an advection–diffusion model that incorporates the mechanism of the horizontal eddy-flux generation is used to simulate how various terms of Equqtion (1) evolve as a function of distance downwind of an abrupt change in the source strength. Recognizing the fact that horizontal sensor arrays are rarely deployed in micrometeorological field studies, a system of equations is presented, in an appendix, to approximate the divergence term with measurements made by a vertical array of sensors on a single tower.

For reasons to be discussed below, the horizontal eddy flux, $\overline{u'c'}$, is large in the surface layer, but because its variations usually occur at a (horizontal) length scale much larger than the (vertical) scale over which the variation of the vertical eddy flux, $\overline{w'c'}$, is measured, its horizontal divergence is often neglected, a simplification that is adopted in model assessments of fetch and footprint requirements (Philip, 1959; Dyer, 1963; Horst and Weil, 1992) and is implied in all observational studies of vegetation–air exchange. Clearly, the flux-divergence term may become non-negligible at times when one or more of the non-ideal conditions are encountered. Furthermore, the information produced by the tower-based eddy-covariance studies is not used to its full potential because analysis of the observational data is almost always limited to the vertical eddy fluxes only. Examinations of the horizontal flux may reveal new constraints on the mechanism of horizontal and vertical advection.

Generation of the Horizontal Eddy Flux

In a scalar field where the mean concentration varies in both the x and the z directions, the fluctuation of c can be expressed, on the simple mixing-length analogy, as:

$$c' = -u'\tau_x{}' \frac{\partial \bar{c}}{\partial x} - w'\tau_z{}' \frac{\partial \bar{c}}{\partial z} \tag{2}$$

where $\tau_x{}'$ and $\tau_z{}'$ are a horizontal and a vertical eddy time-scale, respectively. Multiplying Equation (2) by u' and performing Reynolds averaging, we obtain an expression for the horizontal eddy flux:

$$\overline{u'c'} = -\tau_{ux}\overline{u'^2}\frac{\partial \bar{c}}{\partial x} - \tau_{uz}\overline{u'w'}\frac{\partial \bar{c}}{\partial z} \tag{3}$$

where $\tau_{ux} = \overline{u'^2\tau_x'} / \overline{u'^2}$ and $\tau_{uz} = \overline{u'w'\tau_z{}'} / \overline{u'w'}$

An expression similar to Equation (3) can be written for the vertical eddy flux $\overline{w'c'}$:

$$\overline{w'c'} = -\tau_{wx}\overline{u'w'}\frac{\partial \bar{c}}{\partial x} - \tau_{wz}\overline{w'^2}\frac{\partial \bar{c}}{\partial z} \tag{4}$$

where τ_{wx} and τ_{wz} are two eddy time-scales analogous to τ_{ux} and τ_{uz}. Eliminating $\partial\bar{c}/\partial z$ from Equations (3) and (4), we obtain:

$$\overline{u'c'} \approx -\tau_{ux}\overline{u'^2}\frac{\partial \bar{c}}{\partial x} + \alpha\frac{\overline{u'w'}\;\overline{w'c'}}{\overline{w'^2}} \tag{5}$$

where α ($= \tau_{uz}/\tau_{wz}$) is an empirical coefficient. The asymptotic limit of Equation (5), attained in a flow where the horizontal $\bar{c}$ gradient vanishes, is:

$$\overline{u'c'} = \alpha\frac{\overline{u'w'}\;\overline{w'c'}}{\overline{w'^2}} \tag{6}$$

The coefficient α takes a value of 2.4 for neutral and unstable conditions and 3.3 for stable conditions (Fig. 10.1). Note that Equation (6) differs in form from that of Wyngaard *et al.* (1971), who used dimensional analysis to relate $\overline{u'c'}$ to $\partial\bar{u}/\partial z$ and $\partial\bar{c}/\partial z$.

Several points suggested by Equations (5) and (6) warrant commenting on here. First, the horizontal eddy flux can originate from gradient diffusion in a horizontally heterogeneous scalar concentration field (first term, Equation 5) or from a correlation between the horizontal and vertical velocities (second term, Equation 5). The first term of Equation (5) vanishes in an idealized, horizontally homogeneous scalar field, but the second term is usually large, except under free convection. Since $\overline{u'w'}$ is negative in the surface layer, the horizontal

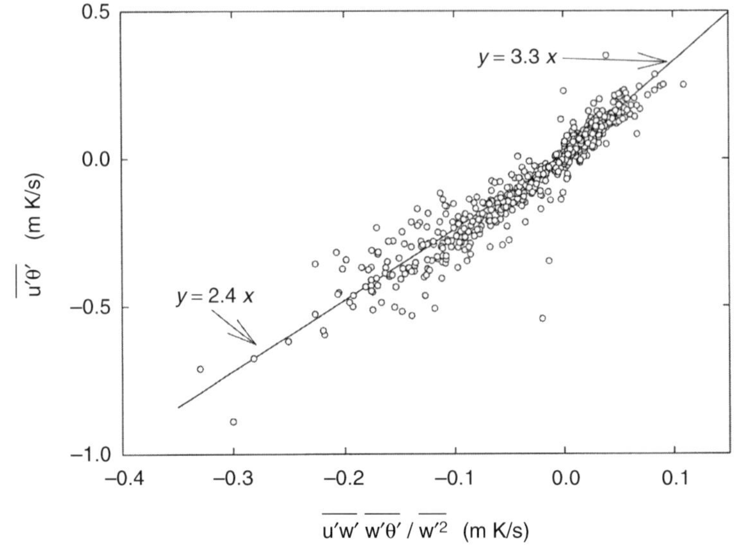

Fig. 10.1. The relationship between $\overline{u'\theta'}$ and $(\overline{u'w'}\ \overline{w'\theta'})/\overline{w'^2}$ at the Great Mountain forest, June–December 1999.

eddy flux is generally opposite in sign to the vertical flux, $\overline{w'c'}$. Secondly, the horizontal flux is often parameterized in the form $\overline{u'c'} = -K_x \partial\bar{c}/\partial x$ in modelling studies of the boundary-layer flow. This is inappropriate in the surface layer, where the horizontal flux is generated mainly by the u–w correlation. Thirdly, Equation (6) shows that, even in a horizontally uniform scalar field ($\partial\bar{c}/\partial x = 0$), a streamwise change in the horizontal flux can exist owing to changes in the shear stress, calling into question the assumption that the streamwise diffusion is negligible in a transitional flow downwind of a change in the surface roughness (Garratt, 1990). Finally, according to Equation (6), it is incorrect to ignore the horizontal-flux divergence in a situation where the surface source strength varies in the streamwise direction, as in modelling assessments of fetch requirement for flux observations (Philip, 1959; Dyer, 1963). The reason for this is that a streamwise variation in the source strength will result in a streamwise gradient of $\overline{w'c'}$, which in turn will result in a streamwise gradient of $\overline{u'c'}$. The last point is discussed further in the Modelling section.

A Field Observation of the Flux Divergence

The experiment

This section presents the result of a field experiment in which the temperature-flux divergence was measured. The experiment was conducted at the Great Mountain Forest on rolling topography (Lee and Hu, 2002). Tree height varied between 16 and 22 m. Three sonic anemometers/thermometers were operated at the same height of 30.4 m above the ground. Two of these were of the same model (model CSAT3, Campbell Scientific, Logan, Utah) and were separated by a distance Δx (= 3.82 m). Their synchronous device for measurement (SDM) signals were recorded at 5 Hz for archiving and for the analysis presented below. The SDM diagnostic flags were used to eliminate those observations that had data-quality problems. The horizontal eddy flux of temperature, $\overline{u'\theta'}$, its horizontal divergence, $\partial\overline{u'\theta'}/\partial x$, and the covariance term, $\overline{u'(\partial u'\theta'/\partial x)}$, were determined with the two sonic units in parallel flow, that is, when wind direction was within $\pm 10°$ of their separation direction (wind direction range 315–335°). All turbulence statistics were expressed in a long-term streamline coordinate.

Results

The horizontal temperature-flux divergence had a tendency to be opposite in sign to the vertical temperature flux, $\overline{w'\theta'}$, and appeared to be of the same order of $-\overline{w'\theta'}/z$ (Fig. 10.2). The midday mean divergence value was -0.008 K/s. The large divergence value is not a real surprise, given that large spatial variations in both the shear stress and the surface heat flux, and hence variations in the horizontal flux (Equation 6), are expected for rolling terrain. Lee and Hu (2002) found that the vertical momentum and heat fluxes were not constant with height in the roughness sublayer over the same forest; this could be interpreted as evidence for spatial heterogeneity.

At first glance, it seems extremely naïve to measure the flux divergence over a short separation distance, as any measurement biases could make the data useless. For example, a slightly lower u sensitivity of the upwind sensor would result in $\partial\overline{u'\theta'}/\partial x$ with an opposite sign to $\overline{w'\theta'}$. Prompted by this possibility, we have computed the flux difference $\Delta\overline{u'\theta'} = (\overline{u'\theta'})_2 - (\overline{u'\theta'})_1$ for periods when the wind was perpendicular to sensor separation (wind direction

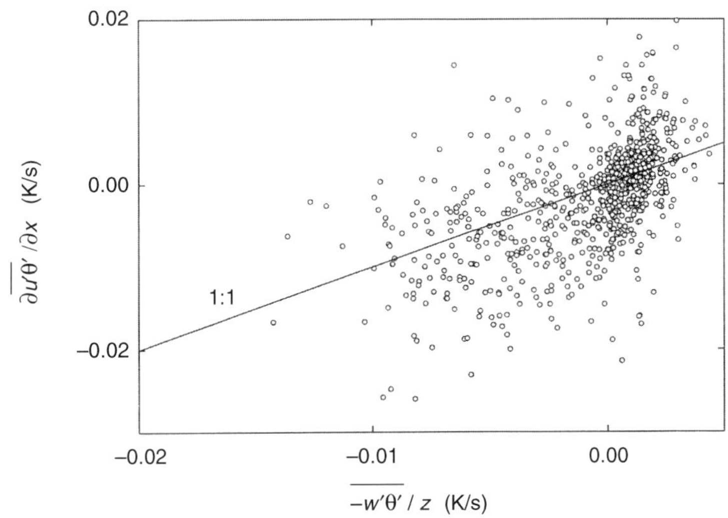

Fig. 10.2. A scatter plot of $\partial\overline{u'\theta'}/\partial x$ against $\overline{w'\theta'}/z$ for Great Mountain, 1999.

235°) and have found no dependence on $\overline{w'\theta'}$. Here subscripts 1 and 2 denote the two CSAT3 sensors. Furthermore, the value of $\Delta\overline{u'\theta'}/\Delta x$ at a wind direction of 235° fell in a range much smaller than the range in Fig. 10.2. Thus sensor bias is unlikely to be a cause of the pattern in Fig. 10.2. A second source of error could arise from the aerodynamic interference of the tower with the flow. The aerodynamic influence on both sensors was found to be negligible at the wind direction of 325° (Lee and Hu, 2002). Had there been a slight reduction in the horizontal velocity at the downwind sensor location, it would have resulted in a divergence term of the same sign as the temperature flux, contrary to the pattern shown in Fig. 10.2.

The implication of the negative correlation between $\partial\overline{u'\theta'}/\partial x$ and $\overline{w'\theta'}$ for the surface-layer energy balance can be explored in the context of the conservation equation (Equation 1). The observed daytime total eddy fluxes (sensible-heat flux plus latent-heat flux) was 8% higher than the available energy flux for the periods with wind parallel to the sensor separation (Lee and Hu, 2002). Having the divergence term with opposite sign to the vertical flux implies that attempts to correct for the divergence would make the estimate of NEE of sensible heat smaller in magnitude than that without the correction, and hence would improve the energy balance. However, such improvement is rather fortuitous. Because a non-zero horizontal-flux divergence indicates that the flow or the scalar concentration field is not horizontally homogeneous, other advection terms must also be non-negligible. In other words, the divergence term cannot be evaluated in isolation for other terms of the mass-conservation equation. Unfortunately, experimental attempts to close the mass/energy balance by measuring every term of the conservation equation are extremely difficult.

Modelling

Basic equations

To overcome the limitation inherent in the point observation, in this section we use an advection–diffusion model to investigate how various terms of the conservation equation may interact with one another. For computational convenience but without losing generality, the source is assumed to exist only at ground level. We consider the classical advective flow, in which a step change in the NEE rate (from zero to F) occurs at position $x = 0$. On the assumption that the flow is at steady state and is convergence-free, Equation (1) is simplified to:

$$\frac{\partial\overline{u'c'}}{\partial x} + \bar{u}\frac{\partial\bar{c}}{\partial x} + \frac{\partial\overline{w'c'}}{\partial z} = 0 \qquad (7)$$

The eddy fluxes are parameterized as:

$$\overline{w'c'} = -K\frac{\partial\bar{c}}{\partial z} \qquad (8)$$

$$\overline{u'c'} = -K\frac{\partial \bar{c}}{\partial x} + \alpha \frac{\overline{u'w'}\,\overline{w'c'}}{\overline{w'^2}} \tag{9}$$

where K is eddy diffusivity, assumed to be the same for both vertical and horizontal diffusion, and $\alpha = 2.4$.

In seeking a numerical solution to Equations (7)–(9), we use the standard surface-layer functions for neutral air:

$$\bar{u} = \frac{u_*}{k}\ln(z/z_o)$$

$$K = ku_*z$$

$$\overline{u'w'} / \overline{w'^2} = -0.64$$

where z_o is surface roughness length, u_* is friction velocity, and k (= 0.4) is the von Karman constant. The boundary conditions are:

$$\bar{c} = 0,\ x = 0$$

$$\overline{w'c'} = F,\ z = z_o$$

$$\frac{\partial \overline{w'c'}}{\partial z} = 0,\ z = H$$

$$\frac{\partial \overline{w'c'}}{\partial x} = \frac{\partial \overline{u'c'}}{\partial x} = 0,\ x = L$$

A full description of the numerical procedure is given by Lee (2003). Briefly, the solution is achieved using the method of Patankar (1980). The vertical domain size (H) is 800 m, and the horizontal domain size (L) is 4000 m. Parameter z_o takes a value of 2 m to represent the surface roughness of a forest of about 20 m in height.

Integration of Equation (7) in respect of z gives:

$$\frac{1}{\Phi}\int_0^{\zeta_r} \frac{\partial \overline{\upsilon'\chi'}}{\partial \xi} + \frac{1}{0}\int_0^{\zeta_r} \bar{\upsilon}\frac{\partial \bar{\chi}}{\partial \xi}d\zeta + \frac{1}{\Phi}\overline{\omega'\chi'}\Big|_{\zeta_r} = 1 \tag{10}$$

where z_r is measurement height relative to the source location. In the following, the numerical solution is integrated according to Equation (10), for a measurement height of 26 m, a typical height of eddy-flux observations in forests. In addition, as shown by Leclerc and Thurtell (1990) and Horst and Weil (1992), the solution for this particular type of advective flow can be used to derive the flux footprint f as:

$$f = \frac{1}{F}\frac{\partial \overline{w'c'}}{\partial x} \tag{11}$$

Results

Figure 10.3 presents the evolution with fetch distance of the three terms on the left of Equation (10) and the flux footprint computed from Equation (11). Also given as a reference is the solution with the horizontal eddy-flux divergence term in Equation (7) set to zero. The vertical eddy flux downwind of the source discontinuity increases with distance to approach the source strength *F*, or the true NEE. The horizontal eddy flux (not shown) approaches the asymptotic limit of $\alpha\, \overline{u'w'}\, F/\overline{w'^2}$. The horizontal divergence of the horizontal eddy flux is negative, that is, with opposite sign to the vertical flux, and is not negligible up to a fetch distance of about 100 m. Within this fetch distance, the inclusion of the horizontal flux divergence in the calculation has a fairly large impact on the

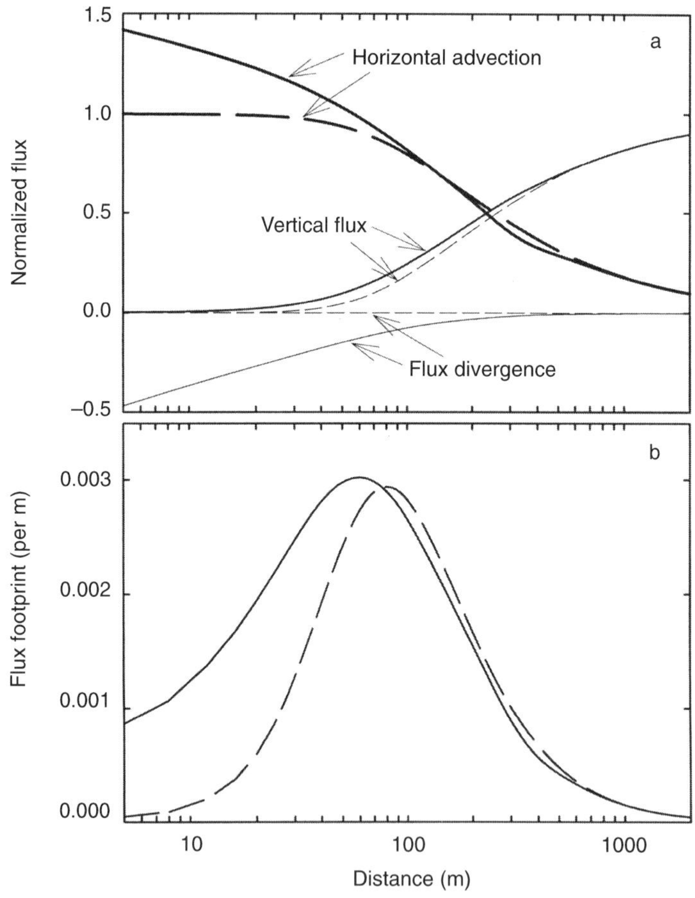

Fig. 10.3. (a) Evolution of various terms of the mass conservation as a function of fetch distance (flux divergence: term 1 on the left of Equation 10; horizontal advection: term 2; vertical flux, term 3). The dashed lines represent the solution with the horizontal flux divergence set to zero. (b) Flux footprint with (solid line) and without (dashed line) the inclusion of the flux divergence in the conservation equation.

horizontal advection term and on the flux footprint and a minor effect on the vertical flux. Interestingly, a compensating mechanism exists between horizontal advection and the horizontal flux divergence: the flux-divergence term is roughly equal in magnitude to the difference in the horizontal advection calculated with and without the inclusion of the horizontal eddy flux in the model. From a modelling viewpoint, omission of the horizontal flux divergence does not introduce serious errors if the objective is to solve for the vertical eddy flux. However, from an experimental viewpoint, because of the compensating mechanism, efforts to correct the vertical eddy flux for either horizontal advection or flux divergence alone do not necessarily lead to a complete surface-layer mass/energy-balance closure.

Conclusions

The horizontal eddy flux is generated by gradient diffusion and by a correlation between the vertical and horizontal velocities. In the roughness sublayer, where the shear stress is large, the shear term (second term on the right, Equation 5) is much larger than the diffusion term (first term), except perhaps in the vicinity of a surface-source discontinuity.

Over a forest on rolling terrain, the horizontal temperature-flux divergence was measurable with two flux sensors separated by a short separation distance. The non-zero horizontal flux divergence suggests streamwise gradients of the shear stress and/or the surface heat flux at the site (cf. Equation 6).

Model simulations of the advective flow downwind of an abrupt change in the surface-source strength suggest a compensating mechanism between horizontal advection and the horizontal flux divergence. Experimental efforts to correct the vertical eddy flux for either horizontal advection or the flux divergence alone do not necessarily lead to a complete surface-layer mass/energy-balance closure. Inclusion of the flux-divergence term in the computation produces a flux footprint that is markedly different from that without this term, up to the distance where the footprint function reaches its peak value.

Acknowledgements

This work was supported by the US National Science Foundation through grant ATM-0072864 and the US Department of Energy National Institute for Global Environmental Change (NIGEC) program (grant 901214-HAR).

References

Dyer, A.A. (1963) The adjustment of profiles and eddy fluxes. *Quarterly Journal of the Royal Meteorological Society* 89, 276–280.

Finnigan, J. (1999) A comment on the paper by Lee (1998): 'On micrometeorological observations of surface–air exchange over tall vegetation'. *Agriculture and Forest Meteorology* 97, 55–64.

Gao, W., Shaw, R.H. and Paw U, K.T. (1989) Observation of organized structure in turbulent flow within and above a forest canopy. *Boundary-Layer Meteorology* 47, 349–377.

Garratt, J.R. (1990) The internal boundary layer – a review. *Boundary-Layer Meteorology* 50, 171–203.

Garratt, J.R. (1992) *The Atmospheric Boundary Layer*. Cambridge University Press, New York.

Horst, T.W. and Weil, J.C. (1992) Footprint estimation for scalar flux measurements in the atmospheric surface layer. *Boundary-Layer Meteorology* 59, 279–296.

Leclerc, M.Y. and Thurtell, G.W. (1990) Footprint prediction of scalar fluxes using a Markovian analysis. *Boundary-Layer Meteorology* 52, 247–258.

Lee, X. (1998) On micrometeorological observations of surface–air exchange over tall vegetation. *Agriculture and Forest Meteorology* 91, 39–49.

Lee, X. (2003) Fetch and footprint of turbulent fluxes over vegetation stands with elevated sources. *Boundary-Layer Meteorology* 107, 561–579.

Lee, X. and Hu, X. (2002) Forest–air fluxes of carbon and energy over non-flat terrain. *Boundary-Layer Meteorology* 103, 277–301.

Massman, W.J. and Lee, X. (2002) Eddy covariance flux corrections and uncertainties in long-term studies of carbon and energy exchanges. *Agriculture and Forest Meteorology* 113, 121–144.

Patankar, S.V. (1980) *Numerical Heat Transfer and Fluid Flow*. Taylor and Francis, London.

Paw U, K.T., Baldocchi, D.D., Meyers, T.P. and Wilson, K.B. (2000) Correction of eddy-covariance measurements incorporating both advective effects and density fluxes. *Boundary-Layer Meteorology* 97, 487–511.

Philip, J.R. (1959) The theory of local advection. *Journal of Meteorology* 16, 535–547.

Wyngaard, J.C., Cote, O.R. and Izumi, Y. (1971) Local free convection, similarity, and the budgets of shear stress and heat flux. *Journal of Atmospheric Sciences* 28, 1171–1182.

Appendix: an Approximation for the Horizontal Eddy-flux Divergence

Recognizing the fact that horizontal sensor arrays are rarely deployed in micrometeorological field studies, here we present a system of equations to approximate the divergence term with measurements that can be made by a vertically separated array of sensors. This 'single-tower' strategy is in line with a previous study, which suggests that additional constraints, developed from first principles, can generate new insights into the advection problem with observations made on one tower (Lee, 1998).

The divergence term can be written as:

$$\frac{\partial \overline{u'c'}}{\partial x} = \overline{u'\frac{\partial c'}{\partial x}} + \overline{c'\frac{\partial u'}{\partial x}} = \overline{u'\frac{\partial c'}{\partial x}} - \overline{c'\frac{\partial w'}{\partial z}} \tag{12}$$

where the following continuity equation for the fluctuating velocities has been used to eliminate $\partial u'/\partial x$:

$$\frac{\partial u'}{\partial x} + \frac{\partial w'}{\partial z} = 0 \tag{13}$$

To derive an expression for the first term on the right of Equation (12), we note the conservation equation for c':

$$\frac{\partial c'}{\partial t} + \frac{\partial}{\partial x}(u'c' - \overline{u'c'} + \bar{u}\,c' + u'\bar{c}\,) + \frac{\partial}{\partial z}(w'c' - \overline{w'c'} + \bar{w}\,c' + w'\bar{c}\,) = K_c\left(\frac{\partial^2 c'}{\partial x^2} + \frac{\partial^2 c'}{\partial z^2}\right) \tag{14}$$

where K_c is molecular diffusivity (Garratt, 1992). Multiplying both sides of Equation (14) by u', making use of the continuity equation for the mean velocities:

$$\frac{\partial \bar{u}}{\partial x} + \frac{\partial \bar{w}}{\partial z} = 0 \tag{15}$$

and performing Reynolds averaging, we obtain:

$$\underset{\text{I}}{\overline{u'\frac{\partial c'}{\partial t}}} + \underset{\text{II}}{\overline{u'\left(\frac{\partial u'c'}{\partial x} + \frac{\partial w'c'}{\partial z}\right)}} + \underset{\text{III}}{\bar{u}\,\overline{\left(u'\frac{\partial c'}{\partial x}\right)}} + \underset{\text{IV}}{\bar{w}\,\overline{\left(u'\frac{\partial c'}{\partial z}\right)}} + \underset{\text{V}}{\overline{u'^2}\left(\frac{\partial \bar{c}}{\partial x}\right)} + \underset{\text{VI}}{\overline{u'w'}\left(\frac{\partial \bar{c}}{\partial z}\right)} = \underset{\text{VII}}{K_c\,\overline{u'\left(\frac{\partial^2 c'}{\partial x^2} + \frac{\partial^2 c'}{\partial z^2}\right)}} \tag{16}$$

Obviously term VI of Equation (16) is usually non-zero. A strong correlation exists between the horizontal velocity and the time rate of change or the horizontal gradient of c' in the roughness sublayer, where coherent motion dominates the flow, and hence terms I and III are relatively important. For example, under unstable conditions, the decrease in temperature due to the passage of a temperature microfront is usually accompanied by an acceleration of the horizontal velocity (Gao *et al.*, 1989). Indeed, our field observations indicate that terms I and III are of the same order of magnitude as term VI. It is reasonable to assume that the molecular term (term VII) and the higher-order term (term II) are negligible in comparison with terms I and III (Wyngaard *et al.*, 1971; Garratt, 1992). Term IV can also be ignored because $\bar{w}/\bar{u} << 1$. Term V arises from a horizontal concentration gradient, which is usually non-zero in advective flow, but it is appropriate to assume:

$$\frac{\partial \bar{c}}{\partial x} / \frac{\partial \bar{c}}{\partial z} << 1 \tag{17}$$

for the roughness sublayer. Equation (17) may be regarded as a weak advection criterion. It follows from the above scale analysis that Equation (16) can be simplified to:

$$\overline{u' \frac{\partial c'}{\partial x}} \approx -\frac{1}{\bar{u}} (\overline{u' \frac{\partial c'}{\partial t}} + \overline{u'w'} \frac{\partial \bar{c}}{\partial z}) \tag{18}$$

Combining Equations (12) and (18) gives an approximation for the divergence of horizontal eddy flux, which can be measured with a vertical array of sensors on one tower:

$$\frac{\partial \overline{u'c'}}{\partial x} = -\frac{1}{\bar{u}} \overline{u' \frac{\partial c'}{\partial t}} - \frac{\overline{u'w'}}{\bar{u}} \frac{\partial \bar{c}}{\partial z} - \overline{c' \frac{\partial w'}{\partial z}} \tag{19}$$

In Fig. 10.A1, the direct measurement of $\overline{u'(\partial\theta'/\partial x)}$ is compared with the Taylor frozen-turbulence approximation, $-u'(\partial\theta'/\partial t)/\bar{u}$, and the approximation that includes the shear term (Equation 18). The frozen-turbulence relation failed to capture the range of variation in $u'(\partial\theta'/\partial x)$. On the other hand, Equation (18) did a reasonable job, considering the fact that the weak advection criterion (Equation 17) may not always be satisfied because the site was on rolling terrain. The reader is also reminded that the coarse precision of the temperature gradient, which was measured with unventilated and unshielded thermocouples, may also have contributed to the scatter. The consistency between the measured and calculated $u'(\partial\theta'/\partial x)$ lends support to the conclusion that the pattern shown in Fig. 10.2 was genuine rather than due to measurement artefacts.

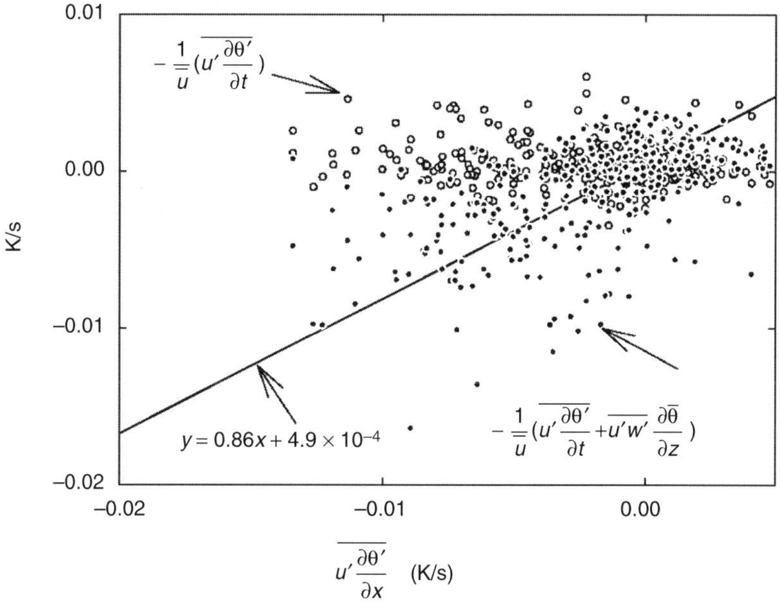

Fig. 10.A1. A comparison of the covariance $\overline{u'(\partial\theta'/\partial x)}$ calculated by the approximation equation (Equation 19) against that measured with the two sonic sensors separated in the direction of the flow at Great Mountain. Line represents a linear regression of the data. Also given is a comparison of the measured $\overline{u'(\partial\theta'/\partial x)}$ with the Taylor frozen-turbulence approximation (open circles).

11

Review of Forest Evaporation[1] Studies, Primarily in the United Kingdom

J.B. Stewart
Department of Geography, University of Southampton, Highfield, Southampton SO17 1BJ, UK

Abstract

The major experimental and modelling studies that have contributed to developing our understanding of the physical principles governing the process of forest evaporation are reviewed with emphasis on developments in the UK over the last 50 years. The importance of considering the two components of forest evaporation separately – transpiration and evaporation of intercepted rainfall – and the compensating mechanisms that control the surface resistance are stressed. Finally, the remaining problems of understanding and modelling forest evaporation are discussed.

Introduction

Forest evaporation has been of interest to foresters and hydrologists for much longer than the last 50 years. For over a century it has been realized that the presence of forests helps to prevent soil erosion, but there has been great controversy about the effect of forests on the runoff from catchments. On the one hand, there was the opinion that forests were bad news for water resources, because they could use more water than shorter vegetation, as demonstrated by Clarke and Newson (1978). On the other hand, early catchment studies in the USA suggested that the presence of forests could smooth out the peaks in runoff, thereby reducing floods and extending the period of runoff when there were droughts (Bates and Henry, 1928). It was also popularly thought that the planting of forests resulted in increased rain-

[1] Note that throughout this review 'evaporation' is used to denote the conversion of liquid water into vapour, the source of liquid water being either within the vegetation (transpiration) or on the surface of the vegetation as a result of precipitation or condensation (e.g. evaporation of intercepted rainfall).

 Forests at the Land–Atmosphere Interface (eds M. Mencuccini, J. Grace, J. Moncrieff and K.G. McNaughton)

fall. So the hydrology of forested areas has been of interest for a long time, but only in the last 50 years has the physical basis of forest evaporation been unravelled.

Initially, interest in forest evaporation was related to land management and water resources, using catchment studies of forests versus grassland or other forms of low status vegetation. With improvements in instrumentation and the development of computers, micrometeorological methods were used to measure evaporation from forests over periods of a day or so. These short-term studies led to an appreciation of the importance of large-scale feedbacks between forests and the atmosphere. Following the Kyoto Agreement, interest is now much greater in the exchange of carbon dioxide between atmosphere and forests. Since the uptake of carbon dioxide takes place through stomata, which allow transpiration to occur, the interest in forest evaporation has continued. However, in relation to the sequestering of carbon by forests, the interest has moved from short-term studies of the physical processes to much longer-term monitoring and modelling.

The aim of this review is to chart the development of our physical understanding of the process of forest evaporation, highlighting in particular the developments in the UK. To a considerable extent, these developments have followed from the availability of improved instrumentation, measurement techniques and processing of data. As our knowledge has increased, it has been realized that the evaporation of rainfall intercepted by forests has to be considered separately from forest transpiration. Also the compensating mechanisms, which are very important in controlling forest evaporation, will be described. Improved physical knowledge has provided the basis for developing models of forest evaporation, which are particularly important for a world undergoing climate change.

Earlier Studies

Towards the end of the 19th century, a research programme aimed at the management and protection of forests with the aim of improving runoff and controlling erosion was started in the USA. As a result the first catchment research studies were set up at Wagon Wheel Gap in 1910 and Great Basin in 1912 and in the mid-1930s further studies were set up, for example, Coweeta (Hoover, 1944). All these studies determined the forest evaporation as the residual of the water balance and so gave values over long periods, i.e. months or a year, and there was considerable uncertainty in the derived evaporation, due to the accumulating errors in the other hydrological components. Also during this period measurements were made of the interception of rainfall by forests by comparing the through fall and stem flow beneath the forest with the rainfall falling above the trees or in a nearby clearing. However, these early measurements of intercepted rainfall were not related to the controlling weather conditions (Kittredge *et al.*, 1941). These studies were able to provide a quantitative value of the total forest evaporation, but not an understanding of the physics involved, and so could not allow reliable predictions of forest evaporation under changed circumstances. It was generally considered that evaporation from water surfaces and vegetation was controlled by 'atmospheric demand'.

To measure this 'atmospheric demand', a network of evaporation pans and tanks was set up in many countries. It was soon discovered that the design of pans or tanks affected the measured evaporation (FAO, 1977).

Penman: Potential Evaporation

Howard Penman began his research into evaporation in 1944 – a spin-off from investigations into the prediction of workable soil surfaces for wheeled traffic for the war effort. From physical principles he developed a formula to estimate pan evaporation from meteorological data for regions where there were no direct measurements. Empirical factors, which were less than unity, were then used to estimate crop evaporation from the values obtained from evaporation pans. In the same year that Penman published his paper (Penman, 1948), Thornthwaite published a paper (Thornthwaite, 1948) in which he adopted the term 'potential evapotranspiration', which Penman preferred to call 'potential evaporation' (Penman, 1949a). Penman, in tune with the times, believed that 'a crop that has an ample supply of water, that is growing in the midst of an extended area of similar plants ... behaves like a sheet of wet green blotting paper that is covered up at night' (Penman, 1949b). So there was no consideration of the effect of the plant stomata on the rate of transpiration.

Howard Penman then moved on to consider crop evaporation directly rather than via pan evaporation. Working with Kenworthy Schofield he reviewed earlier works, particularly those of Brown and Escombe (1900) and Jeffreys (1918), and noted the 'overwhelming importance of external resistances'. They developed a 'big-leaf' canopy model, where the stomata of all the leaves were assumed to conduct in parallel (Penman and Schofield, 1951). This can be considered as a forerunner of the Penman–Monteith equation. Penman and Schofield applied their model to a wheat canopy, now using values of stomatal and aerodynamic 'diffusion lengths' and net radiation appropriate to the crop rather than to a UK Meteorological Office sunken pan. For the wheat crop they calculated a stomatal diffusion length of 0.78 mm – equivalent to a canopy resistance of 52 s/m – which is still considered plausible today. For an aerodynamic diffusion length they only had the empirical value determined for the sunken pan, which gave a value of 200 s/m at a wind speed of 2.2 m/s at a height of 2 m; this is now known to be about five times the typical value for a wheat crop. As a result of this error Penman inferred that stomatal resistance had an insignificant effect in controlling evaporation and decided to neglect stomatal resistance in future. He revised his 1948 formula by replacing net radiation for open water by that for a green crop to give direct estimates of potential evaporation for a crop.

Another consequence of the overestimate of the aerodynamic resistance in relation to the stomatal resistance was the expectation that the evaporation from vegetation when it was wetted by rain would be similar to that when the vegetation was transpiring. So it was predicted that intercepted rain on forests would evaporate at similar rates to that intercepted by short vegetation. Though there were many catchment studies comparing the runoff from forests and that from short vegetation, the evaporation derived from these water balances showed a lot of scatter, since it was computed from the difference between two variables – rainfall and runoff – which were of similar magnitude. So there was no convincing evidence of a significant difference between the evaporation from forests compared with short vegetation.

Other studies, for example, Bange (1953), Milthorpe and Spencer (1957) Monteith *et al.* (1965) and Milthorpe and Penman (1967), showed that canopy resistances were not negligible. Also Businger (1956) developed an accurate method of estimating the aerodynamic resistance. However, Penman's original potential-evaporation equation continued to gain popularity and is still used extensively today.

Law's Experiments

At the British Association for the Advancement of Science meeting in 1956, Frank Law presented the results of his comparison of forest and grassland evaporation (Law, 1958). Law, who was Chief Engineer of the Fylde Water Board, used measurements of the water balance of his main catchment area, which was grass-covered, to determine the evaporation from grassland. He also isolated a small spruce plantation to measure the runoff and measured other components of the water balance. From these measurements he concluded that the forested area evaporated significantly more than the grassed area. There was, however, some uncertainty about extrapolating these results to larger areas of forest, because the plot of forest was not very large and so it was suggested that advection could have caused these unexpected results.

Setting Up the Hydrological Research Unit

During the 1950s the Forestry Commission of Britain was undertaking a major programme of afforestation of the upland areas of the UK. These same upland areas provide most of the water-supplies for the UK. Frank Law's results implied that the runoff would be less if the areas were afforested and so more reservoirs would be required. To study the effects of afforestation on hydrology, the Hydrological Research Unit (renamed the Institute of Hydrology in 1973 and the Centre for Ecology and Hydrology, Wallingford site, in 2000) was set up in 1962. It was proposed to use paired catchments to compare the evaporation from grassland and forest. Since Howard Penman suggested that the major difference in water use between forest and grassland would be due to the greater depth of tree roots, which would maintain potential evaporation rates for a longer period under conditions of developing soil-moisture deficit, the interest was in establishing the research in areas where drought conditions could be expected every few years. Initially the Long Mynd, on the Welsh border with England, was suggested as an ideal site for the catchments. However, the Long Mynd is an area of 'common land', i.e. land historically available for grazing by animals of the local population, so it could not be afforested. Finally, the adjoining headwater catchments of the Severn and Wye Rivers in the Plynlimon Mountains in west Wales were chosen, though the area had a much higher rainfall and was therefore less likely to experience periodic droughts.

Rutter's Experiments

In the 1950s Jack Rutter, a plant physiologist, became interested in forest evaporation. To assess the dependence of the evaporation on environmental conditions, he weighed detached shoots supported in a conifer canopy in a forest near Crowthorne, Berkshire, UK, on dry and rainy days (Rutter, 1959). From these studies he found that on rainless days the transpiration was similar to Penman's estimate of open-water evaporation, E_0, whereas on rainy days the sum of the transpiration and evaporation of intercepted rainfall exceeded E_0 (Rutter, 1966). Rutter (1967) showed that the rate of evaporation of intercepted water was on average about four times as great as the transpiration in the same environmental conditions. These results contradicted the conclusions that Penman drew from his studies. However, again there was uncertainty about extrapolating Rutter's results to extensive canopies. Rutter (1963) also set up plots to measure the interception of rainfall. Based on these measurements he developed a model to estimate rainfall interception from weather data (Rutter *et al.*, 1971, 1975).

Monteith: Estimation of Actual Evaporation

John Monteith, building on the earlier work of Penman and Schofield and incorporating the results of his own field experiments (Monteith *et al.*, 1965), developed what is now known as the Penman–Monteith formula (Monteith, 1965), which specifically included two resistances, called surface and aerodynamic resistances. Thom (1972) showed that, for a vegetated dry surface, the surface resistance in the Penman–Monteith formula was related to the bulk stomatal resistance and therefore was affected by biophysical processes. For bare soil the surface resistance represents the resistance to water-vapour transfer from the soil pores to the atmosphere. For wet vegetation or soil the surface resistance becomes negligible. Monteith (1965) used his formula to demonstrate that the difference between the evaporation from wet and dry grass would be very small, whereas for a typical forest the difference would be a factor of five. The Penman–Monteith formula has been used extensively to analyse experimental measurements, in particular to determine the surface resistance of plant canopies, as an alternative to measurements of stomatal resistance made using porometers. These analyses showed that the surface resistance could vary greatly from tens to thousands of seconds per metre. For forests, where the aerodynamic resistance is small, the transpiration depends strongly on the value of the surface resistance. So large variations in surface resistance imply large variations in the forest transpiration rate.

Early Micrometeorological Measurements of Forest Evaporation

With improvements in instrumentation, a number of projects were set up to measure forest evaporation by micrometeorological methods, using aerodynamic and Bowen-ratio energy-budget methods, for example, Baumgartner (1956), Dzerdzeevskii (1963), Rauner (1966), Mukammal *et al.* (1967), Grulois (1968) and Denmead (1969). These workers found making accurate measurements above forest canopies very difficult, because vertical gradients above the canopy were very small (typically less than 1°C over 10 m, even at midday on a cloudless summer day). In addition there is the need for tall towers and hence long cables, which also increased the instrumental problems. As a result these measurements were obtained over limited periods and had large uncertainties. So they did not greatly further our understanding of the process of forest evaporation and did not significantly conflict with the view held at that time, that evaporation from vegetation was controlled by atmospheric demand and so differences between types of vegetation were insignificant.

Stewart's Thetford Project (1968–1976)

To resolve the conflict between the results of Rutter's and Penman's studies, John Stewart set up an investigation to measure the evaporation components from an extensive area of forest. Following a suggestion of John Monteith, the project was set up in Thetford Forest, a large area of flat forest on the Norfolk–Suffolk border, where the effect of advection would be negligible. The evaporation was measured by the Bowen-ratio energy-budget method. Since measurements were made at heights up to 10 m above the forest canopy, the derived evaporation is representative of areas of hectares, thereby overcoming the criticism of the small scale of Frank Law's and Jack Rutter's experiments. With the technological advances in the 1960s, the project was designed around a computer-based data-acquisition system, which allowed real-time monitoring of the measurements. The 50 sensors were connected to a Hewlett Packard computer, which was about the size of a modern washing-machine but initially had a memory of only 8 kilobytes! So, to convert the raw data into hourly averages in SI units in real-time, the programming had to be done in machine language. The initial analysis by Alastair Thom (Stewart and Thom, 1973) of the Bowen-ratio measurements made at six fixed heights above the canopy confirmed that the daytime surface resistance was an order of magnitude larger than the aerodynamic resistance. From 1974 the significant errors in the measurements of the small gradients, particularly of humidity, were reduced by using a thermometer-exchange system (McNeil and Shuttleworth, 1975), following the pioneering work of McNaughton and Black (1973). Measurements of evaporation of intercepted rainfall during and after rain events showed that they often exceeded the input of energy as net radiation (Stewart, 1977), whereas the transpiration was always less than net radiation. On average, it was found that on days when the canopy was dry, transpiration only accounted for about 40% of net radiation, assuming that soil heat flux was only a few per cent of net radiation (Gash and Stewart, 1975). In contrast, Monteith (1965) quotes studies that showed that transpiration from a number of agricultural crops was about 80% of net radiation when there was sufficient soil water.

Jarvis's Fetteresso Project (1969–1975)

In the late 1960s, Paul Jarvis set up a project to determine the sources and sinks of carbon in Sitka spruce growing in Fetteresso Forest, Kincardineshire (Ludlow and Jarvis, 1971; Watts *et al.*, 1976). In 1971 the instrumentation was rearranged to measure the fluxes of carbon dioxide and water vapour from the forest canopy (Jarvis *et al.*, 1976). Analysis of 50 days of measurements showed that, when the canopy was dry, the transpiration used 25–60% of the net radiation, while, in contrast, when the canopy was at least partially wet, evaporation used 60–70% of the net radiation (Jarvis, 1993). The project also included studies of the responses of stomata to environmental conditions, using controlled-environment chambers (Watts and Neilson, 1978).

Relationship between Evaporation and Net Radiation

The effect of stomata on evaporation rates

In the 1960s, meteorologists were impressed by the strong relationship between net radiation and evaporation, which supported Howard Penman's view that vegetation could be considered as 'green wet blotting paper' as long as there was a plentiful supply of soil water. However, Richard Lee reviewed the literature on the control of transpiration by stomata (Lee, 1967) and drew the conclusion that the evidence supported the view that stomata were important in controlling transpiration. This led to a heated discussion in the literature: for example, Idso (1968), Lee (1968) and van Bavel (1968). With the development of improved instruments to measure stomatal resistance, later studies confirmed that models had to take account of the effect of stomata and leaf-area index on transpiration (Black *et al.*, 1970; Goltz *et al.*, 1971; Brun *et al.*, 1972). The debate continues about the response of transpiration to meteorological and biological variables: see, for example, Idso (1983), Choudhury and Monteith (1986) and Monteith (1995). Useful reviews of surface conductance values have been given by Kelliher *et al.* (1993, 1995).

Compensating mechanisms

The minimum surface resistances of all types of vegetation are similar (Monteith, 1965; Kelliher *et al.*, 1995). However, as a result of the rough surface of forests compared with low vegetation, the aerodynamic resistances of forests are an order of magnitude less than those of low vegetation. So, for forests, the aerodynamic resistance is much smaller than the surface resistance. In this situation, the rate of forest transpiration becomes almost directly proportional to the vapour-pressure deficit. It is well known that the vapour-pressure deficit exhibits a large diurnal and seasonal variation, implying that the forest transpiration should also show a large diurnal and seasonal variation. Some studies did show such a variation (e.g. McNaughton and Black, 1973), whereas others (e.g. Stewart and Thom, 1973) showed that forest transpiration followed the diurnal variation in net radiation rather than that of vapour-pressure deficit.

Analysis of results from controlled-environment chamber studies (e.g. Watts and Neilson, 1978) and field experiments (e.g. Stewart, 1988) showed that the predicted increase in evaporation as the vapour-pressure deficit increased was compensated by an increase in the surface resistance. As a result, the evaporation rate often shows a similar diurnal variation to that of net radiation rather than to that of the vapour-pressure deficit. It was also found that variations in leaf-area index of different tree species, which would be expected to affect the surface resistance and hence the transpiration from the trees, were offset by the presence or absence of an understorey: see, for example, Roberts *et al.* (1982). Roberts (1983) showed that for temperate forests of Western Europe transpiration was quite conservative, giving maximum daily values of 3–4 mm and an average annual value of 335 mm.

Evaporation from wet forests

In the case of evaporation of intercepted rainfall there have been many studies that have shown that the rate of evaporation exceeds that of net radiation; for a review, see Blake (1975). The occurrence of much larger surface resistances than aerodynamic resistances for forests physically explains the observed difference between the rate of evaporation of intercepted rainfall and the rate of transpiration. The former is only controlled by the aerodynamic resistance, whereas the latter is controlled by the sum of the two resistances. McNaughton and Jarvis (1983) suggested three possible sources for the extra energy required to maintain evaporation in excess of net radiation: dry air may be introduced horizontally (advection); heat may be added to the air by downward movement of air from a near cloud base; and down-slope winds (föhn effect). They were unable to predict the relative importance of these possible mechanisms, because of the many factors involved. To assess the importance of horizontal advection would require measurements of evaporation of intercepted rainfall in relation to net radiation over very large areas of flat forest, combined with very accurate measurements of the horizontal divergence of temperature and humidity.

Modelling Rainfall Interception by Forests

The earliest estimates of rainfall interception by forests were based on empirical factors applied to gross precipitation. Jack Rutter was the first to produce a model of rainfall interception that had a physical basis, though it required some empirical parameters, such as canopy capacity (Rutter *et al.*, 1971, 1975). The Rutter model requires hourly meteorological data, which limits its use. John Gash (1979) developed an analytical model that linked Rutter's method with earlier regression approaches. The Gash model is a storm-based method, which requires knowledge of the average rates of evaporation and precipitation during rainfall. Since it can estimate evaporation of intercepted rainfall given daily rainfall data, available from most standard meteorological stations, it can be used more widely. An alternative approach was developed by Ian Calder (1986), using a so-called stochastic model. Gash *et al.* (1995) reformulated the original Gash model so that it has successfully estimated rainfall interception from sparse forests. Roberts (1999) summarizes the successful tests of these models.

Stomatal/Surface-resistance Modelling

On the basis of the experimental results from Fetteresso Forest and from controlled-environment chambers, Paul Jarvis (1976) developed a model to estimate the stomatal conductance (the reciprocal of the resistance) as a function of stress factors, which were assumed to act independently. These stress factors depended on solar radiation, vapour-pressure deficit, temperature and predawn leaf-water potential. Stewart (1988) modified the model to estimate surface resistance with stress factors depending on solar radiation, vapour-pressure deficit, temperature and soil-moisture deficit. Subsequently, a number of alternative models have been developed by others, such as Lohammar *et al.* (1980) and Lindroth (1985).

Dual-source Models

Experimental evidence showed that evaporation from understorey vegetation could make a significant contribution to the total evaporation from a forested area. Waggoner *et al.* (1969) developed a model to simulate the microclimate of a forest canopy. Models were then developed from the Penman–Monteith equation to explicitly deal with two sources of water for evaporation by Shuttleworth (1976), Shuttleworth and Wallace (1985) and Shuttleworth and Gurney (1990). These models have been used particularly for semi-arid areas, where any forests/woodlands have an open rather than a closed canopy and then evaporation from patches of low vegetation and/or bare soil is important.

Coupled Carbon- and Water-flux Models

The purpose of stomata is to allow the absorption of carbon dioxide while controlling the loss of water by transpiration. Therefore, during the last 20 years, there has been an increasing awareness that models needed to consider the exchanges of water and carbon dioxide in concert (e.g. Baldocchi and Meyers,

1998). The modelling of flows of water and carbon also need to account for the flow of nitrogen, since the presence or lack of nitrogen affects the rate of assimilation. Baldocchi and Meyers (1998), in their review and development of a coupled model, built on the results of Ball *et al.* (1987) and Leuning (1990, 1995) among many others. Williams *et al.* (1996) developed an improved model that incorporated a conductance model whose stress factors were allowed to interact, unlike the original Jarvis model (1976).

Remaining Problems

There are still major problems remaining in understanding and predicting forest evaporation.

Control of stomata

The understanding of the biophysical processes that control stomatal functioning is still incomplete, though the work of Mott and Parkhurst (1991) has been a major step forward, since it has demonstrated that the stomata respond, directly or indirectly, to the transpiration flux; previously it was not certain whether stomata responded to transpiration or directly to vapour-pressure deficit.

Dependence of stomatal resistance on soil-moisture deficit

There is still considerable uncertainty about the effect of soil-moisture deficit on stomatal/surface resistance (Lagergren and Lindroth, 2001). In the field it is very difficult to separate the effect of soil-moisture deficit from the effects of other variables on the stomatal/surface/canopy resistance, since soil-moisture deficit only builds up over periods of days, weeks or months. During periods as long as these, other meteorological variables and canopy characteristics, such as leaf-area index and leaf age, can change.

Evaporation from sparse forests

It is very difficult to obtain reliable measurements of evaporation from sparse and/or heterogeneous forests, so modelling forest evaporation – both evaporation of intercepted rainfall and transpiration – is very uncertain. Knowledge of the hydrological budgets of sparse forests is particularly important, because they occur in semi-arid areas, where estimation of forest evaporation is essential to estimate water resources accurately.

Modelling effect of climate change on forest evaporation

To model forest evaporation successfully for climate-change scenarios, it is essential to model the interaction between the planetary boundary layer and the underlying forest. This is particularly critical for predicting the evaporation of precipitation intercepted by forest, since small uncertainties in the estimated canopy vapour-pressure deficit cause large errors in the estimates of evaporation of intercepted precipitation.

Evaporation from forests on a regional scale

To assess the effects of climate change on the water resources of forested areas requires models that are known to be accurate at the regional scale. However, validation of forest evaporation models is very difficult at these larger scales, since the methods available – scintillometer measurements over scales of 1 to 10 km, replicated eddy-correlation measurements, estimation from remotely sensed data – are of uncertain accuracy. Another alternative is to develop models that have a strong biophysical basis with a minimum of empiricism, validate them at smaller scales and then assume that they are also valid at larger scales.

Acknowledgements

The valuable commentary by Keith McNaughton on the development of the Penman formula and the resulting treatment of the biological control of transpiration in the formula is gratefully acknowledged. Also the suggestion from Mike Unsworth for additional references is gratefully acknowledged.

References

Baldocchi, D.D. and Meyers, T.P. (1998) On using eco-physiological and biochemical theory to evaluate carbon dioxide, water vapour and trace gas fluxes over vegetation: a perspective. *Agricultural and Forest Meteorology* 90, 1–25.

Ball, J.T., Woodrow, I.E. and Berry, J.A. (1987) A model predicting stomatal conductance and its contribution to the control of photosynthesis under different environmental conditions. In: Biggens, J. (ed.) *Progress in Photosynthetic Research*. Martinus Nijhoff, The Netherlands, pp. 221–224.

Bange, G.G.J. (1953) On the quantitative explanation of stomatal transpiration. *Acta Botanica Neerlandica* 2, 255–297.

Bates, C.G. and Henry, A.J. (1928) Second phase of streamflow experiment at Wagon Wheel Gap, Colorado. *Monthly Weather Review* 56, 79–85.

Baumgartner, A. (1956) Untersuchungen über den Wärme- und Wasserhaushalt eines jungen Waldes. *Berichte des Deutschen Wetterdienstes* 5, 1–53.

Black, T.A., Tanner, C.B. and Gardner, W.R. (1970) Evapotranspiration from a snap bean crop. *Agronomy Journal* 62, 66–69.

Blake, G.J. (1975) The interception process. In: Chapman, T.G. and Dunin, F.X. (eds) *Prediction in Catchment Hydrology.* Australian Academy of Science, Canberra, pp. 59–81.

Brown, H.T. and Escombe, F. (1900) Static diffusion of gases and liquids in relation to the assimilation of carbon and translocation in plants. *Philosophical Transactions of the Royal Society of London, Series B* 193, 223–291.

Brun, L.J., Kanemasu, E.T. and Powers, W.L. (1972) Evapotranspiration from soybean and sorghum fields. *Agronomy Journal* 64, 145–148.

Businger, J.A. (1956) Some remarks on Penman's equations for the evapotranspiration. *Netherlands Journal of Agricultural Science* 44, 77–80.

Calder, I.R. (1986) A stochastic model of rainfall interception. *Journal of Hydrology* 89, 65–71.

Choudhury, B.J. and Monteith, J.L. (1986) Implications of stomatal response to saturation deficit for the heat balance of vegetation. *Agricultural and Forest Meteorology* 36, 215–225.

Clarke, R.T. and Newson, M.D. (1978) Some detailed water balance studies of research catchments. *Proceedings of the Royal Society, London* A 363, 21–42.

Denmead, O.T. (1969) Comparative micrometeorology of a wheat field and a forest of *Pinus radiata. Agricultural Meteorology* 6, 357–371.

Dzerdzeevskii, B.L. (1963) Study of the heat balance of the forest. *Silvia Fennica* 113, 1–17.

FAO (1977) *Crop Water Requirements.* Irrigation and Drainage Paper 24, Food and Agricultural Organization, Rome, 144 pp.

Gash, J.H.C. (1979) An analytical model of rainfall interception in forests. *Quarterly Journal of the Royal Meteorological Society* 105, 43–55.

Gash, J.H.C. and Stewart, J.B. (1975) The average surface resistance of a pine forest derived from Bowen ratio measurements. *Boundary-Layer Meteorology* 8, 453–464.

Gash, J.H.C., Lloyd, C.R. and Lachaud, G. (1995) Estimating sparse forest rainfall interception with an analytical model. *Journal of Hydrology* 170, 79–86.

Goltz, S.M., Tanner, C.B., Millar, A.A. and Lang, A.R.G. (1971) Water balance of a seed onion field. *Agronomy Journal* 63, 762–765.

Grulois, J. (1968) Heat fluxes and evapotranspiration during a clear summer day. *Bulletin of the Royal Botanical Society of Belgium* 102, 27–41.

Hoover, M.D. (1944) Effect of removal of forest vegetation upon water-yield. *American Geophysical Union Transactions* 23, 969–975.

Idso, S.B. (1968) Comments on paper by Richard Lee, 'The hydrologic importance of transpiration control by stomata'. *Water Resources Research* 4, 665–666.

Idso, S.B. (1983) Stomatal regulation of evaporation from well-watered plant canopies: a new synthesis. *Agricultural Meteorology* 29, 213–217.

Jarvis, P.G. (1976) The interpretation of the variations in leaf water potential and stomatal conductance found in canopies in the field. *Philosophical Transactions of the Royal Society of London, Series B* 273, 593–610.

Jarvis, P.G. (1993) Water losses of crowns, canopies and communities. In: Smith, J.A.C. and Griffiths, H. (eds) *Water Deficits – Plant Responses from Cell to Community.* Environmental Plant Biology Series, BIOS Scientific Publishers, Oxford, pp. 285–315.

Jarvis, P.G., James, G.B. and Landsberg, J.J. (1976) Coniferous forest. In: Monteith, J.L. (ed.) *Vegetation and the Atmosphere*, 2. Academic Press, New York, pp. 171-240.

Jeffreys, H. (1918) Some problems of evaporation. *London, Edinburgh and Dublin Philosophical Magazine Journal of Science* 35, 270–280.

Kelliher, F.M., Leuning, R. and Schulze, E.-D. (1993) Evaporation and canopy characteristics of coniferous forests and grasslands. *Oecologia* 95, 153–163.

Kelliher, F.M., Leuning, R., Raupach, M.R. and Schulze, E.-D. (1995) Maximum conductances for evaporation from global vegetation types. *Agricultural and Forest Meteorology* 73, 1–16.

Kittredge, J., Loughead, H.J. and Mazurak, A. (1941) Interception and streamflow in a pine plantation. *Journal of Forestry* 39, 505–522.

Lagergren, F. and Lindroth, A. (2001) Transpiration response to soil moisture in pine and spruce trees in Sweden. *Agricultural and Forest Meteorology* 112, 67–85.

Law, F. (1958) Measurement of rainfall, interception and evaporation losses in a plantation of Sitka spruce trees. *International Association of Scientific Hydrology, General Assembly, Toronto* 2, 397–411.

Lee, R. (1967) The hydrological importance of transpiration control by stomata. *Water Resources Research* 3, 737–752.

Lee, R. (1968) Reply to Idso, S.B. (1968) Comments on paper by Richard Lee, 'The hydrologic importance of transpiration control by stomata'. *Water Resources Research* 4, 667–669.

Leuning, R. (1990) Modelling stomatal behaviour and photosynthesis of *Eucalyptus grandis*. *Australian Journal of Plant Physiology* 17, 159–175.

Leuning, R. (1995) A critical appraisal of a combined stomatal–photosynthesis model for C_3 plants. *Plant, Cell and Environment* 18, 339–355.

Lindroth, A. (1985) Canopy conductance of coniferous forests related to climate. *Water Resources Research* 21, 297–304.

Lohammar, T., Larsson, S., Linder, S. and Falk, S.O. (1980) FAST – simulation models of gaseous exchange in Scots pine. In: Persson, T. (ed.) *Structure and Function of Northern Coniferous Forests – an Ecosystem Study. Ecology Bulletin (Stockholm)* 32, 505–523.

Ludlow, M.M. and Jarvis, P.G. (1971) Photosynthesis in Sitka spruce [*Picea sitchensis* (Bong.) Carr.]. I. General characteristics. *Journal of Applied Ecology* 8, 925–953.

McNaughton, K.G. and Black, T.A. (1973) A study of evapotranspiration from a Douglas fir forest using the energy budget approach. *Water Resources Research* 9, 1579–1590.

McNaughton, K.G. and Jarvis, P.G. (1983) Predicting effects of vegetation changes on transpiration and evaporation. In: Kozlowski, T.T. (ed.) *Water Deficits and Plant Growth*, VII. Academic Press, New York, pp. 1–47.

McNeil, D.D. and Shuttleworth, W.J. (1975) Comparative measurements of the energy fluxes over a pine forest. *Boundary-Layer Meteorology* 9, 297–313.

Milthorpe, F.L. and Penman, H.L. (1967) Diffusive conductivity of stomata of wheat leaves. *Journal of Experimental Botany* 18, 422–457.

Milthorpe, F.L. and Spencer, H.J. (1957) Experimental studies of the factors controlling evaporation, III. *Journal of Experimental Botany* 8, 413–437.

Monteith, J.L. (1965) Evaporation and environment. *Symposium of the Society of Experimental Biology* 19, 206–234.

Monteith, J.L. (1995) A reinterpretation of stomatal responses to humidity. *Plant, Cell and Environment* 18, 357–364.

Monteith, J.L., Szeicz, G. and Waggoner, P.E. (1965) The measurement and control of stomatal resistance in the field. *Journal of Applied Ecology* 2, 345–355.

Mott, K.A. and Parkhurst, D.F. (1991) Stomatal response to humidity in air and helox. *Plant, Cell and Environment* 14, 509–515.

Mukammal, E.I., MacHattie, L.B. and Turner, V.E. (1967) Petawawa forest meteorology project. In: *Proceedings First Canadian Conference on Micrometeorology, 12–14 April, 1965, Toronto, Ontario*. Department of Transport, Toronto, pp. 499–506.

Penman, H.L. (1948) Evaporation from open water, bare soil and grass. *Proceedings of the Royal Society, London A* 193, 120–145.
Penman, H.L. (1949a) The dependence of transpiration on weather and soil conditions. *Journal of Soil Science* 1, 74–89.
Penman, H.L. (1949b) An account of the physics of irrigation control. *Quarterly Journal of the Royal Meteorological Society* 75, 293–302.
Penman, H.L. and Schofield, R.K. (1951) Some physical aspects of assimilation and transpiration. *Symposium of the Society Experimental Biology* 5, 115–129.
Rauner, J.L. (1966) Methodik und einige Ergebnisse der experimentellen Erforschung des Wärmehaushaltes der Waldbestände. *Zeitschrift für Angewandte Meteorologie* 5, 157–165.
Roberts, J.M. (1983) Forest transpiration: a conservative hydrological process. *Journal of Hydrology* 66, 133–141.
Roberts, J.M. (1999) Plants and water in forests and woodlands. In: Baird, A. and Wilby, R. (eds) *Ecohydrology: Plants and Water in Terrestrial and Aquatic Ecosystems*. Routledge, London, pp. 181–236.
Roberts, J.M., Pitman, R.M. and Wallace, J.S. (1982) Comparison of evaporation from stands of Scots pine and Corsican pine in Thetford Chase, East Anglia. *Journal of Applied Ecology* 19, 859–872.
Rutter, A.J. (1959) Evaporation from a plantation of *Pinus sylvestris* in relation to meteorological and soil conditions. In: *Symposium of Hannoversch-Münden*, I, *Water and Woodlands*. Publication No. 48, International Association of Scientific Hydrology, Gentbrugge pp. 101–110.
Rutter, A.J. (1963) Studies on the water relations of *Pinus sylvestris* in plantation conditions I. Measurement of rainfall and interception. *Journal of Ecology* 51, 191–203.
Rutter, A.J. (1966) Studies on the water relations of *Pinus sylvestris* in plantation conditions, IV. Direct observations on the rates of transpiration, evaporation of intercepted water, and evaporation from the soil surface. *Journal of Applied Ecology* 3, 393–405.
Rutter, A.J. (1967) An analysis of evaporation from a stand of Scots pine. In: Sopper, W.E. and Lull, H.W. (eds) *Forest Hydrology*. Pergamon, Oxford, pp. 403–417.
Rutter, A.J., Kershaw, K.A., Robbins, P.C. and Morton, A.J. (1971) A predictive model of rainfall interception in forest, I. Derivation of the model from observations in a plantation of Corsican pine. *Agricultural Meteorology* 9, 367–384.
Rutter, A.J., Morton, A.J. and Robbins, P.C. (1975) A predictive model of rainfall interception in forest, II. Generalisation of the model and comparison with observations in some coniferous and hardwood stands. *Journal of Applied Ecology* 12, 367–380.
Shuttleworth, W.J. (1976) A one-dimensional theoretical description of the vegetation–atmosphere interaction. *Boundary-Layer Meteorology* 10, 273–302.
Shuttleworth, W.J. and Gurney, R.J. (1990) The theoretical relationship between foliage temperature and canopy resistance in sparse crops. *Quarterly Journal of the Royal Meteorological Society* 116, 497–519.
Shuttleworth, W.J. and Wallace, J.S. (1985) Evaporation from sparse crops – an energy combination theory. *Quarterly Journal of the Royal Meteorological Society* 111, 839–855.
Stewart, J.B. (1977) Evaporation from the wet canopy of a pine forest. *Water Resources Research* 13, 915–921.
Stewart, J.B. (1988) Modelling surface conductance of pine forest. *Agricultural and Forest Meteorology* 43, 19–35.
Stewart, J.B. and Thom, A.S. (1973) Energy budgets in pine forest. *Quarterly Journal of the Royal Meteorological Society* 99, 154–170.

Thom, A.S. (1972) Momentum, mass and heat exchange of vegetation. *Quarterly Journal of the Royal Meteorological Society* 98, 124–134.

Thornthwaite, C.W. (1948) An approach toward a rational classification of climate. *Geographical Review* 38, 55–94.

van Bavel, C.H.M. (1968) Further to the hydrologic importance of transpiration control by stomata. *Water Resources Research* 4, 1387–1388.

Waggoner, P.E., Furnival, G.M. and Reifsnyder, W.E. (1969) Simulation of the microclimate in a forest. *Forest Science* 15, 37–45.

Watts, W.R. and Neilson, R.E. (1978) Photosynthesis in Sitka spruce [*Picea sitchensis* (Bong.) Carr.]. VII. Measurements of stomatal conductance and $^{14}CO_2$ uptake in controlled environments. *Journal of Applied Ecology* 15, 245–255.

Watts, W.R., Neilson, R.E. and Jarvis, P.G. (1976) Photosynthesis in Sitka spruce [*Picea sitchensis* (Bong.) Carr.]. VII. Measurements of stomatal conductance and $^{14}CO_2$ uptake in a forest canopy. *Journal of Applied Ecology* 13, 623–638.

Williams, M.W., Rastetter, E.B., Fernandes, D.N., Goulden, M.L., Wofsk, S.C., Shaver, G.R., Melillo, J.M., Munger, J.W., Fan, S.-M. and Nadelhoffer, K.J. (1996) Modelling the soil–plant–atmosphere continuum in a *Quercus–Acer* stand at Harvard Forest: the regulation of stomatal conductance by light, nitrogen and soil/plant hydraulic properties. *Plant, Cell and Environment* 19, 911–927.

12

Scaling the Estimate of Maximum Canopy Conductance from Patch to Region and Comparison of Aircraft Measurements

Ying-Ping Wang,[1] Ray Leuning,[2] Peter Isaac[3] and Guoyi Zhou[4]

[1]*CSIRO Atmospheric Research, Private Bag No. 1, Aspendale, Victoria 3195, Australia;* [2]*CSIRO Land and Water, FC Pye Laboratory, PO Box 1666, Canberra, ACT 2601, Australia;* [3]*53 Brewer Road, Bentleigh, Victoria 3204, Australia;* [4]*South China Institute of Botany, Chinese Academy of Sciences, Guang Zhou, China*

Introduction

Recognition of the role of vegetation in partitioning energy at the earth's surface and its feedback on climate has led to the inclusion of stomatal-conductance models in the surface–atmosphere exchange component of climate models. One of the earliest models for leaf stomatal conductance that incorporated responses to multiple environmental factors was developed by Professor Paul Jarvis, and his approach has been used in models at scales ranging from single leaves to the globe (Jarvis, 1976; Sellers *et al.*, 1986; Dickinson and Henderson-Sellers, 1988). However, the Jarvis stomatal model was developed to describe and predict the variation of stomatal conductance in the field, and was not initially envisaged for use at a regional or global scale. In compiling global data sets, Sellers and Dorman (1987) estimated parameters of the Jarvis model using field measurements at the point or patch scale. Concerns have thus been raised about applying estimates of model parameters obtained from leaf- or patch-scale measurements to regional or global scales, because the stomatal-conductance model itself is non-linear in its parameters and because simple averages of surface conductances for constituent patches will not lead to correct estimates of regional evaporation over heterogeneous surfaces (McNaughton, 1994; Raupach, 1995). This is because there is generally a non-linear dependence between model parameters and fluxes of mass and energy between the biosphere and the atmosphere. Model parameters may thus need to be estimated at the scale of the desired predictions or suitable parameter-scaling rules need to be devised.

 Forests at the Land–Atmosphere Interface (eds M. Mencuccini, J. Grace, J. Moncrieff and K.G. McNaughton)

Several studies have explored theoretically how to scale parameters estimated from patch-scale measurements to regional scales (McNaughton, 1994; Raupach, 1995). Using the principle of flux conservation, the authors of these studies concluded that regional evaporation is calculated correctly when the regional surface conductance is calculated as the mean of surface conductances of each homogeneous patch within the region, weighted by the net available energy and area fractions of each patch. Unfortunately, there are some difficulties in applying this averaging scheme. First, available energy is usually not known at subgrid scales of climate models. Secondly, the available energy of a patch is a function of surface temperature, which in turn depends on surface conductance. We thus require methods to estimate surface conductances at the regional scale that correctly predict regional evaporation. Surface conductance is often modelled as a function of environmental variables and the maximum stomatal conductance (Jarvis, 1976; Kelliher *et al.*, 1995). Estimates of the parameters at the patch scale may be made using field measurements such as those available from the global-flux station network (Baldocchi *et al.*, 1996), and the challenge is to scale these parameters from patch to region consistent with flux conservation.

Wang *et al.* (2001) performed a simulation study using a coupled model for heat, water vapour and CO_2 fluxes from vegetation to examine the consequences of using linear combinations of model parameters when estimating regional fluxes for a heterogeneous surface consisting of variable fractions of well-watered and water-stressed pastures. They found that a linear combination of parameters, weighted by the area fractions of the two different land-cover types, gave satisfactory estimates of latent heat and CO_2 fluxes, but was less satisfactory for sensible heat. Best results were obtained when the number of interpolated parameters was ≤ 2, with all other parameters being held constant across both patch types.

In this chapter, we develop a simple model of surface conductance for evaporation and estimate a key parameter in the model, maximum canopy conductance (G_{cmax}), for six patches along a 100 km transect in south-east Australia during the spring of 1995 (Leuning *et al.*, 2004). We then estimate the latent-heat flux of the region, using G_{cmax} for the different patches, weighted by their area fractions in the region, and compare the results with latent-heat fluxes measured using instrumented aircraft flights.

To apply the simple averaging scheme, we need to know the area and G_{cmax} of each homogeneous patch in the region. Determining the boundaries between patches can be quite difficult if the parameters for a patch vary with time or if differing parameters do not have the same distribution. We utilize satellite measurements of the normalized differences of vegetation index (NDVI) to differentiate between patches and to estimate the distribution of G_{cmax} at subgrid scales. We also estimate the spatial variability of G_{cmax}, using the correlations between G_{cmax} and rainfall in the 3 months before the OASIS (stands for observation at several interacting scales) field measurements (R_3: July, August and September, 1995), as a surrogate for the available soil water. This correlation is then used to calculate the spatial average of canopy conductance, G_c, a function of G_{cmax}, absorbed light and atmospheric water-vapour-pressure deficit.

The simple averaging using area fraction is practical but not as rigorous as the averaging schemes developed by McNaughton (1994) and Raupach (1995), as it does not always conserve flux. Using flux measurements from instrumented towers and an aircraft during the OASIS field study, we assess the errors introduced into calculations of latent-heat fluxes using the simple averaging scheme for surface conductance. Specifically, we examine: (i) the number of parameters in the surface-conductance model that can be estimated independently using measured latent-heat fluxes; (ii) whether measurements of NDVI or R_3 can be used to represent the spatial variability of G_{cmax} over the OASIS region; and (iii) if estimates of spatial averages of G_{cmax}, G_c and λE from the aircraft measurements can be reliably predicted using the area-weighted G_{cmax}.

Issac *et al.* (2004a,b) have used the same measurements to analyse the spatial variation of G_{cmax} and its correlation with NDVI. This study complements their analysis, but with a stronger emphasis on scaling up from patch to region the estimate of G_{cmax}, an input parameter required by many land-surface schemes in climate models.

The Model

Latent-heat flux from a land surface, λE, is related to surface conductance (G_s) through the Penman–Monteith equation. That is:

$$\lambda E = \frac{\varepsilon A + C_p D G_a / \gamma}{\varepsilon + 1 + G_a / G_s} \tag{1}$$

where λ is the latent heat of vaporization (J/mol), E is evapotranspiration from the land surface (mol/m^2/s), ε is the change in latent-heat content of saturated air with change in sensible-heat content (dimensionless), A is available energy at the surface (W/m^2), G_a is the aerodynamic conductance (mol/m^2/s), G_s is the surface conductance (mol/m^2/s), C_p is the molar specific heat of dry air at constant pressure (J/mol/K), γ is the psychrometric constant in Pa/K, and D is the saturation deficit of the ambient air (Pa). The thermodynamic coefficients, λ, ε and γ, vary with air temperature, T_a.

Rearranging Equation (1) for G_s we obtain:

$$G_s = \frac{\lambda E G_a}{\varepsilon A + C_p D G_a/\gamma - \lambda E(\varepsilon + 1)} \tag{2}$$

Evaporation occurs at both leaf and soil surfaces, so we assume that G_s is the sum of a canopy conductance (G_c) and a conductance for evaporation from the soil (G_g). That is:

$$G_s = G_c + G_g \tag{3}$$

We also assume that G_c is calculated as the parallel sum of conductances for individual leaves on a ground-area basis:

$$G_c = \int_0^L g_c d\xi \tag{4}$$

where L is canopy leaf-area index, g_c is leaf stomatal conductance and ξ is the cumulated leaf-area index from the top of the canopy. Following Kelliher *et al.* (1995) and Issac *et al.* (2004b), g_c is taken to be a function of absorbed photosynthetically active radiation (PAR) (Q) and saturation deficit (D) according to:

$$g_c = \frac{g_{cmax} Q}{Q + Q_{0.5}} \left(\frac{1}{1 + D / D_{0.5}} \right) \tag{5}$$

where g_{cmax} is the maximum stomatal conductance of the leaves, Q is absorbed PAR, and where $Q_{0.5}$ and $D_{0.5}$ are the values of Q and D when g_c is half its maximum value due to limitations by light and saturation deficit, respectively. Because measurements of soil water contents were not available, the effects of plant water stress on stomatal conductance are included in daily variation of g_{cmax} in this model.

For a horizontally homogeneous canopy, Q can be estimated as:

$$Q = ak_1 Q_0 \exp(-k_1 \xi) \tag{6}$$

where a is leaf absorptance for PAR (= 0.8), k_1 is the extinction coefficient of PAR within the canopy (= 0.64) and Q_0 is incident PAR flux density above the canopy (μmol/m^2/s). We also assume that:

$$Q_{0.5} = Q_{0.5,0} \exp(-k_1 \xi) \tag{7}$$

and:

$$g_{cmax} = g_{cmax,0} \exp(-k_1 \xi) \tag{8}$$

where $Q_{0.5,0}$ and $g_{cmax,0}$ are the values of parameter $Q_{0.5}$ and g_{cmax} for the leaves at the top of the canopy (at $\xi = 0$), respectively. There is ample evidence in the literature that leaves acclimatize to their mean light environment and hence that both $Q_{0.5}$ and g_{cmax} decrease with depth in the canopy (Evans, 1995; Meir *et al.*, 2002).

Substituting Equations (6), (7) and (8) into (5), and Equation (5) into (4) we obtain:

$$G_c = G_{cmax} \frac{Q_0}{Q_0 + Q_{0.5,0} / ak_1} \frac{1}{1 + D / D_{0.5}} \tag{9}$$

and:

$$G_{cmax} = \frac{g_{cmax,0} (1 - \exp(-k_1 L))}{k_1} \tag{10}$$

Therefore the total surface conductance, G_s, is modelled as:

$$G_s = G_{cmax} \frac{Q_0}{Q_0 + Q_{0.5,0} / ak_1} \frac{1}{1 + D / D_{0.5}} + G_g \tag{11}$$

There are four unknowns in the above equations, $g_{cmax,0}$, $Q_{0.5,0}$, $D_{0.5}$ and G_g, which are to be estimated using the measurements of λE at six sites at patch scale or from aircraft measurements for the region.

The expression for G_c developed here differs significantly from that of Kelliher *et al.* (1995) and Issac *et al.* (2004b), because these authors assumed that the half-saturation light intensity, $Q_{0.5}$, is a constant, whereas here it is allowed to decrease with depth in the canopy. The main difference between this and the earlier model for G_c is the expected behaviour of the ratio $G_c/g_{cmax,0}$ as a function of L, the leaf-area index. Stomatal conductance decreases less rapidly with depth in the canopy in the earlier model, so at any given value of L, $G_c/g_{cmax,0}$ is approximately double that predicted by the current model.

The Data

Data for the analysis in this chapter were obtained from instruments mounted on fixed towers and on an aircraft. Fluxes of latent and sensible heat were measured at three paired sites (crop and pasture) located at approximately 50 km intervals along a 100 km transect in south-east Australia during October 1995 (Leuning *et al.*, 2004). Locations of the three sites, Urana (U), Browning (B) and Wagga Wagga (W) are shown in Fig. 12.1. Annual rainfall generally decreases from Wagga Wagga to Browning and Urana (see Fig. 12.1). The sites are denoted as UP, UW, BP, BO, WP and WT, where the first letter represents the location and the second letter represents pasture (P), or the crops wheat (W), triticale (T) or oats (O). Leaf-area indices were 1.0, 2.5, 1.5, 3.5, 2.0 and 4.0 for UP, UW, BP, BO, WP and WT, respectively. Fluxes of latent and sensible heat were measured using eddy-flux techniques at Urana and Browning, while the Bowen-ratio method was used to estimate these fluxes at Wagga Wagga. As described by Issac *et al.* (2004a), 11 direct flights of an instrumented aircraft

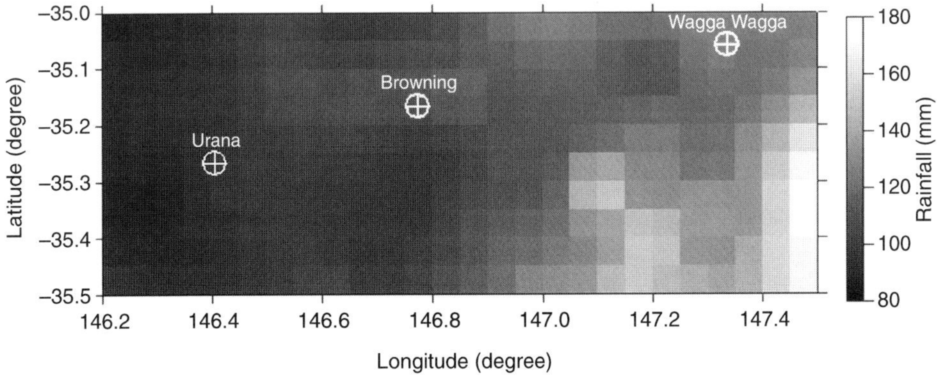

Fig. 12.1. Locations of three paired sites of the OASIS field experiment in south-east Australia. Different shading represents the total amount of rainfall (mm) in July, August and September 1995 (R_3) at each cell.

were made between Wagga Wagga and Urana between 16 and 30 October 1995. The return flights from Urana to Wagga Wagga were interrupted by gridded flights near Browning to examine in more detail relationships between fluxes measured on the towers and by the instrumented aircraft (Issac *et al.*, 2004b). We shall refer to the transect flight between Wagga Wagga and Urana via Browning as transect I, between Browning and Wagga Wagga as transect II and between Urana and Browning as transect III.

Measurements of incoming short-wave radiation, relative humidity, wind speed and air temperature were made at the three paired sites and on the transect flights. Friction velocity, u_*, was also measured at each site and by the aircraft, and aerodynamic conductance (G_a) was estimated from the measurements of u_*, roughness length and zero-plane displacement, using surface-layer similarity theory (Garratt, 1994).

Mean values of all meteorological variables and surface fluxes were calculated every 15 min for the two paired sites at Browning and Urana and every 30 min for the pasture and crop sites at Wagga Wagga. To ensure consistency in comparing data from the towers and the aircraft, mean values of all meteorological variables and surface fluxes were calculated for the period 16–30 October 1995. We excluded some flux measurements when the canopy was wet due to either rain or dew, when there was strong advection ($H + \lambda E > R_n$) and when net radiation was negative. The number of flux measurements included in the study are 12, 88, 107, 103, 59 and 62 for UP, UW, BP, BO, WP and WT, respectively.

The total amount of rainfall for the 3 months prior to the field campaign (July, August and September) was 175.4 mm at Wagga Wagga, 100.4 mm at Browning and 90.1 mm at Urana.

Results

Parameter estimation

The parameters in Equation (11) (G_g, $D_{0.5}$, G_{cmax} and $Q_{0.5,0}$) were estimated using the non-linear curve-fitting routine in SIGMA PLOT (version 5.0, SPSS Inc., 2001), which uses the Marquardt–Levenberg algorithm (Press *et al.*, 1992). Output from SIGMA PLOT 5.0 includes the estimated mean and standard error of each parameter and the correlation matrix. Initial estimates of the parameters were $G_g = 0.01$ mol/m^2/s, $D_{0.5} = 1.5$ kPa, $G_{cmax} = 0.2$ mol/m^2/s and $Q_{0.5,0} = 250$ μmol/m^2/s.

Non-linear parameter estimation involves minimizing the so-called cost function, $\phi = \sum_{i=1}^{m}(P_i - O_i)^2$, where P_i are model predictions and O_i are the 1 … *m* observations.

Figure 12.2 shows a contour plot for ϕ as a function of two parameters, $D_{0.5}$ and G_{cmax} for the BP site. We see that there is no unique minimum for ϕ and that estimates of two parameters are strongly and negatively correlated. Similarly, we found a strong correlation ($r > 0.9$) between estimates of the parameters for all data sets until the number of parameters to be estimated

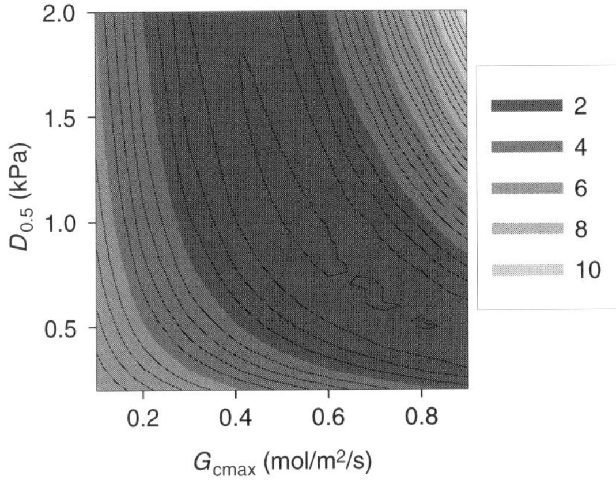

Fig. 12.2. Contour plot of the cost function ϕ (dimensionless), for the sum of squared deviations between observed and modelled λE, as a function of parameters $D_{0.5}$ and G_{cmax}.

was reduced to one. All parameters except G_{cmax} were thus fixed to their initial values, and the optimization was repeated. The resultant estimates of G_{cmax} for the three paired sites are presented in Table 12.1. Considering the relatively small value of G_g, our estimates of G_{cmax} are within the range of maximum surface conductance found by Kelliher *et al.* (1995) for different vegetation types. The relatively poor fit of the model to data for the UP and WP sites is from different causes. For the UP site, only 12 measurements were valid because of instrument malfunctioning and the measured responses of G_s to Q_0 or D were quite variable. For the WP site, the poor fit results from the considerable scatter in the response of G_s to D in the data. Whereas pastures at Urana and Browning consisted of annual grasses, the pasture at Wagga Wagga was dominated by deep-rooted lucerne with access to water at depth, and this may have caused differences in model performance between the various pastures.

Table 12.1. Mean and standard error (SE) of maximum canopy conductance (G_{cmax}) for the three paired sites.

	UP	UW	BP	BO	WP	WT
G_{cmax}	0.073	0.164	0.370	0.393	1.201	1.053
SE	0.011	0.003	0.010	0.010	0.039	0.031
L	1.0	2.5	1.5	3.5	2.0	4.0
$\frac{1-\exp/(-k_1 L)}{k_1}$	0.74	1.25	0.96	1.40	1.13	1.44
r^2	0.34	0.68	0.57	0.66	0.33	0.62
n	12	88	107	103	59	62

The unit for the mean and SE of G_{cmax} is mol/m²/s; n is the number of measurements used in the estimation, and r^2 is the correlation coefficient of the non-linear regression; the term(1 − exp(k_1L))/k_1 is the ratio of maximum leaf at the top stomatal conductance of the canopy and maximum canopy conductance. L is canopy leaf-area index.

To obtain independent estimates of a greater number of parameters requires extra information. Wang *et al.* (2001) applied non-linear parameter inversion to the measurements of CO_2, latent-heat, sensible-heat and ground-heat fluxes during OASIS 1995 to estimate parameters in a more complex model of surface conductance than the one adopted here. They found that different types of flux measurements provided differing constraints on estimates of model parameters, but, despite the extra fluxes measured, only three parameters could be estimated independently.

Estimation of $\langle G_{cmax} \rangle$ for three different transects

We next compare estimates of $\langle G_{cmax} \rangle$, the spatially averaged values of G_{cmax} over the transect derived from tower-based and aircraft measurements. The first approach is to calculate $\langle G_{cmax} \rangle$ as the mean from estimates of G_{cmax} of different patches weighted by their area fraction. This gives the correct $\langle G_{cmax} \rangle$ but not $\langle \lambda E \rangle$ for the region because of the non-linear relationship between G_{cmax} and λE. The second approach is to estimate mean $\langle G_{cmax} \rangle$ using Equations (2) and (11) and spatial averages of all independent variables in both equations as measured by the aircraft. This approach gives the correct spatial average of $\langle \lambda E \rangle$ but not $\langle G_{cmax} \rangle$. If all other variables are spatially uniform, the first approach will overestimate mean $\langle \lambda E \rangle$ and the second approach will underestimate mean $\langle G_{cmax} \rangle$ because $dE/dG_{cmax} > 0$ and $\frac{d^2E}{dG^2_{cmax}} < 0$.

Applying non-linear parameter estimation to the measured latent-heat fluxes, we obtained the estimates of G_{cmax} for each of the three paired sites. Results in Table 12.1 show that the estimates of G_{cmax} can vary by over an order of magnitude between pastures and crops at different locations, and those differences cannot be accounted for solely by differences in leaf-area index. To estimate G_{cmax} for all patches along the three transects, we seek to interpolate values of G_{cmax} between the tower sites using a surrogate, spatially explicit variable that correlates with G_{cmax}. For this purpose we examined the NDVI, obtained using remote sensing, and total rainfall in the 3 months prior to the OASIS campaign in 1995 (denoted R_3). The latter was chosen because plant growth and canopy conductance are strongly influenced by winter–spring rainfall. A strong correlation was observed between G_{cmax} and R_3 ($r^2 = 0.97$, Fig. 12.3a), whereas the correlation between G_{cmax} and NDVI was non-linear and the relationships between crops and pasture were different (Fig. 12.3b). These results suggest that R_3 is a suitable surrogate variable for interpolating G_{cmax} between the tower sites, allowing us to estimate $\langle G_{cmax} \rangle$ for each transect. The standard error of mean R_3 was estimated to be 10% of its mean.

In the second approach, we calculated the mean $\langle G_s \rangle$ using Equation (2) and the spatial means of incident solar radiation ($\langle Q_0 \rangle$), water-vapour-pressure deficit ($\langle D \rangle$), aerodynamic conductance ($\langle G_a \rangle$) and latent-heat flux ($\langle \lambda E \rangle$) of each aircraft flight. We then estimated mean $\langle G_{cmax} \rangle$ for each transect by fitting Equation (11) using the estimates of mean $\langle G_s \rangle$ and spatial averages of $\langle Q_0 \rangle$ and $\langle D \rangle$ of all flights.

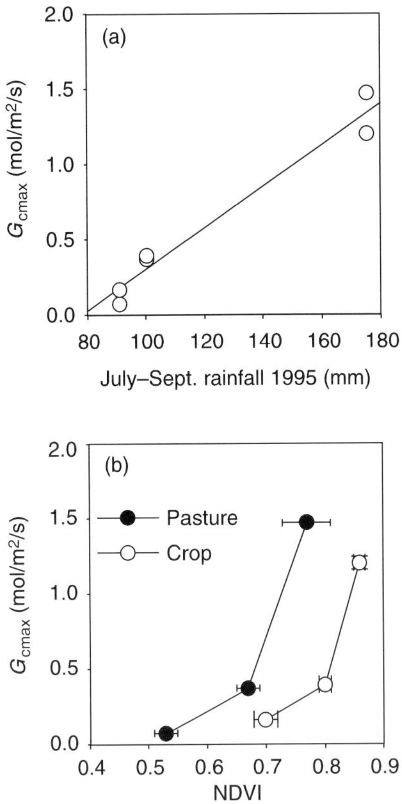

Fig. 12.3. (a) Correlation between G_{cmax} and the amount of rainfall in the 3 months before October 1995 (R_3). The best fitted regression equation is $G_{cmax} = -1.0737 + 0.0138R_3$, $r^2 = 0.97$. (b) Correlation between G_{cmax} and NDVI at each of three paired sites.

Figure 12.4 compares the two estimates of mean $\langle G_{cmax} \rangle$. The second approach underestimates $\langle G_{cmax} \rangle$, but the differences are not significant for any of the transects. Note that $\langle G_{cmax} \rangle$ for transect I is not the mean of transects II and III, even though they cover the same distance. This is because rainfall was not distributed linearly along the transect.

Comparison of the predicted spatial averages of G_s and λE using the area-weighted mean of G_{cmax} for all three transects

Using the estimates of $\langle G_{cmax} \rangle$ interpolated by rainfall and $\langle Q_0 \rangle$, $\langle D \rangle$ and $\langle G_a \rangle$ calculated over a transect, i.e. approach I, we calculated $\langle G_s \rangle$ using Equation (11) and $\langle \lambda E \rangle$ using Equation (1). These estimates were then compared with values of $\langle G_s \rangle$ calculated using Equation (2) and $\langle \lambda E \rangle$ from aircraft flux measurements. Figure 12.5 shows that calculated $\langle G_s \rangle$ and $\langle \lambda E \rangle$ agree well with estimates from aircraft measurements for all transects except $\langle G_s \rangle$ for

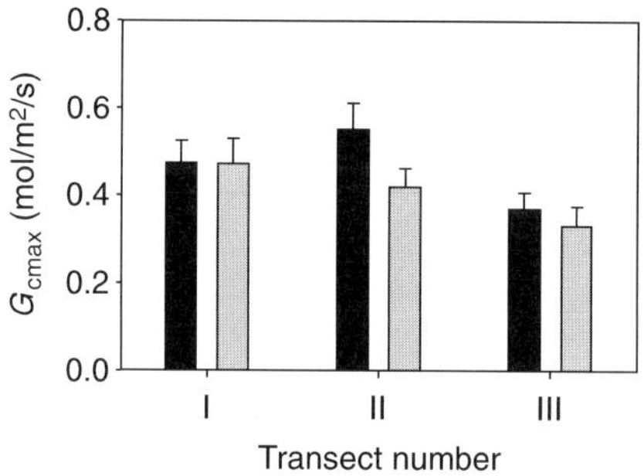

Fig. 12.4. Comparison of the estimates of maximum canopy conductance ($<G_{cmax}>$) using the regression between $<G_{cmax}>$ and the amount of rainfall in July, August and September 1995 (R_3) (black bar) with the estimates from non-linear inversion and the measurements of latent-heat flux by aircraft over three different transects (grey bar). The error bar represents one standard error of $<G_{cmax}>$.

transect II. We significantly overestimate $<G_s>$ for transect II. This suggests that we may have overestimated the mean rainfall for transect II. As shown in Fig. 12.6, we overestimate both $<G_s>$ and $<\lambda E>$ when they are relatively low and underestimate them when they are relatively high. The intercept and slope of the linear regression are significantly greater than zero and less than one, respectively, for either $<G_s>$ or $<\lambda E>$. However, the slope of the linear regression of two estimates is not significantly different from one if the intercept is forced to zero for either $<G_s>$ or $<\lambda E>$.

To assess the agreement between two estimates more objectively, we calculated the agreement index (Willmott, 1981), which is defined as the degree to which model predictions approach their observed counterparts. The agreement index is equal to 1 for a perfect agreement and 0 for no agreement at all. We found the agreement indices to be 0.75 and 0.85 for $<G_s>$ and $<\lambda E>$, respectively. Therefore we concluded that the agreement is satisfactory between the two estimates.

Issac *et al.* (2004b) estimated the maximum canopy conductance for every 10 km along transects I and II, and they also found that $<G_{cmax}>$ estimated using aircraft data near Wagga Wagga was significantly lower than $<G_{cmax}>$ obtained using flux measurements at the fixed towers. The difference in maximum canopy conductance between Wagga Wagga and its surrounding patches could not be explained by the NDVI data. They concluded that the deep-rooted lucerne pasture at Wagga Wagga had access to deeper soil water than the surrounding vegetation and therefore had a higher maximum canopy conductance (Leuning *et al.*, 2004).

Even though the estimate of $<G_s>$ obtained using R_3 is significantly higher than that calculated from the aircraft measurements of $<\lambda E>$ (Equation 2), $<\lambda E>$ values calculated using $<G_s>$ (Equation 1) are not significantly different from the measured $<\lambda E>$ values for transect II (Fig. 12.5). These results show that the fractional error in $<\lambda E>$ is less than in $<G_s>$. Sensitivity of λE to G_{cmax} is given as:

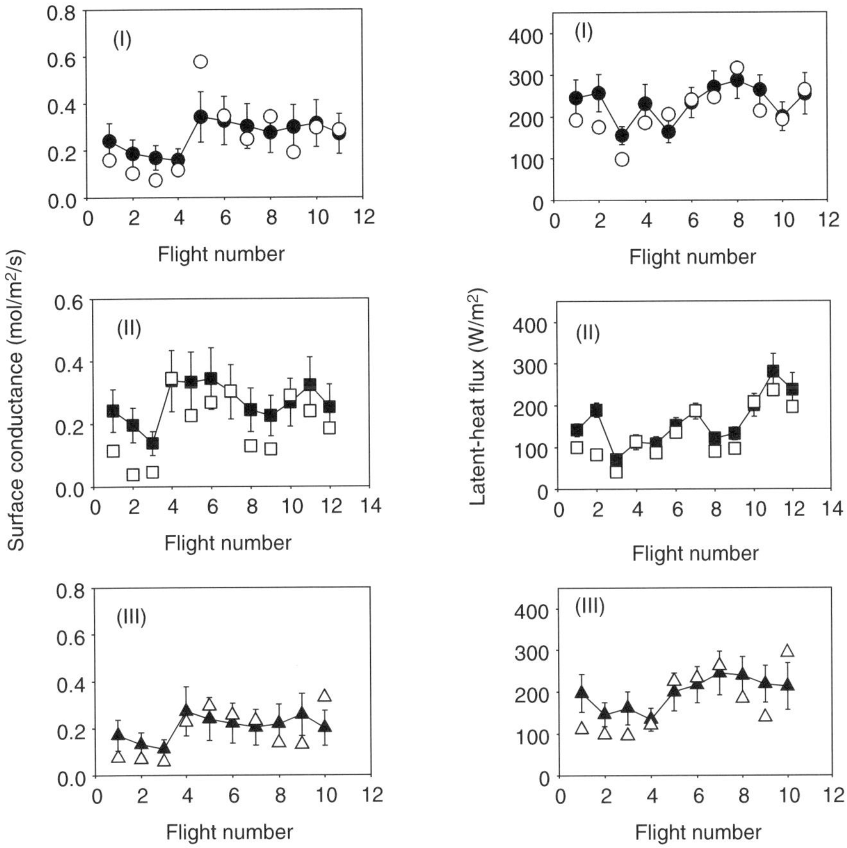

Fig. 12.5. Left panels: Mean surface conductance $<G_s>$ calculated using Equation (2) and aircraft measurements of $<\lambda E>$ (open symbols) compared with $<G_s>$ calculated from interpolation of tower-based estimates using the relationship between rainfall (R_3) and G_{cmax} shown in Fig. 12.3a (closed symbols). Right panels: Mean $<\lambda E>$ from aircraft measurements (open symbols) compared with estimates of $<\lambda E>$ calculated using Equation (1) and $<G_{cmax}>$ deduced from the relationship between G_{cmax} and R_3. All calculations use $<D>$, $<G_a>$ and $<A>$ from the aircraft flights, which are numbered on the *x* axis.

$$\frac{d(\lambda E)\,/\,(\lambda E)}{dG_{cmax}\,/\,G_{cmax}} = \frac{d(\lambda E)/(\lambda E)}{dG_c\,/\,G_c} = 1 - \Omega \tag{12}$$

where we assume that $dG_g/dG_{cmax} = 0$ and $G_g << G_s$. The decoupling coefficient (Ω) was defined by Jarvis and McNaughton (1986) as:

$$\Omega = \frac{\varepsilon + 1}{\varepsilon + 1 + G_a\,/\,G_s} \tag{13}$$

Using the aircraft measurements, we estimated the mean values of Ω to be 0.48 $\pm$ 0.08, 0.67 $\pm$ 0.08 and 0.32 $\pm$ 0.06 for transects I, II and III, respectively. Equation (12) thus shows that the fractional sensitivity of λE to G_{cmax} or G_c lies

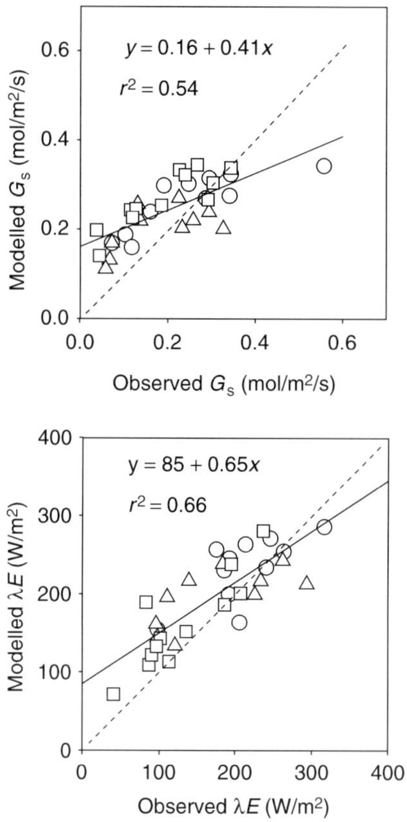

Fig. 12.6. Comparison of the modelled $\langle G_s \rangle$ (upper) or $\langle \lambda E \rangle$ (lower panel) calculated from interpolation of tower-based estimates using the relationship between rainfall (R_3) and G_{cmax} with those estimated from the aircraft measurements (observed). The dotted line represents the 1:1 relationship.

between 0.68 and 0.33 for the three transects. Transect II has the lowest sensitivity and therefore the smallest influence of the modelled $\langle G_{cmax} \rangle$ on the predicted $\langle \lambda E \rangle$.

Discussion and Conclusions

In this study, we developed a simple model of stomatal conductance and estimated one of the key parameters, G_{cmax}, in the model using measurements from flux towers at three paired sites during the OASIS 1995 field campaign. We found that only one parameter can be estimated independently using the measurements of latent-heat flux at each site. Using the flux measurements from flux towers and aircraft, we found a practical and reasonably accurate method for scaling maximum canopy conductance from patch to region for calculating latent-heat flux. This is significant as several flux networks have been operational continuously (Baldocchi *et al.*, 1996; Running *et al.*, 1999) and given that the footprint of a flux tower is only a few km^2, about three to four orders of magnitude smaller than the size of the typical grid box in a global climate model.

Because the relationship between λE and G_{cmax} is non-linear, using area-weighted $<G_{cmax}>$ within a grid cell of a climate model will overestimate the latent-heat flux of that cell. The averaging scheme developed by McNaughton (1994) and Raupach (1995) that conserves latent-heat flux should be used to calculate the mean surface conductance of the region, but is difficult to apply because the spatial variation in R_n is not available and because corrections to G_c depend on G_c itself. In this study, we examined the error in estimating latent-heat flux of a region using the area-weighted average of G_{cmax}. This approach is less rigorous than the averaging scheme of McNaughton (1994) and Raupach (1995), because flux is usually not conserved. However, our approach is a practical one, amenable to field measurements. The linear relationship between G_{cmax} and R_3 may be specific to semi-arid terrestrial ecosystems in south-east Australia. The difficult task is to find an independent variable that is linearly related to the model parameter and is measured spatially.

We found that the relationship between G_{cmax} and NDVI is non-linear and varies between crop and pasture, but that G_{cmax} is linearly related to R_3, the total amount of rainfall in the 3 months prior to the field study. Only linear relationships give a spatial average that is independent of the spatial resolution of the independent variables, i.e. R_3, in our case. We thus used the linear relationship between R_3 and G_{cmax} to calculate $<G_{cmax}>$ and hence to estimate $<G_s>$ and $<\lambda E>$ for each of three transects. Good agreement was obtained with the corresponding estimates from the aircraft measurements for all three transects.

This study has found that simple area-weighted averaging can be used to scale the estimates of G_{cmax} from patch to region for calculating latent-heat flux. More studies are required to determine whether area-weighted averaging of G_{cmax} can be used to calculate other fluxes at regional scale.

Acknowledgements

Some of the ideas presented in this chapter were conceived during a visit by YPW and RL to the South China Institute of Botany, Chinese Academy of Sciences. For their financial support, we thank the Natural Science Foundation of China (grant number 39928007), the Chinese Academy of Sciences' Funds for Cooperation with Overseas Chinese Talents and CSIRO International.

References

Baldocchi, D.D., Valentini, R., Running, S.R., Oechel, W. and Dahlman, R. (1996) Strategies for measuring and modelling carbon dioxide and water fluxes over terrestrial ecosystems. *Global Change Biology* 2, 159–168.

Dickinson, R.E and Henderson-Sellers, A. (1988) Modelling tropical deforestation: a study of GCM land-surface parameterization. *Quarterly Journal of Royal Meteorological Society* 114, 439–462.

Evans, J.R. (1995) Carbon fixation profiles do reflect light absorption profiles in leaves. *Australian Journal of Plant Physiology* 22, 865–873.

Garratt, J.R. (1994) *The Atmospheric Boundary Layer*. Cambridge University Press, Cambridge, 316 pp.

Issac, P.R., McAneney, J., Leuning, R. and Hacker, J.M. (2004a) Comparison of aircraft and ground based flux measurements during OASIS 95. *Boundary-Layer Meteorology* (in press).

Issac, P.R., Leuning, R., Hacker, J.M., Cleugh, H.A., Coppin, P.A., Denmead, O.T. and Raupach, M.R. (2004b) Estimation of regional evapotranspiration by combining aircraft and ground based measurements. *Boundary-Layer Meteorology* (in press).

Jarvis, P.G. (1976) The interpretation of the variation in leaf water potential and stomatal conductance found in canopies in the field. *Philosophical Transactions of the Royal Society of London* 273B, 593–610.

Jarvis, P.G. and McNaughton, K.G. (1986) Stomatal control of transpiration. *Advances in Ecological Research* 15, 1–49.

Kelliher, F.M., Leuning, R., Raupach, M.R. and Schulze, E.-D. (1995) Maximum conductances for evaporation from global vegetation types. *Agricultural and Forest Meteorology* 73, 1–16.

Leuning, R., Coppin, P.A., Cleugh, H.A., Issac, P.R., Denmead, O.T., Raupach, M.R., Zegelin, S. and Dunin, F.X. (2004) Spatial and temporal variation in fluxes of energy, water vapour and carbon dioxide during OASIS 1994 and 1995. *Boundary-Layer Meteorology* (in press).

McNaughton, K.G. (1994) Effective stomatal and boundary-layer resistance of heterogeneous surfaces. *Plant, Cell and Environment* 17, 1061–1068.

Meir, P., Kruijt, B., Broadmeadow, M., Barbosa, E., Kull, O., Carswell, F., Bobre, A. and Jarvis, P.G. (2002) Acclimation of photosynthetic capacity to irradiance in tree canopies in relation to leaf nitrogen concentration and leaf mass per unit area. *Plant, Cell and Environment* 25, 343–357.

Press, W.H., Teukolsky, S.A, Vetterling, W.T. and Flannery, B.P. (1992) *Numerical Recipes: the Art of Scientific Computing*, 2nd edn. Cambridge University Press, Cambridge, 933 pp.

Raupach, M.R. (1995) Vegetation–atmosphere interaction and surface conductance at leaf, canopy and regional scales. *Agricultural and Forest Meteorology* 73, 151–180.

Sellers, P.J. and Dorman, J.L. (1987) Testing the simple biosphere model (SiB) using point micrometeorological and biophysical data. *Journal of Applied Meteorology* 26, 622–651.

Sellers, P.J., Mintz, Y., Sud, Y.C. and Dalcher, A. (1986) A simple biosphere model (SiB) for use within general circulation models. *Journal of Atmospheric Science* 43, 505–531.

Wang, Y.-P., Leuning, R., Cleugh, H.A. and Coppin, P.A. (2001) Parameter estimation in surface exchange models using non-linear inversion: how many parameters can we estimate and which measurements are most useful? *Global Change Biology* 7, 495–510.

Willmott, C.J. (1981) On the validation of models. *Physical Geography* 106, 85–100.

13

Land Sinks: the Kyoto Process and Scientific Implications

Melvin G.R. Cannell

Centre for Ecology and Hydrology, Bush Estate, Penicuik, Midlothian EH26 0QB, UK

Abstract

This chapter outlines the process of negotiating land sinks in the Kyoto Protocol up to the agreement in Marrakesh in 2001. Before Kyoto, land sinks were not conceived as part of the Protocol, but they turned out to be crucial in getting an agreement. This overview begins by stating the principles of the Protocol and the flaws perceived by the USA. An outline is then given of the land-sink problems that contributed to the failure of the talks in The Hague in 2000 and the way they were subsequently resolved. A more detailed account is then given of nine specific land-sink issues, namely: definition of a forest, unintended outcomes, additionality and baseline, 'directly human-induced', measurement and verification, permanence and reversibility, leakage, equity and collateral damage or benefits. Finally, the main scientific challenges arising from the agreement are specified and a reflection is given on the process so far.

Introduction

Sinks were never intended to be part of the Kyoto Protocol. It was originally conceived as setting targets to curb the use of fossil fuels. It was only a few weeks before Parties of the Framework Convention on Climate Change (FCCC) met in Kyoto in November 1997 that some countries raised the question of offsetting emissions by increasing carbon uptake by forests. The negotiators had not thought through the complications and were poorly prepared to deal with the sink issue. It is not surprising, therefore, that sinks caused some confusion. Devils soon appeared in the detail (Schlamadinger and Marland, 1998; Schneider, 1998). What began as an apparently simple, additional low-cost mechanism turned out to be a complex, controversial issue involving 4 years of difficult negotiation.

 Forests at the Land–Atmosphere Interface (eds M. Mencuccini, J. Grace, J. Moncrieff and K.G. McNaughton)

This chapter briefly tells the story of those negotiations, focusing on the issues raised by including land sinks. It is a fascinating story, because sinks – not thought of originally – became, in the end, a means of rescuing the Protocol. In fact, in 2000, the sink issues became entangled with misgivings about the value of the Protocol itself, culminating in the withdrawal of the USA in 2001. Consequently, we need to re-examine the basic principles underlying the whole debate.

First, an outline is given of events so far – the institutions that are involved, the main meetings, the US withdrawal and progress to a compromise agreement. Secondly, an overview is given of the principles on which the Kyoto Protocol was founded, the claims by the USA that it is fatally flawed and the alternatives. Thirdly, an account is given of the crucial role that land sinks played in getting agreement during the make-or-break negotiations in Bonn and subsequently in Marrakesh in 2001. Fourthly, an account is given of nine specific issues that were introduced into the negotiations either directly or indirectly as a result of including land sinks. Finally, a summary is given of the challenges that are now presented to the scientific community to enable sinks to be included in ratified agreements by Parties to the Protocol.

The Institutions and Process from 1992 to 2002

There are two global institutions involved in the international debate: (i) the Intergovernment Panel on Climate Change (IPCC); and (ii) the Conference of Parties (COP) to the United Nations FCCC (UNFCCC).

The IPCC was set up by the World Meteorological Organization and the UN Environment Programme and has produced three assessments of the science of climate change and its projected impacts, in 1990/91, 1995/96 and 2000/01. It is a scientific body that influences the political process, but is not part of it.

The COP (and the Meeting of Parties (MOP)) was set up by the UNFCCC to negotiate agreements to prevent 'dangerous' climatic change (Article 2 of the Convention). It has a bureau in Bonn and meets annually (but met twice in 2001). It is served by a Subsidiary Body for Scientific and Technological Advice (SBSTA) and a Subsidiary Body for Implementation (SBI), which meet three to four times a year. Financial matters are dealt with through the Global Environment Facility of the World Bank. Details are available on the UNFCCC website, along with all the major documents produced by the COP and subsidiary bodies.

The UNFCCC was one of several Conventions set up in 1992 at the Earth Summit in Rio. It was signed by 87 Parties. In the 10 years to the second Earth Summit in Johannesburg in 2002, the COP met seven times (Fig. 13.1). It took 3 years after 1992 to establish the negotiating machinery and two meetings (COP1 and 2) to develop the principles for the Kyoto Protocol, which was signed by 84 countries at COP3 in 1997. The following year, at COP4, the Parties outlined the Buenos Aires Plan of Action for 'operationalizing' the Kyoto Protocol.

1992 Rio Earth Summit	**UNFCCC (87 Parties)**
1995 COP1 Berlin	Berlin Mandate
1996 COP2 Geneva	Ministerial Declaration
1997 COP3 Kyoto	Kyoto Protocol (84 signatories; 38 Annex I Parties)
1998 COP/MOP4 Buenos Aires	Buenos Aires Plan of Action (Operationalizing Kyoto)
1999 COP/MOP5 Bonn	Revised reporting guidelines, non-Annex I reports
2000 COP/MOP6 The Hague	Plan of Action not agreed
2001 March – USA withdraws from the Kyoto Protocol	
2001 April – Pronk proposal on sinks	
2001 July COP6bis Bonn	Modified Plan of Action agreed
2001 Nov. COP/MOP7 Marrakesh	Marrakesh Accords and Declaration
2002 Johannesburg Earth Summit	**Way Forward**

Fig. 13.1. The timetable of meetings of the Conference and Meeting of Parties (COP/MOP) to the UN Framework Convention of Climate Change (UNFCCC) from the first Earth Summit in Rio in 1992 to the second one in Johannesburg in 2002.

It was at this point that serious problems arose. We shall concentrate here on those concerning land sinks. At Kyoto, four articles had been agreed in principle on sinks, without knowing how they would be implemented. They were as follows:

- Article 3.3, to include sinks and sources during the commitment period of 2008–2012 that had been created by afforestation, reafforestation and deforestation (ARD) since the base year 1990.
- Article 3.4, to consider other forestry and agricultural sinks.
- Article 3.7, to include land sources in 1990 baseline emissions (the larger these are, the easier it is to achieve a reduction).
- Article 12, to include sinks in the Clean Development Mechanism (CDM) (actions in non-Annex I countries used to offset emission-reduction targets in Annex I countries – Annex I being those Parties with agreed emission-reduction targets in the Protocol).

The COP referred these issues to SBSTA, which then requested a Special Report on Land Use, Land-use Change and Forestry from IPCC. A comprehensive report was produced in 2000 (Watson *et al.*, 2000) and was considered at SBSTA meetings that year (notably in Lyons). The representatives at SBSTA agreed how to implement Articles 3.3 (ARD) and 3.7 (to include deforestation sources in base-year emissions) but disagreed on Articles 3.4 and 12.

Consequently, COP6 was convened in The Hague in November 2000 with no pre-agreement on how to deal with many of the difficult issues concerning sinks (as well as other issues (see below)). The meeting was also overloaded with work. The outcome was that the Plan of Action was not agreed. Depending on the viewpoint, the talks were said to have ended without time to get an agreement or ended with disagreement. Problems for the Protocol were then exacerbated with the announcement, in March 2001, by the new US President that the USA was withdrawing from the Protocol, claiming that it was 'fatally flawed'. Without the USA, all remaining Parties had to stay in the Protocol for it to survive – highlighting the need to agree on sinks (see below).

The Protocol was rescued with a variety of compromise deals. Notably, in April 2001, Jan Pronk, the Dutch chairman of COP6, put forward an innovative proposal on how to include Article 3.4 sinks (see below). COP6 was then reconvened as COP6bis in Bonn in July 2001 and a modified Plan of Action was agreed. This plan was concluded in Accords and a Declaration at Marrakesh in November 2001 (COP7).

The Kyoto Protocol Principles: Are They Flawed?

The Kyoto Protocol was formulated with four principles in mind (Grubb *et al.*, 2001). These are as follows:

- Fairness. At the outset, it was recognized that developed countries (Annex I Parties) should take the lead. They are, at present, the major emitters of greenhouse gases from fossil fuels and, historically, they also emitted most from land-use change (mainly deforestation). Developing countries need to concentrate resources on poverty alleviation, food security and development and need assistance regardless of climate change. They will also need assistance from any adverse effects of climate change, especially sea-level rise and shifts in rainfall.
- Successive targets. It was decided to set a series of binding targets for each Party, to be achieved over a series of commitment periods. In the first period, Annex I countries (strictly Parties – the European Union (EU) operates as one) agreed emission reductions amounting, overall, to 5.2% below their 1990 emissions by 2008–2012, measured in CO_2 equivalents, but comprising a 'basket' of six greenhouse gases. All countries would report emissions (including non-Annex I Parties) and Annex I Parties would be subject to sanctions for non-compliance.
- Lowest cost. In order to reduce the cost, three Flexible Mechanisms were agreed: (i) Joint Implementation, in which Annex I Parties could meet part of their emissions-reduction targets jointly, e.g. by one providing clean technology for another; (ii) Emissions Trading, where Parties with surpluses or emission credits (i.e. below their 1990 emissions) could sell them to Parties with emission debits; and (iii) the CDM, where Annex I countries could meet part of their targets by contributing to emission reductions in non-Annex I countries. Land sinks were not conceived as a fourth Flexible Mechanism, although they turned out to be one of the principal means by which some Parties lowered the cost of meeting their commitments.
- Meaningful. In order to be globally meaningful, the Protocol must be rati-

fied by 55% of the Parties and by Parties that make up 55% of the Annex I country emissions in 1990: that is, there must be a majority among the major emitters (Annex I Parties).

The USA questioned the first three of these four principles when they claimed that the Kyoto Protocol was 'fatally flawed'. The three US concerns were as follows:

- No developing-country commitment. The US position is that climate change is a global problem and a truly meaningful agreement needs to include everyone who is a significant emitter of greenhouse gases. The atmosphere makes no distinction between countries like China, Brazil and India, which are major emitters, and Annex I countries. This argument strikes at the first principle of fairness and of richer countries providing leadership. It also ignores the fact that non-Annex I Parties are required to report their emissions and can be given emission-reduction targets in subsequent commitment periods. It is true that emissions from non-Annex I countries are increasing, but they are unlikely to exceed Annex I emissions before 2050. Also, we may note that China is taking independent action to cut emissions anyway, while rapidly expanding its economy.
- No long-term goal. The USA would like there to be an ultimate target for greenhouse-gas emissions that would solve the problem, analogous to the Montreal Protocol, which cut chlorofluorocarbon (CFC) emissions to slow stratospheric ozone depletion by a meaningful amount to prevent ultravoilet (UV) damage. This argument opposes the principle of setting successive targets. But it ignores the fact that: (i) the science on what constitutes an ultimate target is immature (i.e. there is still debate on the level at which greenhouse-gas concentrations should be stabilized); and (ii) cutting greenhouse-gas emissions involves restructuring economies and lifestyles and should be approached cautiously. Although climate change is a pollution problem like stratospheric ozone, the solution is a development problem with many dimensions and trade-offs. Viewed in that way, it may be sensible to progress slowly.
- Too costly. The current US Senate will not agree to any commitment that has an adverse effect on the US economy. The Kyoto target for the USA was an 8% cut in emission below 1990 levels. Some commentators claim that actions to meet this target would lower the US gross domestic product (GDP) by 4%. On the other hand, the IPCC Working Group III Third Assessment estimated that the Kyoto targets of Organization for Economic Cooperation and Development (OECD) countries could be met at a cost of 0.2–2.0% of GDP, or 0.1–1.0% with Emissions Trading. Nevertheless, it is probably true that the US Kyoto target was set too high, given the anticipated growth in US emissions. Had it stayed in the agreement, the USA would certainly have needed to negotiate special terms, like Canada, Japan and Australia (see below).

In brief, the USA recognized that it was facing an agreement that would not solve the climate-change problem and that it could not keep. There is an obvious contradiction here, in that much larger targets required to be sure of solving the problem (e.g. 60% reductions in emissions) would be much harder to keep. Nevertheless, the withdrawal of the USA has led to a constructive re-examination of the Kyoto principles. Are there alternatives?

Four schemes have been proposed as alternatives to the Kyoto method of reducing emissions of greenhouse gases (Muller *et al.*, 2001).

1. Contraction and convergence schemes advocate a progressive move towards equal emissions per capita worldwide. This ideal can, in fact, be accommodated within a Kyoto scheme of progressive targets by assigning Parties different targets.
2. Others have advocated coordinated policies and measures to reduce greenhouse-gas emissions rather than setting targets. This approach was considered in the run-up to Kyoto and was rejected as being weak and unworkable.
3. Concerns about the effects of emission controls on GDP have led to the idea of setting intensity targets – such as an amount of greenhouse-gas emission per unit of GDP. The idea is that intensity targets would permit economic growth. However, there are problems in agreeing common measures of GDP and of determining what happens in recession. Also, such schemes could not deliver greenhouse-gas stabilization and would exacerbate inequities between countries.
4. A fourth proposal has been to develop a price-capped emissions-permit market, in which each country has an allocation of tradable permits, with more permits issued when the price rises beyond a threshold. Most analysts consider such schemes to be unstable and unlikely to be agreed.

Overall, it may be concluded that the Kyoto scheme of successive binding targets is a good one and is capable of being improved over time to accommodate flaws highlighted by the US withdrawal. It would be tragic to abandon the consensus that has been established between such diverse parties. There has been almost a decade of intellectual effort in the Kyoto process, which should not be wasted.

Getting Agreement in Bonn and Marrakesh: Role of Land Sinks

The main problems resolved

The overarching challenge in Bonn was to save the Kyoto Protocol, which meant keeping all major emitters on board (Giles, 2001). The USA represented 36.1% of the Annex I emissions in 1990, so the Parties in Bonn represented 63.9%. Thus, it was technically possible to get agreement by Parties totalling 55% of 1990 emissions without including everyone, but in practice the Protocol would be seriously weakened without agreement by all the remaining major emitters. All Europe (32.4%) and Russia (17.4%) together represented 49.8% of 1990 emissions. The 55% target could be just reached by adding Canada (3.3%) and Australia/New Zealand (2.3%) or, alternatively, Japan (8.5%). But, to be convincing, Canada, Australia/New Zealand and Japan should all be included.

The problem was that Canada, Australia and Japan were the countries that were having most difficulty in attaining their targets. The EU was favoured by the 1990 baseline date, when their emissions were high, before the shift from coal- to gas-powered electricity generation in the UK and the clean-up of industry in former East Germany. In contrast, from 1990 to 1998, fossil-fuel emissions in Canada had increased by 11%, in Australia by 13% and in Japan by 10%. (US emissions increased by 15% from 1990 to 1998.) Canada, Australia and Japan needed the Flexible Mechanisms and sink offsets.

Consequently, in the end, all three Flexible Mechanisms were allowed with relaxed conditions. This included Emissions Trading, even though Russia, Ukraine and the 'economies in transition' are likely to have tradeable surpluses totalling over twice the emissions-reduction targets of the Annex I countries minus the USA (Victor *et al.*, 2001). Thus, there is a risk of 'hot-air' trading.

In addition, land sinks were allowed under Articles 3.4 and 12, as well as Articles 3.3 and 3.7: that is, sinks attributable to forest, cropland and grassland management, plus 'revegetation', were allowed (see below, Article 3.4) and afforestation and reafforestation (A/R) were allowed under the CDM up to a maximum of 1% of a Party's emission in 1990. It may be noted that curbing deforestation in developing countries, which could hugely benefit the atmosphere, was not allowed, because it was considered unworkable to define the baseline of prevented actions (i.e. to avoid pessimistic projections or threatened deforestation).

The other major challenges at Bonn and Marrakesh were: (i) how to meet the needs of developing and vulnerable countries; and (ii) how to enforce compliance. The former was resolved with the establishment of four funds, totalling $530 million/year, three of which are for developing countries, with the EU contributing about half. The latter was solved by agreeing that, for every 1.0 Mt C/year by which a Party misses its target in 2010, it would get 1.3 Mt C/year added to its target in the second commitment period.

Overall, the Protocol was considerably weakened, with an expected overall emission reduction by Annex I countries of only 1.5% below their 1990 emission (rather than 5.2%), although perhaps over 10% less than business-as-usual emission in 2008–2010 relative to 1990.

Resolving the land-sink issues

As mentioned above, in April 2001, Jan Pronk put forward a specific proposal on how to calculate land-sink allowances (credits). The proposal can be divided into five stages, which are illustrated below, using the example of Canada:

1. ARD credit. A positive figure (i.e. a sink or 'credit') for ARD can be allowed (i.e. can be used to offset an emissions-reduction target).

2. ARD debit compensation. If deforestation sources exceed A/R sinks, then the resulting negative figure for ARD (the ARD 'debit') can be subtracted from any forest-management sink claimed under Article 3.4. In the case of Canada, ARD was −4.30 Mt C/year and the forest-management sink claimed was 9.11 Mt C/year, so the 'debit compensation' was 9.11 − 4.30 = 4.81 Mt C/year.

3. Forest-management 85% discount. Bearing in mind uncertainties in the estimation of forest-management sinks, and more particularly in the fraction of the sink due to direct human intervention (as opposed to CO_2/N fertilization, etc. (see below)) only 15% of the claimed sink should be allowed (i.e. 85% discounted). In the case of Canada, $4.81 \times 0.15 = 0.72$ Mt C/year.
4. Net–net calculation of agricultural-management sink. Unlike the forest-management sink, sinks due to agricultural management (of crop and grasslands) should be calculated on a net–net basis: that is, the sink is calculated as the change in carbon stock in the commitment period 2008–2012 (divided by five to get an annual stock change) minus the carbon-stock change in 1990. Canada claimed a net–net agricultural-management sink of 5.00 Mt C/year.
5. Cap on total sink allowance. Some countries, particularly those in the EU, considered that all Parties should begin to revise their energy policies and restructure their economies to cut fossil-fuel use. Restructuring land-use policies alone was not enough. There was talk of simply watching trees grow. Consequently, Pronk proposed that no Party should meet over 50% of its emissions-reduction target using sinks. To avoid this, he suggested that sink allowances should be capped at between 2.5% and 4.0% of 1990 emissions (depending on the country). In the case of Canada, 1990 emissions were 166.17 Mt C /year and the cap was set at 3%, so the maximum sink allowance was $116.17 \times 0.03 = 4.99$ Mt C/year. The sum of the forest- and agricultural-management sinks calculated above is 5.72 Mt C/year, so only 4.99 Mt C/year would be allowed.

In Bonn and Marrakesh (COP6bis and COP7), the Pronk proposals were debated and the outcome was as follows:

1. ARD credit. This was accepted.
2. ARD debit compensation. This was accepted, but the level of compensation was capped at 9.0 Mt C/year.
3. Forest-management 85% discount. This was not agreed. Instead, forest-management sinks were set Party by Party 'according to country circumstances' but guided by the Pronk principles. Some countries (those that had difficulties meeting their Kyoto targets) were allowed relatively large forest-management sinks. The forest-management sink allowed for Canada was 12.0 Mt C/year (7.2% of Canada's 1990 emission) rather than the 4.99 Mt C/year in the Pronk proposal. For Japan, the allowance was 13.0 Mt C/year (3.9% of Japan's 1990 emission) rather than the 8.8 Mt C/year in the Pronk proposal.
4. Net–net calculation of agricultural-management sink. This was accepted, and 'revegetation' sinks were also allowed.
5. Cap on total sink allowance. This principle was recognized but relaxed, so that some countries, notably Canada and Japan, will be able to meet over 50% of their emissions-reduction targets using sink allowances. Countries with large forests and agricultural industries argued that the creation and preservation of land sinks did, in fact, involve a significant restructuring of their economies.

Overall, the outcome of the Marrakesh Declaration was that about 169 Mt C/year of the total emission-reduction target of Annex I countries (excluding the

USA) could be met using land sinks. Excluding sinks, Annex I countries will now reduce emissions below 1990 levels by about 1.8% rather than 5.2% by 2008–2012. This seems a small amount, but remember that the reduction in emissions below projected business-as-usual emissions in 2008–2012 (without the Kyoto Protocol) is possibly over 10%.

Nine Specific Land-sink Issues

It is possible to distinguish nine specific land-sink issues that caused difficulty in the Kyoto Protocol. All of them were addressed in the IPCC Special Report (Watson *et al.*, 2000). Below, these issues are briefly explained and the Bonn/Marrakesh 'solution' is given.

Definition of a forest

'Kyoto forests' need to be defined areas of land with specific characteristics that distinguish them from non-forest areas. This is not a trivial problem. The Special Report considered the merits of a land-use approach (i.e. areas designated as forests) as opposed to a land-cover approach (i.e. areas with a physical cover of trees) and showed that very different areas and sink claims would be made, depending on the definition. Furthermore, within each of these approaches there are problems. For instance, if forests were defined as areas with, say, over 70% tree canopy cover, this would omit large areas of sparsely forested land, which may then be cleared without penalty. On the other hand, if the forests were areas with over 20% tree-canopy cover, forests can be severely degraded (thinned) without it being called deforestation. The areas and sinks would also depend on the spatial scale of definition – whether areas were defined, for example, to the nearest 100 ha or 0.1 ha. At Marrakesh, the agreement was that Parties can define forests within the bounds of 0.05–1.0 ha, 10–30% tree canopy cover and 2–5 m minimum height at maturity.

Unintended outcomes

Within the first commitment period, the IPCC Special Report identified two specific instances where Parties might be unfairly penalized or rewarded within the rules of Article 3.3 (ARD). One would occur when deforestation after 1990 was followed by reafforestation, which created a sink in 2008–2012, but, in fact, the total carbon stock had decreased since 1990. The deforestation–reafforestation (D-R) rule proposed was that credits be awarded for reafforestation only when the carbon stock in the commitment period was above that in 1990. The second was when areas that were afforested or reafforested shortly after 1990 were thinned or felled in 2008–2012, so there would be a debit (emission) in the commitment period, although beneficial A/R had taken place. The A/R rule proposed was that an A/R debit be counted only if it exceeded a credit in a previous commitment period – which means that it would not be counted in the first commitment period. Both of these rules were accepted at Bonn and Marrakesh.

Additionality and baseline

The Kyoto Protocol states that sinks should be created by direct human activities that are additional to those that existed in 1990 and additional to those that would exist anyway in the absence of the Protocol. Gross–net accounting would assume that there was no sink in 1990, so that the full sink in 2008–2012 could be counted. This was, in fact, agreed for forest-management sinks, but as explained, the allowances were set country by country. In contrast, as mentioned above, agricultural sinks were calculated on a net–net basis (annual stock change in 2008–2012 minus stock change in 1990). The Marrakesh Declaration also says that the sinks must be shown to be above baseline, but details on how this is done and verified were left unclear.

Directly human-induced

It was agreed in Kyoto that land sinks must be shown to be the result of directly human-induced actions since 1990. Negotiators were well aware that there is currently a large global terrestrial carbon sink of about 2 ± 1 Gt C/year, most of which is probably due to non-directly human-induced activities, such as CO_2 and N fertilization, or to activities prior to 1990, like the recovery of forests from historical heavy felling or clear-cutting in Europe and North America (Houghton and Hackler, 2001; Royal Society, 2001). This sink is not trivial: if 2 Gt C/year is spread evenly over the global land area (excluding Antarctica) it is 0.15 t C/ha/year and, if spread over that global forest area, it is 0.58 t C/ha/year. This is the primary reason why Pronk proposed that 85% of the forest-management sink claimed by Parties should be discounted. In the end, this crude mechanism was adopted, along with capping, but with Canada, Japan and Russia agreeing to major exceptions to the Pronk rules because of their 'country circumstances'.

Measurement and verification

The original terms on which sinks were allowed into the Kyoto Protocol were that they be transparent, accurate, consistent, comparable and verifiable. The Protocol covered the 'basket' of six greenhouse gases, so account has to be taken of land-management effects on methane and nitrous oxide sources and sinks. Within the agreement, Parties can enumerate carbon-stock changes in every area of Kyoto land or use models or statistical methods. Biomass, litter and soil carbon pools should all be counted, but the problem of accounting for wood products has been deferred. (Briefly, the wood-product options are to count them as zero, measure stock changes in each country or attribute all to the exporting countries or all to the importing countries.) The Special Report elaborated the methods of carbon accounting that are to be formally considered by SBSTA to provide guidance to the Parties.

Permanence and reversibility

All land sinks are inherently reversible, so carbon stocks in vegetation and soils can never be guaranteed in perpetuity. Superficially, this seemed, to some, to be a reason for not including sinks in the agreement at all. However, temporary removal of carbon from the atmosphere into terrestrial carbon stores is equally as valuable as slowing the rate of fossil-fuel use. In other words, removing carbon into organic matter now and emitting it later is the converse of not emitting it from fossil fuels now and emitting it later. The benefit in the period when terrestrial carbon stores exist could be measured in carbon tonne years, which could be rented – the Colombian proposal. Alternatively, forests created to store carbon could subsequently be used as a biofuel, or insurance could be taken to cover the risk of forest destruction by fire, insects or climate change. Whatever risk-management strategy is adopted, the Marrakesh Declaration was clear that geographically defined land areas claimed as sinks in the first commitment period must be counted in subsequent commitment periods. Once a Kyoto forest, always a Kyoto forest, so that any reversal of sinks is accounted for.

Leakage

An action to create a carbon sink has negative leakage when it promotes an activity elsewhere that creates a carbon source (emission) and has positive leakage when it promotes an activity elsewhere that creates an additional sink. Thus, tree planting may oversupply a timber market and undermine incentives for tree planting elsewhere (negative leakage) or provide timber that takes pressure off deforestation (positive leakage). Curbing deforestation may displace it elsewhere (negative leakage) or stimulate tree planting to provide timber elsewhere (positive leakage). Basically, leakage is about shifting the demand for agricultural products, fuel wood or timber, all of which require land to produce. Leakage is greatest when the demand for these products is inelastic (i.e. the consumers must have them), as is the case for staple foods and possibly tropical timbers without regulation. The solution to the leakage problem is to consider activities outside the boundary of the Kyoto lands and preferably within the whole country or region. At Bonn and Marrakesh, leakage was recognized as a particular problem when claiming sinks in non-Annex I countries. This was one reason why sinks in the CDM were limited to 1% of 1990 emissions.

Equity

There was a political desire by many Parties that all Annex I countries should share the burden of cutting fossil-fuel use, as originally conceived. Allowing very large sink offsets was seen as undermining the incentive to restructure economies and the principle of burden sharing. Pronk's solution was to impose caps on sinks, as explained above, but in the end these caps were relaxed. Parties will now meet their targets in a variety of ways, some by largely cutting fossil-fuel use (e.g. the EU), some largely by creating or maintaining land sinks (e.g. Japan) and some by a mixture of the two (e.g. Canada).

Collateral damages or benefits

Any change in land use or management obviously has impacts well beyond any one objective, such as carbon storage. First, environmental impacts need to be considered – on biodiversity, soil quality, soil erosion, water resources, water quality, acidification, aesthetics and climate feedbacks through changes in albedo. Secondly, socio-economic impacts need to be considered – on agricultural markets, timber markets and recreation/tourism and knock-on effects on employment and wealth/poverty. Thirdly, formal account must be taken of other international agreements, such as the Convention on Biological Diversity, the Rio Forest Principle and Agenda 21 on Sustainability. The Marrakesh Declaration specifically states that land use, land-use change and forestry activities must conserve biodiversity, be consistent with the sustainable use of resources and avoid adverse impacts on developing countries.

The Scientific Challenges Regarding Sinks

Many of the challenges presented to the IPCC in 1998 still exist. However, they have now been focused by the negotiations and agreements since the Special Report was completed in 2000. In Marrakesh, the Parties were given 2 years to agree how (and whether) they would claim land sinks. To assist this process, the COP made specific requests for assistance on the following four issues.

- Consistent reporting. The difficulties originally identified to implement ('operationalize') Article 3.3 have been magnified by the liberal inclusion of sinks under Article 3.4. Forest and agricultural land-management sinks are not only difficult to quantify, but they are also uncertain, particularly changes in soil carbon stocks. There is now greater urgency to define good practice in the measurement and modelling of sinks and to provide specific guidance on how to report Article 3.4 sinks and manage uncertainty.
- Distinguishing human-induced sinks. The agreement now includes sinks created by a number of land-management practices, all of which are ill-defined and encompass carbon-stock changes due to a mixture of directly human-induced and non-directly human-induced factors. Forest management, cropland management, grazing management, degradation and revegetation all need to be defined more closely as activities that cause human-induced carbon-stock changes. Most of all, ways need to be found to separate out that part of the sink that is due to CO_2 and N fertilization and other drivers not directly attributable to human activities since 1990. In the absence of any scientific guidance, the agreement will continue to be based on arbitrary discounts, caps and country claims.

- Conserving biodiversity and sustaining resources. Now that large areas of land are to be designated as Kyoto lands, with incentives to manage them to sequester carbon, it becomes all the more important to quantify the trade-offs between carbon storage, biodiversity and resource use (particularly water).
- Impacts on developing countries. Any additional driver for land-use change in developing counties has the potential to exacerbate already complex problems of resource use and poverty. Now that afforestation and reafforestation have been included in the CDM, complex problems need to be addressed, particularly those concerning leakage, permanence and environmental and socio-economic impacts.

Reflection

Land sinks have clearly been very important in the Kyoto Protocol. They have, simultaneously, been both the bane and the saviour of the agreement. Without them, the negotiations would have been simpler but agreement might not have been possible at all. Some might argue that sinks have remained in the negotiations to escape from the uncomfortable economic and political decisions required to curb fossil-fuel emissions. But, as the negotiations have progressed, it has become increasingly recognized that land sinks have a valid and important role to play in slowing the rate of rise in atmospheric CO_2 – a view reinforced by the IPCC Special Report.

The agreement struck in Marrakesh is a very small step towards stabilizing atmospheric greenhouse-gas concentrations, but it is a large step for the global community to get so far in addressing such a complex and vital development issue. It is not surprising that there was a measure of short-term political expediency on all sides. The USA presented its abandonment of the Protocol as an honest withdrawal. Canada, Japan and Australia used land sinks to extract large concessions. The EU claimed the moral high ground, but was favoured by the 1990 baseline. Russia maximized its opportunity to sell carbon credits, while the non-Annex I countries obtained much of the financial support they sought.

Nevertheless, the view that global warming is a serious threat that needs to be addressed has prevailed over the view held by some US commentators that it is a slow process that we do not fully understand and should ignore (Topfer, 2001).

It will be necessary for Annex I countries to show meaningful progress towards achieving their targets by 2005, by which time thought will have been given to the second commitment period, the role of non-Annex I countries and, hopefully, the inclusion of the USA. Over time, responses to the threat of climate change could well be melded with other threats, most importantly energy security. If projections are correct, the need to cut greenhouse-gas emissions will become increasingly evident as a result of climate change itself, and mitigation actions will become increasingly coincident with actions to secure energy supplies, such as investing in new energy-generation technologies (renewables, hydrogen, fuel cells, micropower systems, etc.), meeting fuel-efficiency standards and geological carbon storage.

Acknowledgements

This chapter benefits from support from the UK Department for Environment, Food and Rural Affairs (contract EPG1/1/160) and is a contribution to the Tyndall Centre for Climate Change Research.

References

Giles, J. (2001) Political fix saves Kyoto deal from collapse. *Nature* 412, 365.

Grubb, M., Hourcade, J.-C. and Oberthur, S. (2001) *Keeping Kyoto. A Study of Approaches to Maintaining the Kyoto Protocol on Climate Change*. International Network for Climate Policy Analysis, Sustainable Energy Programme, Shell Foundation, London, UK.

Houghton, R.A. and Hackler, J.L. (2001) Changes in terrestrial carbon storage in the United States. I. The roles of agriculture and forestry. *Global Ecology and Biogeography* 9, 125–144.

Muller, B., Michaelova, A. and Vrolijk, C. (2001) *Rejecting Kyoto. A Study of Proposed Alternatives to the Kyoto Protocol. Climate Strategies*. International Network for Climate Policy Analysis, Sustainable Energy Programme, Shell Foundation, London, UK.

Royal Society (2001) *The Role of Land Sinks in Mitigating Global Climate Change*. Policy Document 10/01, Royal Society, London.

Schlamadinger, B. and Marland, G. (1998) The Kyoto Protocol: provisions and unresolved issues relevant to land-use change and forestry. *Environmental Science and Policy* 1, 1–15.

Schneider, S.H. (1998) Kyoto Protocol: the unfinished agenda. *Climatic Change* 39, 1–21.

Topfer, K. (2001) Whither after the Hague? *Science* 291, 2095–2096.

Victor, D.G., Nakicenovic, N. and Victor, N. (2001) The Kyoto Protocol emission allocations: windfall surpluses for Russia and Ukraine. *Climatic Change* 49, 263–277.

Watson, R.T., Noble, I.R., Bolin, B., Ravindranath, N.H., Verardo, D.J. and Dokken, D.J. (2000) *Land Use, Land-use Change and Forestry*. A Special Report of the IPCC, Cambridge University Press, Cambridge, UK.

14

Spatial and Temporal Assessment of Biospheric Carbon Fluxes at a Continental Scale by Neural-network Optimization

Dario Papale and Riccardo Valentini
University of Tuscia, Department of Forest Science and Environment, via C. de Lellis 01100 Viterbo, Italy

Introduction

In recent years, questions related to the distribution and spatial–temporal dynamics of the terrestrial carbon fluxes have been at the core of many scientific and policy debates. Although driven primarily by the concerns about the increasing CO_2 content in the atmosphere and the associated role of terrestrial ecosystems in mitigating the growth rate in atmospheric CO_2 concentration, this issue has also been discussed in relation to biodiversity, land degradation, and other pressing policy and science questions. Yet current global models and observation are unable to discern carbon sources and sinks with reasonable accuracy at a regional scale; indeed, budgets at regional, continental or basin scales are not in agreement with the global picture. Temporal patterns of major carbon fluxes are poorly understood at seasonal to decadal time-scales. Uncertainty and scientific controversy prevail concerning the location of the land sink, i.e. in the northern USA and Canada, in Europe or in the Siberian forests of Eurasia (Ciais and Meijer, 1998; Fan *et al.*, 1998).

This question is most often addressed using atmospheric inversion techniques, which provide estimates of net surface–atmosphere exchanges of CO_2 using an optimization process.

The atmospheric CO_2 concentrations from a number of measurement sites constrain the source/sink areas in the presence of estimated atmospheric transport and known emissions. Since the typical atmospheric gradient of CO_2 from North to South Pole is only 3 p.p.m., the accuracy of the network is critical. Bousquet *et al.* (2000) estimate a sink of 0.5 ± 0.6 Gt C/year for North America, a sink of 0.3 ± 0.8 Gt C/year for Europe and 1.3 Gt C/year for Siberia. In contrast, Fan *et al.* (1998) concludes that a northern-hemisphere sink of 1.7 Gt C/year may exist entirely over North America. The difference between these two studies is partly caused by the inclusion of more measurement sites in the Bousquet *et al.* study. Although inversion studies can provide useful estimates of the net land–atmosphere CO_2 exchange they are unable to separate this into contributions from accidental effects and land management. In Table 14.1 we present the current best estimates of European sink strength obtained by various methods.

All methods indicate a land carbon sink but there is a factor of 5 between the highest and lowest estimates. A likely reason for these discrepancies is that the various methods use different scaling techniques to transform information obtained at one spatial and temporal scale to another. Several international efforts are today envisaged to construct a 'carbon data assimilation system' where models and observations can be included in an optimization scheme to provide robust estimates at high resolution in both space and time-scales.

Within the carbon-cycle research community, two quite distinct modelling techniques are used to estimate net land–atmosphere CO_2 fluxes. In 'inverse studies', atmospheric CO_2 measurements are used in conjunction with atmospheric transport modelling. These studies produce useful estimates of the large-scale geographical distribution of sources and sinks (Bousquet *et al.*, 2000), but they are unconstrained by what is already understood about physiological and ecological processes and cannot currently provide good estimates at the policy-relevant scale of countries. In 'forward modelling', models of land carbon uptake and release are developed based on an understanding of the relevant processes (e.g. the response of photosynthesis to CO_2, the response of microbial respiration to climate), and then these models are integrated forward in time to produce predictions of the temporal and spatial variability of land carbon sinks (Cramer *et al.*, 2001). Estimates of the land carbon balance produced by forward modelling to the current day from the past are constrained by the understanding of the system embodied in the model (e.g. conservation of carbon and nitrogen), but not constrained by any direct observations of the carbon cycle (e.g. flux measurements, forest inventories, CO_2 flask measurements). Thus traditionally observations and models have been completely separated fields, often reflecting the differences in the scientific communities responsible for such activities. The only areas of overlapping were: the generation of model algorithms by observations and (but not often) the validation of models through observations, but there were no feedbacks between the validation results and the model itself. A representation of this approach is presented in Fig. 14.1.

Table 14.1. Current best estimates of European sink strength obtained by various methods.

Method	Estimated C sink (Gt/year)	Reference
C Pools	0.5 (forests only)	Schulze *et al.* (1999)
Flux measurements	0.3 (forests only)	Martin *et al.* (1998)
Inversion	0.3	Bousquet *et al.* (2000)
Forest inventory	0.1 (forests only)	Nabuurs *et al.* (1997)

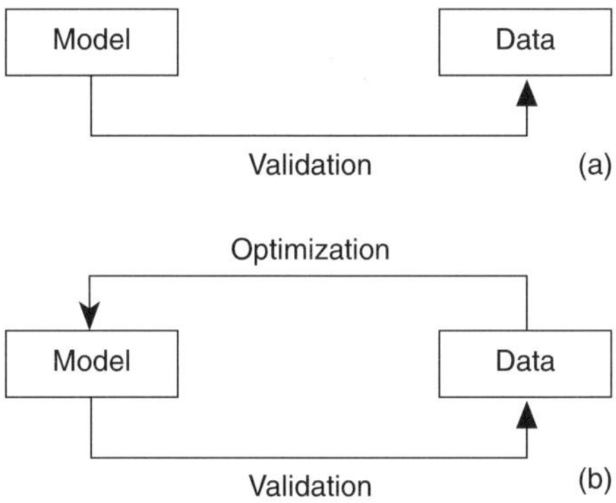

Fig. 14.1. (a) Traditional use of data to validate models; (b) new generation of models with embedded data assimilation and optimization of parameters.

However, a third emerging direction is the multiple-constraint approach. Observations constitute data sets, which challenge the model to produce predictions consistent with all available data. A variety of approaches are available for optimization schemes of this kind, including the use of adjoined model, gradient-based search methods and Kalman filters. Some applications of this type are: the use of combined atmospheric concentrations and surface data at a continental scale (Wang and Barrett, 2003), investigations on combinations of atmospheric concentrations and eddy-flux data at a global scale (Rayner *et al.*, 2001) and applications at the scale of vegetation canopies (Styles *et al.*, 2002). Only recently, neural networks have been used as optimization schemes for predictions of the terrestrial carbon cycle (Barrett *et al.*, 2001).

Artificial neural networks (ANNs) constitute a field in fast expansion that finds application in many disciplines, from properly scientific (physics, biology, chemistry, etc.) to typically humanistic (literature, language, social studies, etc.) (Bishop, 1995; Demuth and Beale, 2000). In biology and ecology applications, the neural methods are able to represent in a very effective way the complexity of phenomena under examination (Lek and Guégan, 2000). Normally, in such a complex system, the manifestation of a function or a property of the organism represents the arranged effect of multiple non-linear factors, which become partially or completely activated only in response to particular external circumstances.

The reproduction of the behaviour of an organism with such a complexity of responses is enormously difficult in traditional systems of simulation, based on mechanistic models or statistical relationships. In this chapter we present a preliminary work where biospheric flux data collected in the EUROFLUX project are used to train a neural-network simulator to provide spatial (1 km $\times$ 1 km) and temporal (weekly) estimates of carbon fluxes of European forests. The novelty of the approach is that flux data are used to constrain and parameterize in real time the neural-network structure with relatively few input driving variables. Since data are collected at high frequency (every half-hour) in 16 different forest ecosystems, the time domain can be a good replacement for the space variability. This means that variability in response to climate at a single climate can reproduce the effects of climate diversity within the European continent.

The work so far is at a preliminary stage; thus results have to be viewed in the light of developing a new methodology rather than as providing a final answer to the European continental carbon balance, although results are very encouraging and consistent with other independent estimates.

Methodology

What is an artificial neural network (ANN)?

Introduction

Much of the inspiration for the field of ANNs came from the desire to produce artificial systems capable of sophisticated, perhaps 'intelligent', computations similar to those the human brain routinely performs, and thereby possibly to enhance our understanding of the human brain. There is no universally accepted definition of an ANN. Most people in the field would agree that an ANN is a network of many simple processing elements ('units' or 'nodes'), each possibly having a small amount of local memory. The units, often organized in layers, are connected by communication channels ('connections') and operate only on their local data and on the inputs they receive via the connections (Fig. 14.2).

Most ANNs have some sort of 'training' rule whereby the weights of connections are adjusted on the basis of data. In other words, ANNs 'learn' from

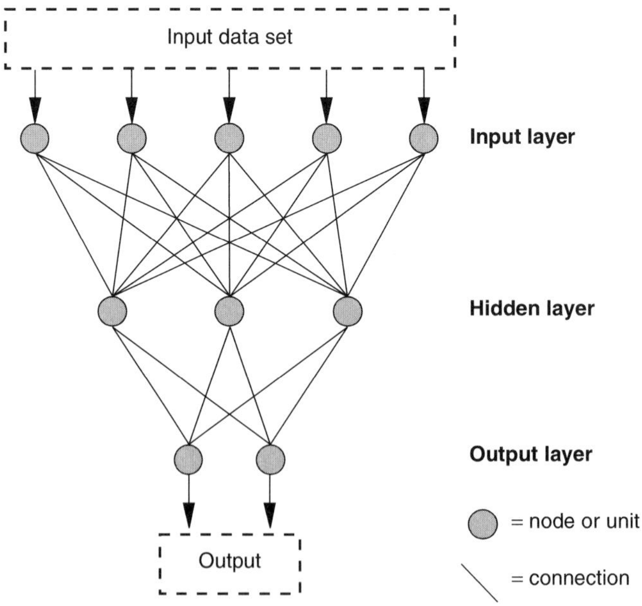

Fig. 14.2. Example of neural network with three layers; this structure can be indicated as 5-3-2, which means five input nodes, three hidden nodes (in only one hidden layer) and two output nodes. A network with two hidden layers with three nodes each would be written as 5-3-3-2.

examples and exhibit some capability for generalization beyond the training data. To summarize, we can say that a neural network is a massively parallel-distributed processor with a natural propensity for storing experiential knowledge and making it available for use. It resembles the brain principally in two respects (Haykin, 1994):

1. Knowledge is acquired by the network through a learning process.
2. Interneuron connection strengths, known as synaptic weights, are used to store the knowledge.

There are several types of neural networks, which differ for some main characteristics:

- Learning process.
- Use.
- Learning algorithm.
- Network architecture.

The learning process can be supervised or unsupervised. In the former case, the network is trained using a set of examples containing both the input and the required associated output; in the latter, the network is able to get information from input data for classification of the examples into categories.

The neural networks can be divided into three main categories according to use:

- Simulators of associative memories: they are made of networks that simulate a system of associating memory – that is, information is stored not by address but by assonance or contents (similar to the human brain). In other words, given a pattern A, the output will be pattern B even if A is imprecise or partial. From a practical point of view, it is often used for letters identification.
- Simulators of processes: they can understand the function that links inputs and outputs on the basis of the examples given during the training phase. Once the training phase is over the network can give outputs in reply to new inputs that have not been used during the training; that is, they interpolate and extrapolate, starting from the training data set. These networks are useful to simulate phenomena that are highly non-linear and whose mechanistic descriptions are not well defined. These are the networks we shall use in the work presented in this chapter.
- Classifiers: these networks enable data classification into categories on the basis of a similarity criterion. The training phase is carried out in unsupervised mode, and inputs are divided into non-predefined categories.

The network architecture refers to the structural organization of all constitutive elements: the node type, number of layers, number of nodes per layer and direction and type of connection among the nodes. As for the learning algorithm, the architecture varies in function accoding to the use of the network. In this work we have used a multilayer feed-forward neural network trained by a back-propagation algorithm (BPN). This kind of network, also called multilayer perception, is at present the ANN used most frequently (Lek and Guégan, 1999).

BACK-PROPAGATION NEURAL NETWORK (BPN) The BPN is a powerful system, often capable of modelling complex relationships between variables, based on the supervised procedure. The architecture of the BPN is a layered feed-forward neural network, in which the neurons are arranged in successive layers and the information flows unidirectionally, from input layer to output layer, through the hidden layer(s). The nodes are connected to all the nodes of the successive layer without connections between nodes of the same layer (lateral connection) and feedback connections. So the outputs are not fed back into the inputs, as is the case with some types of networks known as recurrent networks. What is propagated from the outputs back to the inputs is information about the errors that the network has made with regard to the current training example. Error information is propagated back through the network and this error is used to modify the strengths of the interconnections between the network nodes. The training rule that accomplishes this process is often called error back-propagation. These features will be discussed later in more detail.

The number of input nodes will be equal to the number of independent variables and the number of output nodes equal to the number of dependent variables. Between these two layers there is often a set of hidden nodes, arranged in one or more layers, which are called hidden because they do not directly interact with the outside world.

Functioning of a neural network

In order to simplify and schematize, we can say that each node of the network receives signals coming from the nodes it is connected to through the connections. Each input signal is multiplied by the weight related to its connection and the weighed signal arrives at the node, where it is summed to all the others.

At the node (n_i) all the input weighted signals are summed and the result modified by a transfer function (S), also called an activation function. The result of the activation function is the node output, which can eventually enter into a new node or the final result.

With reference to Fig. 14.3, we suppose that there are n inputs with signal $x_1, x_2, x_3, \ldots x_n$ and weights $w_1, w_2, w_3, \ldots w_n$. The weighted sum a, will be:

$$a = x_1 w_1 + x_2 w_2 + x_3 w_3 + \ldots\ x_n w_n = \sum_{x=1}^{n} x_i w_i$$

The a value will enter into a transfer function S, and is transformed according to different mathematical functions delivering the output y. The more simple one is the hard-limit transfer function (Fig. 14.4), where the output y will be equal to zero if the sum of weighted signals is less than value T, otherwise it will be 1. Thus the output y will be:

$$\begin{cases} y = 1 & \text{if} \quad a \geq T \\ y = 0 & \text{if} \quad a < T \end{cases}$$

There are a lot of transfer functions, but the most common are the linear and especially the sigmoid transfer functions (Fig. 14.5). For the sigmoid transfer function, the output will be:

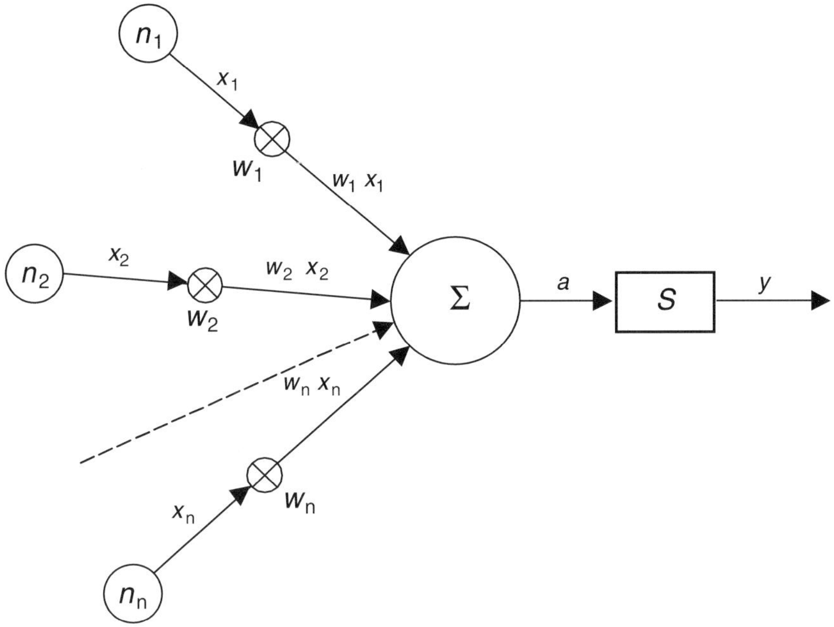

Fig.14.3. Block diagram of an artificial neuron.

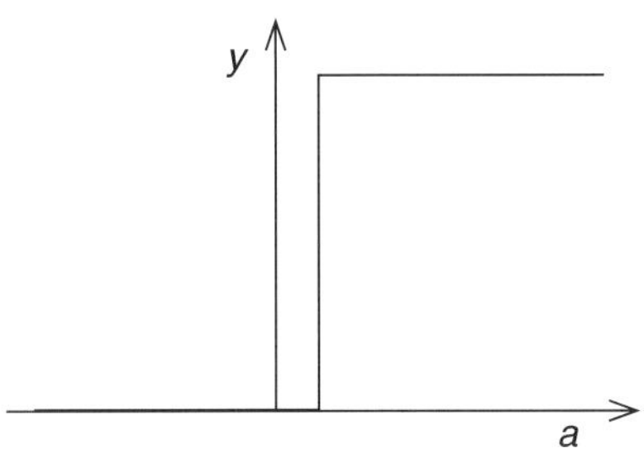

Fig. 14.4. Hard-limit transfer function.

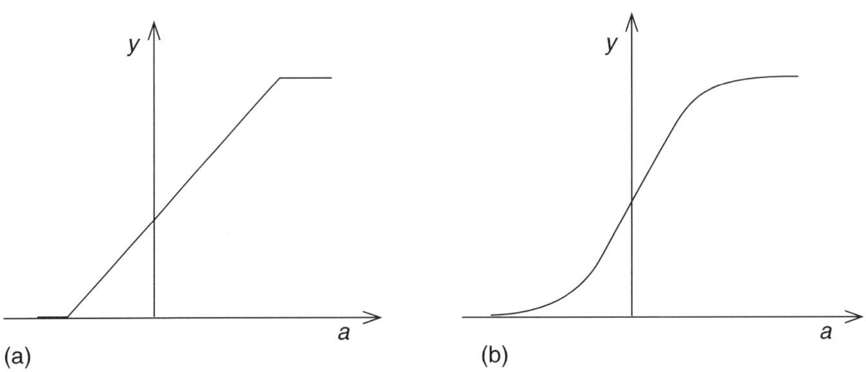

Fig. 14.5. (a) Linear transfer function; (b) sigmoid transfer function.

$$y = \frac{1}{1 + e^{-a/\rho}}$$

The ρ coefficient determines the shape of the sigmoid, a larger value making the curve flatter. In many cases, this parameter is omitted so that it is implicitly assigned the value 1.

Training a neural network

To find the functional relationship between input and output values, the neural network must develop an internal representation of this relationship. This internal representation is coded in terms of the values of the network weights, which are the strengths of the connections between the nodes.

A network is trained by providing a set of training examples, one at a time. Each example is composed of all the input and output values; thus, each input node receives the value of one of the independent variables and the network generates a set of output values by the output nodes. These outputs are the estimates that the network makes of the dependent variables associated with a particular input example.

At the start of network training, the output of the network (actual output) will probably be very different from the real output (target) of each example. These two outputs (actual and target) are compared and the difference is used as the basis for a scheme that modifies the network weights with the goal of minimizing this difference. The weights (whose values were randomly preset at the beginning) are modified according to a rule, typically the error back-propagation rule.

THE ERROR BACK-PROPAGATION ALGORITHM For each example *p*, we have an error E_p that is a function of the weights. Typically this is defined by the square difference between the actual output *y* and the target *t*. Thus, for a single node:

$$E_p = \frac{1}{2}(t - y)^2$$

where *y* is a function of the weights. The total error *E* is then just the sum of the pattern errors:

$$E = \sum_p E_p$$

The idea is to calculate an error each time the net is presented with a training vector and to perform a gradient descent on the error considered as a function of the weights.

Let us now consider the weight of the connection that links the output node *k* to the intermediate node *j* (w_{kj}). Let us suppose that the E_{kj} error varies in response to the weight w_{kj} as in Fig. 14.6. The optimal value of the w_{kj} weight is at the minimum of the error function (point *a*). If a minimum point for each connection can be found, the network is then trained. An additional difficulty lies in the impossibility of analysing each single connection sequentially

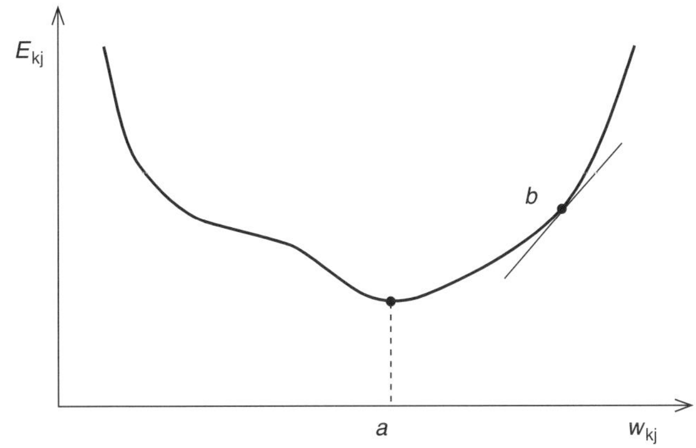

Fig. 14.6. Weight–error function for a single node.

because the function shape that links each weight to the error varies with the varying of the other weights.

This problem could be avoided by the gradient-descent rule technique. This technique is based on little movements of opposite sign and is proportional to the first derivative of the function in that point. In this way we get closer and closer to the minimum of the function and, once it has been reached, the first derivative in that point equal to zero, so we do not move any more. Thus the function shape slowly modifies (because of the modifications of the other weights), but we move on it with movements small enough to consider such movements negligible. This could be mathematically expressed as:

$$\Delta w_{kj} = -\eta \frac{dE_{kj}}{dw_{kj}}$$

η is an important factor named the 'learning rate'. It is the factor that determines the size of the steps that the network takes in navigating through the weight-error function. The size of the learning rate can be set by the user. As a rule, the higher the value of the learning rate, the faster the network will locate the point in the weight space that corresponds to the low training error. But there is a penalty for using a value of the learning rate that is too large: the network may continually overshoot the point of minimum error, never reaching it, but bouncing back and forth around it, proceeding chaotically.

One way of overcoming the problem related to the size of η is to alter the training rule from 'pure' gradient descent to include a term that includes a portion of the last weight change. The new rule is:

$$\Delta w_{kj}(t) = -\eta \frac{dE_{kj}}{dw_{kj}} + \mu \Delta w_{kj}(t-1)$$

where t is the time. The new term on the right-hand side (μ) is called momentum and its function is to increase the size of the step when the direction, in the weight space, is the same as the direction of the previous step. Conversely, momentum decreases the size of the step when the directions of the current and previous steps are not the same. The idea is that, if the network is making good progress in the current direction, then it should keep going forward with an even larger step. If, however, conditions have changed between the previous and current step, then it should take a correspondingly small step.

Among the many enhancements proposed for the error back-propagation algorithm are the variable learning rate and variable momentum. The two factors η and μ are changed during the training in a way such that they are increased if the step is in the same direction of the previous one and decreased if the step directions are not the same. Usually it is possible to define an increasing rate and a decreasing rate for both learning rate and momentum.

LOCAL MINIMA Let us consider the error function shown in Fig. 14.7. Suppose we start with a weight set to point *P*. If we perform the gradient descent, the minimum we encounter is the one at M_l, not that at M_g. M_l is called a local minimum and corresponds to a partial solution for the network in response to the training data. The potential problem of local minima is, however, quite rare with real data. With a one-dimensional weight space, as in Fig. 14.7, local minima are easy to construct. However, the more dimensions there are in the weight space, the more opportunities the network has to escape from the local minimum.

Data sets and errors

At the base of the network training is a data set made of a series of examples, each one containing the observed inputs and outputs. Starting from these examples, the network is trained with the aim of foreseeing the output values

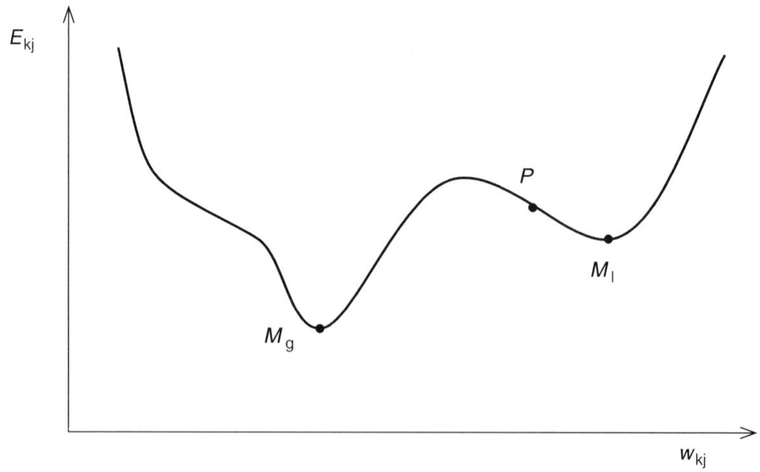

Fig. 14.7. Example of error function with local minima.

starting from input values that have never been seen before (generalization). In order to achieve the goal, the network must not learn too much from the examples given during the training, thus losing the ability to generalize on the basis of new examples (overfitting). The training data set is divided into three groups: training set, used for determination of the weights during neural-network training; test set, used during network training to calculate the errors to prevent overtraining; and validation set, which has not been used at all during training, but is used to assess the network's performance with 'new' data.

In practice, once the three sets of data have been defined, the training-set examples are given to the network and the error between the obtained and the real result is calculated. The modification of the weights occurs after the passage of an epoch – that is, a number of examples that are set before starting the training and which can vary from one to the number of cases in the training set.

Another parameter that has to be defined before training is the number of epochs that have to be processed before the training process stops and the performance of the network, with the current set of weights, is measured using the test set. The training ends either when an acceptable error level is achieved on the test set or when there is no improvement on the network performances after a certain quantity of epochs has passed (this has also been defined at the beginning). Since this procedure can itself lead to some overfitting to the test set, the performance of the selected network is confirmed by measuring its performance on a third independent set of data, called the validation set.

The evaluation of the error made by the network, on both test and validation set, is usually done through the calculation of different error typologies, in particular:

- Pearson correlation coefficient (r)

$$r = \frac{\sum_{n}\left(p - mean(p)\right)\left(a - mean(a)\right)}{\sqrt{\sum_{n}\left(p - mean(p)\right)^{2}\sum_{n}\left(a - mean(a)\right)^{2}}}$$

- Root mean squared error (RMSE)

$$\mathrm{RMSE} = \sqrt{\frac{\sum_{n}\left(p - a\right)^{2}}{n}}$$

- Mean absolute error (MAE)

$$\mathrm{MAE} = \frac{1}{n}\sum_{n}\left|p - a\right|$$

where:

n = number of examples
p = predicted values
a = real values

Neural network and biospheric carbon eddy-flux data

Neural networks have been used in this work with two different aims, even if one is related to the other. As previously shown, we need a data set as close as possible to reality with examples containing inputs and outputs to train a network in a supervised way. In a network for carbon-flux estimates at a continental scale, it is necessary to set examples where, besides the input variables, which will be analysed in detail later on, the output value, which in this case is the net ecosystem exchange (NEE), is also given.

In this chapter we investigate the performance of a neural-network scheme for predicting carbon fluxes at continental scales. The data set is obtained from the measurements of CO_2 fluxes taken at European sites using the eddy-covariance technique (Aubinet *et al.*, 2000). Such measurements could, under some circumstances, be affected by low night flux data, although there is still an intense debate as to whether these data need to be corrected and amended before being used for the NEE estimate. Furthermore, at some times of the year these experimental sites have stored micrometeorological data but, because of some technical problems, they have not been able to store the flux data: when this happens we must fill in the missing information in the best possible way. The data correction and gap filling have already been done on most sites that we have taken into consideration, using multiple regression techniques (Falge *et al.*, 2001), and data are directly available for elaboration. However, there are experimental sites where this work has not been done yet and the data correction and gap filling have been done using some neural networks.

Data filtering and gap filling for each specific site

There are four sites where a network for the flux data correction and gap filling has been trained:

Site 2 – Castelporziano (Italy): Mediterranean climate, evergreen broad-leaf forest (*Quercus ilex*).
Site 4 – Bordeaux (France): oceanic climate, coniferous plantation (*Pinus pinaster*).
Site 6 – Flakaliden (Sweden): boreal climate, coniferous forest (*Picea abies*).
Site 7 – Norunda (Sweden): temperate–boreal transition climate, mixed coniferous forest (*Pinus sylvestris*, *P. abies*).

Data collected every half-hour from the EUROFLUX network before their use for spatialization at a European scale were pre-elaborated and organized in a useful format. The first of these transformations involved the representation of temporal changes (month of the year and time) in such a way as to reduce the linear cumulative numerical weight of time in relation to other variables. Indeed, the progressive increase of time in terms of numbers (from day 1 to day 365) does not have a meaningful trend in terms of information to the network; therefore a transformation of this input into a 'fuzzy' variable was needed.

The inputs of the network trained for each site were air temperature, air moisture, photosynthetically active radiation (PAR) and two series of four fuzzy sets, each series considering the season of the year and the time of day. The fuzzy sets derived from a fuzzy transformation applied to the month of the year and the hour of the day. Each fuzzy set can take on any value between 0 and 1, which represents a degree or percentage membership of the original variable value in the fuzzy set.

The fuzzy transformation for the month of the year (Fig. 14.8) has been done in the following way:

- Winter: start in October, maximum in January, end in April.
- Spring: start in January, maximum in April, end in July.
- Summer: start in April, maximum in July, end in October.
- Autumn: start in July, maximum in October, end in January.

So, for example, May will be 0 winter, 0.667 spring (67%), 0.333 summer (33%) and 0 autumn.

A similar transformation has been done for the time of day (Fig. 14.9)

- Morning: start at 3, maximum at 9, end at 15.
- Afternoon: start at 9, maximum at 15, end at 21.
- Evening: start at 15, maximum at 21, end at 3.
- Night: start at 21, maximum at 3, end at 9.

The original data from each site consist of about 17,500 samples (one measurement every half an hour) but the data set was composed only of those examples with complete input and output data. In addition a further

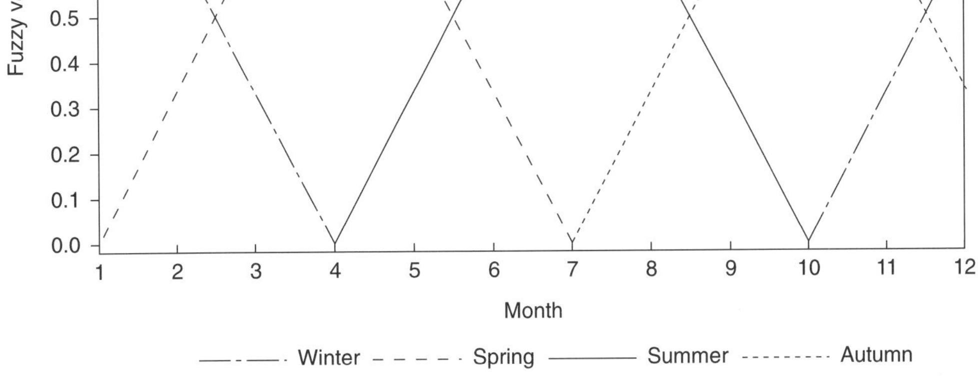

Fig. 14.8. Fuzzy transformation of the month of the year.

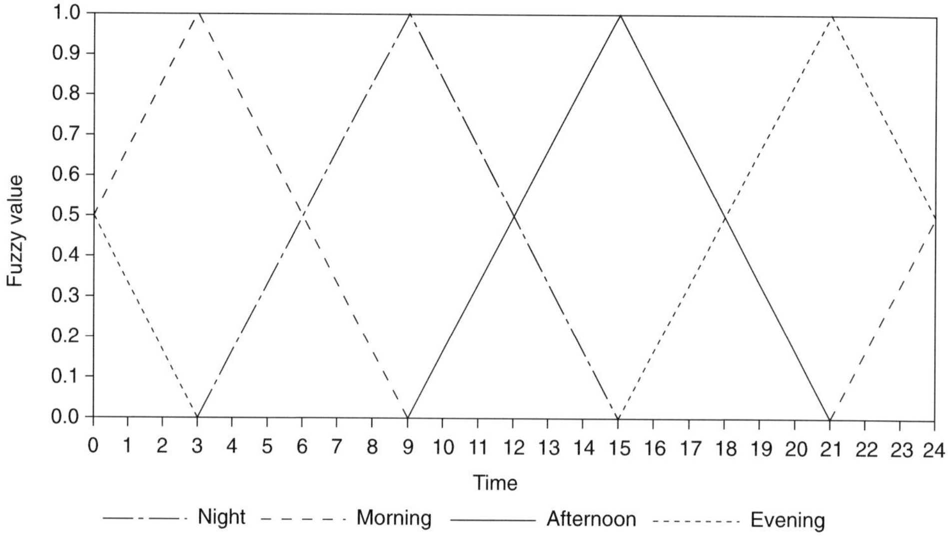

Fig. 14.9. Fuzzy transformation of the time of day.

problem common to all eddy-flux data treatment is the possible presence of unwanted biases originating from underestimation of nocturnal respiration. Although the ANN are very efficient in recognizing 'noise' in the data, they can be seriously affected by systematic biases. Thus a 'gap-filling' procedure, also using neural-network methodologies, was applied to the original data to filter out the nocturnal values below a certain threshold for turbulence intensity. To estimate this critical threshold use is made of the u^* value, defined as:

$$u^* = \sqrt{\frac{\tau}{\rho}}$$

where:

τ = momentum flux density (estimated by the Eddy system)
ρ = air density

The u^* critical value is estimated for each site (see Falge *et al.*, 2001, for the methodology) and it is often around 0.2. The u^* value was calculated for all the measurements taken every half an hour on each site with complete input and output data; only examples with u^* greater than 0.2 were added to the data set. For each site a data set of this kind was then created and used to train the networks to fill the data when u^* was below the threshold.

In those cases where input and output data were not available (failure of data-acquisition technology), flux data were put equal to the mean of the available data of the month at the same time of day. After the corrections and gap filling, it was possible to estimate the daily NEE for each site and these values were used for the second part of the work: the assessment of European NEE.

Assessment of European NEE by means of neural networks

A single network has been used to assess the NEE of Europe in 1997. The year was split up into 48 periods, four for each month, and the assessment concerned the mean NEE of each period. The network, with structure 12-5-1 (Fig. 14.10), was trained with 760 examples (400 training set, 180 test set, 180 validation set) and gave as output the mean value of NEE for the period, expressed as 'g C/day'. The 12 input nodes are:

- one for the dew point temperature;
- one for the maximum temperature;
- one for the minimum temperature;
- one for the mean temperature;
- one for the maximum normalized-difference vegetation index (NDVI);
- three for the land-cover type;
- four for the season (fuzzy).

The data set for the training originated from the sites of the EUROFLUX project and another two Italian sites. In Table 14.2 you can see the list of the stations and their main characteristics. For each site, the NEE was calculated for each period from the CO_2 flux measured.

INPUT DATA To train the network and to apply it to Europe, a set of input data concerning all the continent, in the form of a set of maps, was needed. The values concerning the EUROFLUX site positions were extracted from these maps,

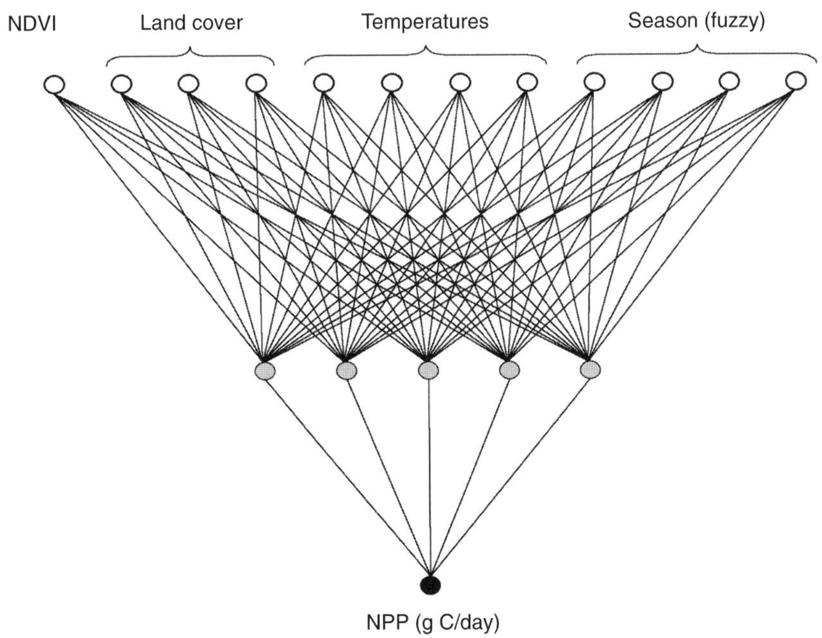

Fig. 14.10. Scheme of the network used to assess the European NEE. NPP, net primary production.

Table 14.2. Main characteristics of the sites used to estimate the NEE with the eddy-covariance system.

Country	Site name	Latitude	Longitude	Vegetation type	Dominant spp.	Climate
BE	Brasschaat	51°18′ N	04°31′ E	Mixed forest	*Pinus sylvestris* *Quercus robur*	T/O
BE	Vielsalm	50°18′ N	06°00′ E	Mixed forest	*Fagus sylvatica* *Pseudotsuga menziensii*	T
DK	Soroe	55°29′ N	11°38′ E	Broad-leaf deciduous	*F. sylvatica*	T/O
FI	Hyytiälä	61°51′ N	24°17′ E	Coniferous evergreen	*P. sylvestris*	B
FR	Bordeaux	44°42′ N	00°46′ W	Coniferous evergreen	*Pinus pinaster*	T/O
FR	Hesse	48°40′ N	07°05′ E	Broad-leaf deciduous	*F. sylvatica*	T/S
DE	Tharandt	50°58′ N	13°34′ E	Coniferous evergreen	*Picea abies*	T/C
IT	San Rossore	43°43′ N	10°17′ E	Coniferous evergreen	*P. pinaster*	M
IT	Collelongo	41°50′ N	13°35′ E	Broad-leaf deciduous	*F. sylvatica*	M/Mt
IT	Roccarespampani	42°23′ N	11°51′ E	Broad-leaf deciduous	*Quercus cerris*	M
IT	Castelporziano	41°45′ N	12°22′ E	Broad-leaf evergreen	*Quercus ilex*	M
NL	Loobos	52°10′ N	05°44′ E	Coniferous evergreen	*P. sylvestris*	T/O
SW	Flakaliden	64°14′ N	19°46′ E	Coniferous evergreen	*P. abies*	B/O
SW	Norunda	60°05′ N	17°28′ E	Coniferous evergreen	*P. abies* *P. sylvestris*	B
UK	Griffin	56°37′ N	03°48′ E	Coniferous evergreen	*Picea sitchensis*	T/O
DE	Bayreuth	50°09′ N	11°52′ E	Coniferous evergreen	*P. abies*	T/C

T, temperate; O, oceanic; B, boreal; C, continental; M, Mediterranean; Mt, montane; S, suboceanic.
BE, Belgium; DE, Germany; DK, Denmark; FI, Finland; FR, France; IT, Italy; NL, The Netherlands; SW, Sweden.

and these data were used as inputs for the training process. In Fig. 14.11 you can see the block scheme for preprocessing the input data to obtain the maps.

MAXIMUM NDVI The NDVI is very often used in remote sensing, because it is correlated to important characteristics of the vegetation (e.g. Gamon *et al.*, 1995). It is based on the green vegetation reflectance in the red and near-infrared (NIR). The equation is:

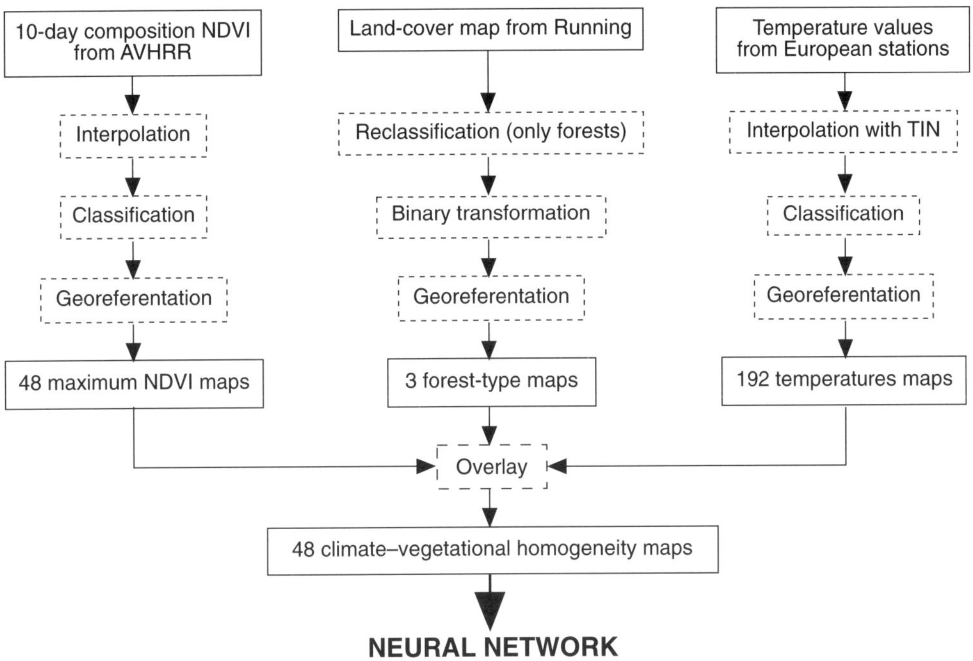

Fig. 14.11. Block scheme of the preprocessing of the input data. TIN, triangulated irregular network.

$$NDVI = \frac{\rho_{NIR} - \rho_R}{\rho_{NIR} + \rho_R}$$

where:

ρ_R = reflectance in the red
ρ_{NIR} = reflectance in the near-infrared

The NDVI maps used are derived from images acquired by the National Oceanic and Atmospheric Administration–Advanced Very High Resolution Radiometer (NOAA-AVHRR) sensor (Cracknell, 1997), and precisely from the first (red) and second (NIR) bands, which are set to 580–680 nm and 720–1000 nm, respectively. This satellite acquires, at the same hour, one image per day of the same area and it has a geometric resolution of about 1.1 km at nadir.

The Earth Resources Observation System (EROS) Data Center Distributed Active Archive Center (DAAC) (http://edcdaac.usgs.gov) provides 10-day image compositions from the AVHRR sensor. These compositions are obtained by overlaying ten images of 10 consecutive days; for each pixel the higher value is taken, which forms the resulting image. Thus, the 10-day composition image is a measure of the maximum value expressed by each pixel in this 10-day period in each acquisition band. This composition image is also provided for the NDVI, so each pixel is the maximum NDVI value expressed in the 10-day period.

The NDVI maps used in this work were acquired and elaborated in 1993. For the purpose of the demonstration of the application of the methodology, this should not have a major impact on the validity of the present work; however, in the future, updated remote-sensing products will be needed for a better quantification of the source and sink distribution.

On a 10-day basis, there were 36 images related to 1 year (three each month), but we needed 48 images to complete our data set, as the images were interpolated. Finally, each image was classified in five classes of NDVI distribution.

TEMPERATURE VALUES The climate data used in this work were extracted from the global-surface summary of day data produced by the National Climatic Data Center (NCDC) in Asheville, USA. This is a database with daily summaries obtained from hourly climate data registered by over 8000 georeferenced worldwide stations, about 3000 of these in Europe (Fig. 14.12). From this database we have extracted the dew point, the mean, the maximum and the minimum temperatures of the European stations with their geographical coordinates. Then, the mean of each temperature was calculated for each station during all the 48 periods. In total we obtained 192 values for each station (four temperatures × 48 periods).

Fig. 14.12. Map of the NCDC micrometeorological stations in Europe.

Through the interpolation method of the Triangulated Irregular Networks (TINs) climatic data were interpolated across Europe. TIN is one of the most commonly used structures for modelling continuous surfaces, using a vector data model. With triangulation, data points with known attribute values (e.g. maximum temperature) are used as the vertices (i.e. corner points) of a generated set of triangles. The result is a triangular tessellation of the entire area that falls within the outer boundary of the data points.

Once a TIN vector image for each period and temperature has been created, raster surfaces may be generated from it. For each raster pixel in the output image, an attribute value was calculated. This calculation is based on the positions and attributes of the three vertex points of the triangular facet within which the pixel centre falls and the position of the pixel centre.

With this procedure, we have created 192 maps of European temperatures, four for each of the 48 periods. These temperature maps were also further reclassified in 12 classes.

LAND COVER The choice of the land-cover map to use was not simple. The most accurate map for Europe is the Corine Land Cover but, unfortunately, this map does not cover all Europe and for this reason Corine was discarded. Other Europe cover maps are available on the EDC DAAC. There are six different maps:

- Global ecosystems (Olson, 1994a,b).
- International Geosphere–Biosphere Program (IGBP) land-cover classification (Belward, 1996).
- US Geological Survey land-use/land-cover system (Anderson *et al.*, 1976).
- Simple biosphere model (Sellers *et al.*, 1986).
- Simple biosphere 2 model (Sellers *et al.*, 1996).
- Biosphere–atmosphere transfer scheme (Dickinson *et al.*, 1986).
- Vegetation life-forms (Running *et al.*, 1994).

All of them are obtained by a classification of satellite images acquired by the AVHRR sensor. We have chosen the Running map because, although overestimating a little the lands covered by forest, it seemed to be the more correct one for forest classes. The Running land-cover map has eight cover types:

1. Evergreen needle-leaf vegetation.
2. Evergreen broad-leaf vegetation.
3. Deciduous needle-leaf vegetation.
4. Deciduous broad-leaf vegetation.
5. Annual broad-leaf vegetation.
6. Annual grass vegetation.
7. Non-vegetated land.
8. Water bodies.

Our work focuses only on forests, so the map was reclassified in a new way, with only three classes: evergreen broad-leaf, deciduous broad-leaf and evergreen needle-leaf. The deciduous needle-leaf class was added to the deciduous broad-leaf because there are no eddy-flux stations in this forest type. The problem, in any case, is marginal because the presence of deciduous needle-leaf forest in Europe is not very large. The land-cover map thus modified was converted into a set of three binary images, with value 0 for false and 1 for true, each of these concerning the presence or absence of one of the three forest cover types.

GEOREFERENCING In order to work with all the eight layers at the same time, the eight maps created needed to be georeferenced. The original maps were, indeed, in a different projection and so it was impossible to overlay them. A set of about 70 ground control points spread over Europe, a third-order mapping function and the latitude–longitude reference system were used for georeferencing. The error in the output maps was less then 1%, due principally to the different geometric resolution of the input images.

LAYER OVERLAY For all 48 periods, the eight layers were overlaid to create a new map where the classes are the result of each combination among the input map classes, like a climate–vegetation homogeneity classification. So each class of each of the new maps (48 new maps, one for each period) is characterized by 12 values: eight from the maps (four for the temperatures, three for the forest type and one for the NDVI) and four from the fuzzy values of the season of the period. The application of the network was carried out on these input data to reproduce the NEE value for each class.

Results and Discussion

Gap filling at each flux tower site

Night-time fluxes and gaps in data acquisitions were filled with a specific neural-network application. Figure 14.13 presents an example of the application of neural-network-generated data for site 4 (Bordeaux), where the model data are compared with the real eddy-flux data set not used during the training phase. Thus this corresponds to the real performance of the generalization. In Fig. 14.14 an example of the generation of daily trends, comparing real and ANN data, is presented. It may be noted that during calm nights ($u^* < 0.2$) the ANN outputs are larger than the real ones, due to the correction for the night-time values. Also some spikes in the real data are smoothed out from the ANN filtering.

The synthesis of the gap-filling procedure for four sites is presented in Table 14.3, where an error analysis is also carried out. In general, for all the sites, the Pearson coefficient is above 0.81, indicating a good agreement of the ANN with real data.

The examples given above indicate the efficiency of ANN methodology for data recovering and gap filling, sometimes superior to the normal regression or modelling approaches (Van Wijk and Bouten, 1999; Falge *et al.*, 2001).

NEE SPATIALIZATION WITH ANN In order to spatialize the single flux-tower data to the European continent, a single ANN was created using all 16 flux-site data

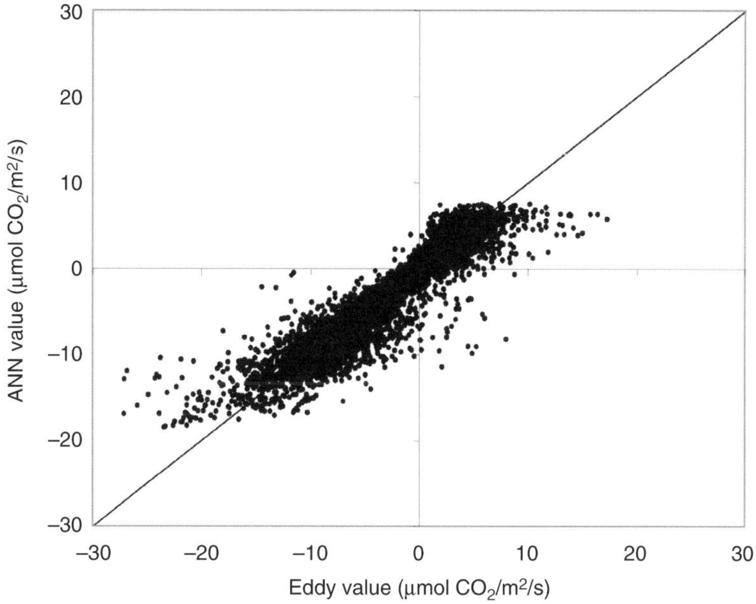

Fig. 14.13. Eddy vs. ANN flux values for validation data set. Site number 4 (Bordeaux). Also shown is the 1:1 line.

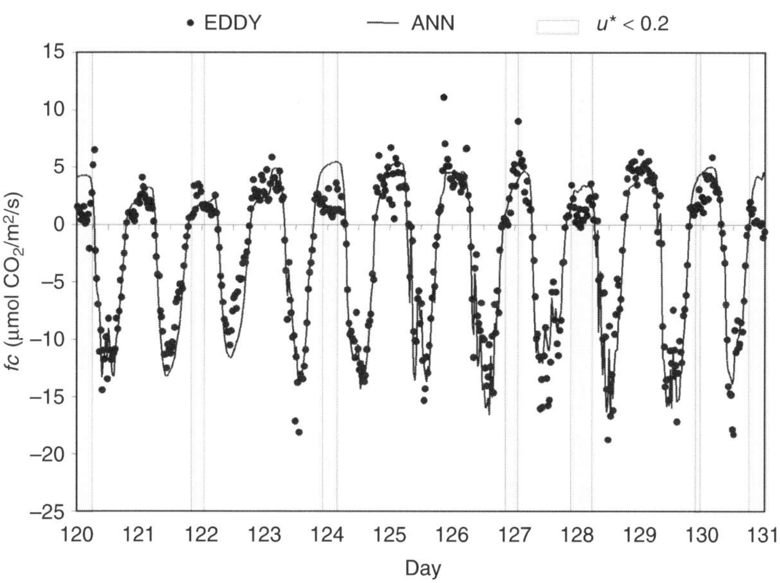

Fig. 14.14. Eddy vs. ANN daily trend for a 10-day period. *Fc*, CO_2 flux.

sets. In this case, the training and validation were created with weekly data. In Fig. 14.15 the comparison between the ANN generalization and the real data, not used during the training phase, is presented.

Table 14.3. Gap-filling summary for four sites. Data and errors analysis. Values in μmol $CO_2/m^2/s$.

	Castelporziano	Bordeaux	Flakaliden	Norunda
Examples with $u^* > 0.2$	9,045	11,372	13,573	13,441
Validation dataset	3,729	5,221	5,459	5,374
Data type				
Mean	−3.611	−2.241	−0.641	0.181
SD	5.925	6.550	3.532	8.659
Minimum value	−21.35	−33.45	−28.68	−49.85
Maximum value	14.53	44.58	24.76	65.38
Errors				
Pearson	0.887	0.916	0.845	0.813
RMSE	2.766	2.489	1.897	5.000
MAE	2.029	1.662	1.135	3.358

SD, standard deviation; MAE, mean absolute error; RMSE, root mean square error.

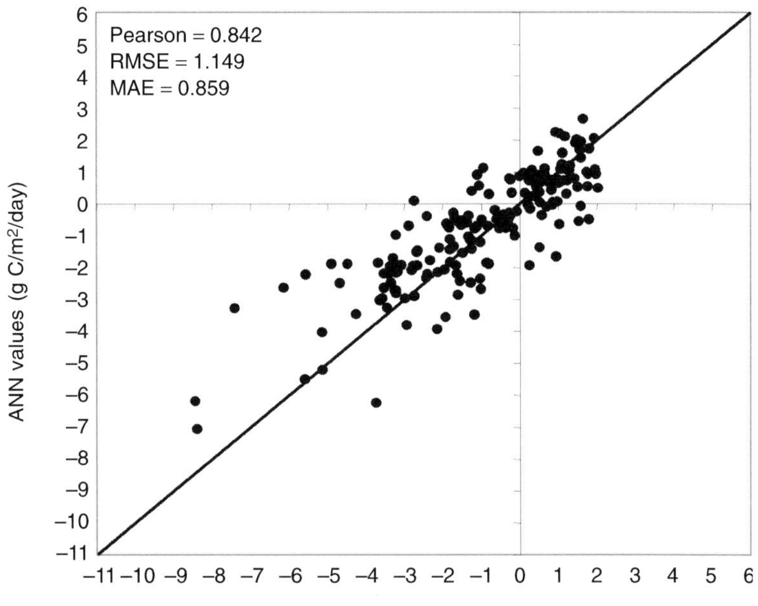

Fig. 14.15. Eddy vs. ANN NEE values for validation data set. Also shown is the 1:1 line.

In Fig. 14.16 the result of the same network but for annual sums is presented. In this case, a certain degree of degradation of the predictive power of ANN is clear. However, an in-depth look at the site-specific data shows evident outliers for sites 7 (Norunda), 8 (Bayreuth), 9 (Tharandt) and 11 (Aberfeldy). Site 7 is indeed a particular site for the network since it is the only slight source of carbon showing a high interannual variability. Site 11 is

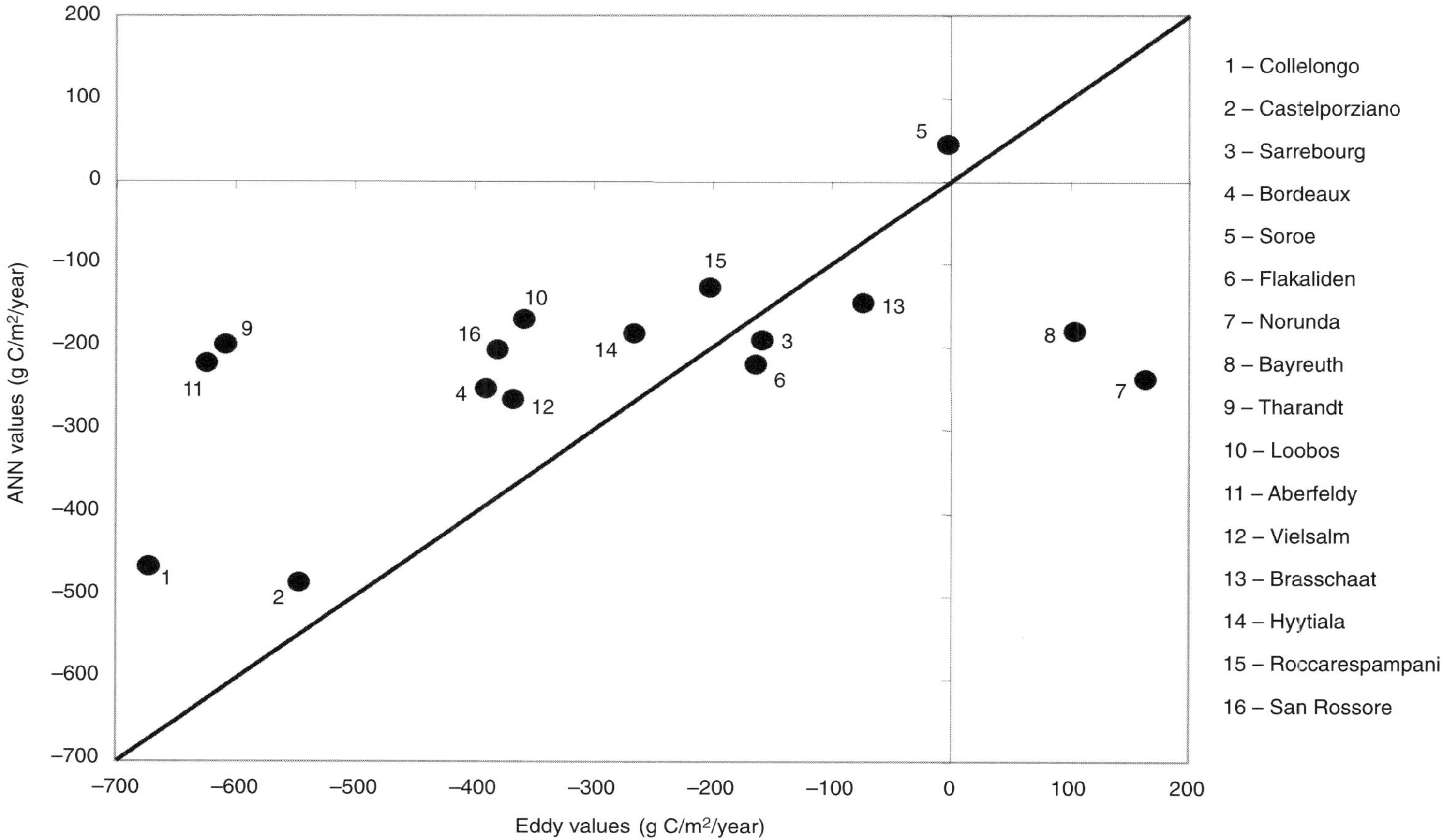

Fig. 14.16. Eddy *vs* ANN NEE annual sum values for the 16 training sites. Also shown is the 1:1 line.

an artificial plantation of fast-growing species and is also an outlier in the latitudinal trend of this data set (Valentini *et al*., 2000). Site 8 has recently been reanalysed and shows a quite heterogeneous canopy with lack of fetch requirements (T. Foken, Budapest, 2002, personal communication). For site 9 we do not have a clear explanation: we can only notice that a software intercomparison for this site showed a greater departure from the mean (EUROFLUX, 1999).

The larger discrepancy for the ANN generalization on annual sums can be attributed to the fact that in this case we have used a single network for all the sites and thus some of the outliers that have been effectively depicted could not be well reproduced. This again speaks in favour of the ability of ANN to detect noise from the underlying signal.

In Fig. 14.17 the final spatialization of NEE data is presented at a European scale for each month. In this case the second week of the month is presented as an example. The overall European carbon uptake following this analysis is 0.47 Gt C/year. It is quite evident that the onset of a sink at a European scale appears in the month of May, particularly due to the northern boreal forests switching from respiration to photosynthesis.

These analyses have to be considered rather preliminary at this stage but they show a consistent trend. Indeed, recent estimates of European carbon balances by several techniques range from 0.2 to 0.7 Gt C/year (Schulze *et al*., 2001).

Conclusions

We have presented a new methodological approach to the estimation of forest-ecosystem fluxes at a European scale by using ANN. The presented approach showed significant advantages and a change in the philosophy of model development for biospheric fluxes, with a strong emphasis on feedback with observational data as a 'constraint' for optimizing the predictions.

The ANN examples of the current study show an interesting application at a single-site level in the general problem of the optimization of gap-filling procedures. The network at the specific site presents the best performance and can be used as a routine system for gap filling.

In the case of weekly data spatialization (one general ANN for all the sites), predictions are still good and can be greatly improved with additional and better-quality input data. In particular we feel that the land-cover representation is rather coarse, the NDVI can be obtained in real time like the other input data, more sites are available today for ANN training and the lack of solar-radiation input could have affected the generalization. The ANN outputs for annual sums were not of the same quality as the other high-temporal-frequency data. However, the discrepancies were found in those sites with known problems of data interpretation. This example, in particular, speaks in favour of ANN methods for their capacity to extract outliers and specific behaviours that can be used for the improvement of process description.

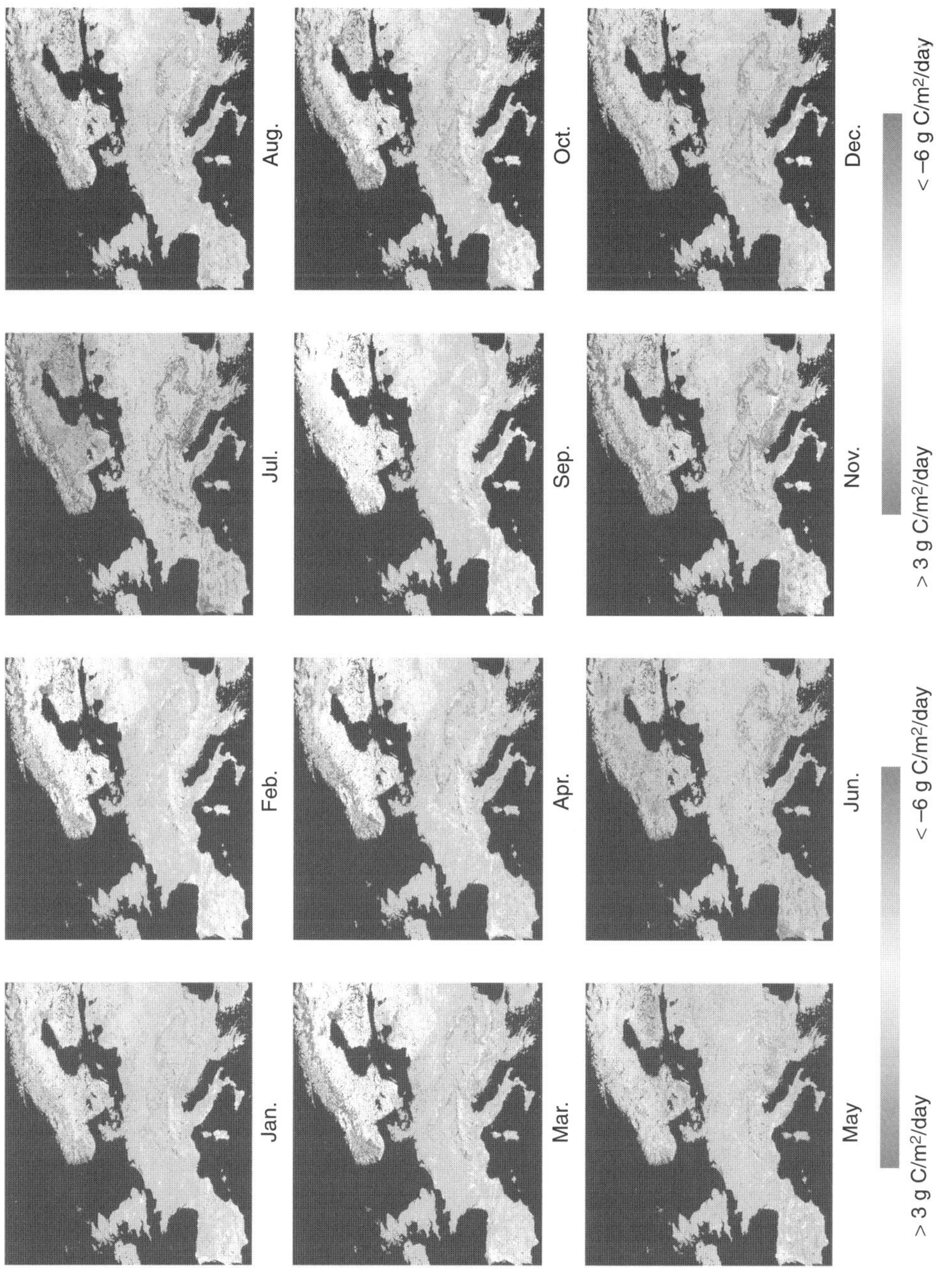

Fig. 14.17. Examples of NEE maps generated. Second week of each month. See frontispiece for colour versions.

References

Anderson, J.R., Hardy, E.E., Roach, J.T. and Witmer, R.E. (1976) *A Land Use and Land Cover Classification System for Use with Remote Sensor Data*. Professional Paper 964, US Geological Survey, Denver, Colorado, 28 pp.

Aubinet, M., Grelle, A., Ibrom, A., Rannik, Ü., Moncrieff, J., Foken, T., Kowalski, A.S., Martin, P.H., Berbigier, P., Bernhofer, C., Clement, R., Elbers, J., Granier, A., Grünwald, T., Morgenstern, K., Pilegaard, K., Rebmann, C., Snijders, W., Valentini, R. and Vesala, T. (2000) Estimates of the annual net carbon and water exchange of forests: the EUROFLUX methodology. *Advances in Ecological Research* 30, 113–175.

Barrett, D.J., Galbally, I.E. and Graetz, R.D. (2001) Quantifying uncertainty in estimates of C emissions from above-ground biomass due to historic land-use change to cropping in Australia. *Global Change Biology* 7, 883–902.

Belward, A.S. (1996) *The IGBP-DIS Global 1 km Land Cover Data Set (DISCover) – Proposal and Implementation Plans*. Working Paper 13, IGBP-DIS, Toulouse, France 61 pp.

Bishop, C.M. (1995) *Neural Networks for Pattern Recognition*. Oxford University Press, New York, 482 pp.

Bousquet, P., Peylin, P., Ciais, P., Le Quere, C., Friedlingstein, P. and Tans, P. (2000) Regional changes in carbon dioxide fluxes of land and oceans since 1980. *Science* 290, 1342–1346.

Ciais, P. and Meijer, H.A.J. (1998) The $^{18}O/^{16}O$ isotope ratio of atmospheric CO_2 and its role in global carbon cycle research. In: Griffiths, H. (ed.) *Stable Isotopes: Integration of Biological, Ecological and Geochemical Processes*. BIOS Scientific Publishers, Oxford, pp. 409–431.

Cracknell, A.P. (1997) *The Advanced Very High Resolution Radiometer*. Taylor & Francis, London.

Cramer, W., Bondeau, A., Woodward, F.I., Prentice, I.C., Betts, R.A., Brovkin, V., Cox, P.M., Fisher, V., Foley, J.A., Friend, A.D., Kucharik, C., Lomas, M.R., Ramankutty, N., Sitch, S., Smith, B., White, A. and Young-Molling, C. (2001) Global response of terrestrial ecosystem structure and function to CO_2 and climate change: results from six dynamic global vegetation models. *Global Change Biology* 7, 357–373.

Demuth, H. and Beale, M. (2000) *Neural Network Toolbox – For Use with MATLAB*, version 4. Math Works, Natick, Massachusetts.

Dickinson, R.E., Henderson-Sellers, A., Kennedy, P.J. and Wilson, M.F. (1986) *Biosphere–Atmosphere Transfer Scheme (BATS) for the NCAR Community Climate Model*. Technical Note NCAR/TN275+STR, NCAR, Boulder, Colorado, 69 pp.

EUROFLUX (1999) *ENV4-CT95-0078, Final Report*. Directorate-General for Research, European Commission, Brussels, Belgium.

Falge, E., Baldocchi, D., Olso, R., Anthoni, P., Aubinet, M., Bernhofer, C., Burba, G., Ceulemans, R., Clement, R., Dolman, H., Granier, A., Gross, P., Grünwald, T., Hollinger, D., Jensen, N.O., Katul, G., Keronen, P., Kowalski, A., Lai, C.L., Law, B.E., Meyers, T., Moncrieff, J., Moors, E., Munger, J.W., Pilegaard, K., Rannik, Ü., Rebmann, C., Suyker, A., Tenhunen, J., Tu, K., Verma, S., Vesala, T., Wilson, K. and Wofsy, S. (2001) Gap filling strategies for defensible annual sums of net ecosystem exchange. *Agricultural and Forest Meteorology* 107, 43–69.

Fan, S., Gloor, M., Mahlman, J., Pacala, S., Sarmiento, J., Takahashi, T. and Tans, P. (1998) A large terrestrial carbon sink in North America implied by atmospheric and oceanic carbon dioxide data and models. *Science* 282, 442–446.

Gamon, J.A., Field, C.B., Goulden, M.L., Griffin, K.L., Hartley, A.E., Joel, G., Peñuelas, J. and Valentini, R. (1995) Relationships between NDVI, canopy structure, and photosynthesis in three Californian vegetation types. *Ecological Applications* 5(1), 28–41.

Haykin, S. (1994) *Neural Networks: a Comprehensive Foundation*. Macmillan, New York.

Lek, S. and Guégan, J.F. (1999) Artificial neural networks as a tool in ecological modelling, an introduction. *Ecological Modelling* 120, 65–73.

Lek, S. and Guégan, J.F. (2000) *Artificial Neuronal Networks – Application to Ecology and Evolution*. Springer-Verlag, Berlin, Heidelberg, 262 pp.

Martin, P., Valentini, R., Jacque, M., Fabbri, K., Galati, D., Quaratino, R., Kennedy, P., Folving, S., Matteucci, G., Granier, A., Berbigier, P., Loustau, D., Jensen, N.O., Lindroth, A., Tenhunen, J., Dolman, A.J., Jarvis, P., Bernhofer, B., Aubinet, M., Ceulemans, R. and Vesala, T. (1998) A new estimate of the carbon sink strength of EU forests integrating flux measurements, field surveys and space observations: 0.17–0.35 Gt(C). *AMBIO* 27(7) 582–584.

Nabuurs, G.J., Päivinen, R., Sikkema, R. and Mohren, G.M.J. (1997) The role of European forests in the global carbon cycle – a review. *Biomass and Bioenergy* 13(6), 345–358.

Olson, J.S. (1994a) *Global Ecosystem Framework – Definitions*. Internal Report, USGS EROS Data Center, Sioux Falls, South Dakota, 37 pp.

Olson, J.S. (1994b) *Global Ecosystem Framework-translation Strategy*. USGS EROS Data Center Internal Report, Sioux Falls, South Dakota, 39 pp.

Rayner, P.J., Knorr, W., Scholze, M., Giering, R., Heimann, M. and Le Quere, C. (2001) Inferring terrestrial biosphere carbon fluxes from combined inversions of atmospheric transport and process-based terrestrial ecosystem models. In: *Extended Abstracts of the 6th International Carbon Dioxide Conference, Sendai, Japan, October 2001*, pp. 1015–1017.

Running, S.W., Loveland, T.R. and Pierce, L.L. (1994) A vegetation classification logic based on remote sensing for use in global biogeochemical models. *Ambio* 23(1), 77–81.

Schulze, E.D., Lloyd, J., Kelliher, F.M., Wirth, C., Rebmann, C., Lukher, B., Mund, M., Milykova, I., Schulze, W., Ziegler, W., Varlagin, A., Valentini, R., Dore, S., Grigoriev, S., Kolle, O. and Vygodskaya, N.N. (1999) Productivity of forests in the Eurosiberian boreal region and their potential to act as a carbon sink. *Global Change Biology* 5, 703–722.

Schulze, E.D., Dolman, A.J., Jarvis, P., Valentini, R., Smith, P., Ciais, P., Grace, J., Linder, S. and Bruning, C. (2001) *The Carbon Sink: Absorption Capacity of the European Terrestrial Biosphere.* EUR19883, Directorate-General for Research, European Commission, Luxemburg.

Sellers, P.J., Mintz, Y., Sud, Y.C. and Dalcher, A. (1986) A simple biosphere model (SiB) for use within general circulation models. *Journal of Atmospheric Science* 43, 505–531.

Sellers, P.J., Randall, D.A., Collatz, G.J., Berry, J.A., Field, C.B., Dazlich, D.A., Zhang, C., Collelo, G.D. and Bounoua, L. (1996) A revised land surface parameterisation (SiB2) for atmospheric GCMs – Part I – model formulation. *Journal of Climate* 9, 676–705.

Styles, J.M., Raupach, M.R., Farquhar, G.D., Kolle, O., Lawton, K.A., Brand, W.A., Werner, R.A. Jordan, A., Schulze, E.D., Shibistova, O. and Lloyd, J. (2002) Soil and canopy CO_2, $^{13}CO_2$, H_2O and sensible heat flux partitions in a forest canopy inferred from concentration measurements. *Tellus B* 54(5), 655–676.

Valentini, R., Matteucci, G., Dolman, A.J., Schulze, E.D., Rebmann, C., Moors, E.J., Granier, A., Gross, P., Jensen, N.O., Pilegaard, K., Lindroth, A., Grelle, A., Bernhofer, C., Grunwald, T., Aubinet, M., Ceulemans, R., Kowalski, A.S., Vesala, T., Rannik, U., Berbigier, P., Loustau, D., Guomundsson, J., Thorgeirsson, H., Ibrom, A., Morgenstern, K., Clement, R., Moncrieff, J., Montagnani, L., Minerbi, S. and Jarvis, P.G. (2000) Respiration as the main determinant of carbon balance in European forests. *Nature* 404, 861–865.

Van Wijk, M.T. and Bouten, W. (1999) Water and carbon fluxes above European coniferous forests modelled with artificial neural networks. *Ecological Modelling* 120, 181–197.

Wang, Y.-P. and Barrett, D.J. (2003) Estimating regional terrestrial carbon fluxes for the Australian continent using a multiple-constraint approach: I. Using remotely sensed data and ecological observations of net primary production. *Tellus B* 55(2), 270–289.

15

Scaling Carbon Uptake from Leaves to Canopies: Insights from Two Forests with Contrasting Properties

David Whitehead,[1] Adrian S. Walcroft,[2] Kevin L. Griffin,[3] David T. Tissue,[4] Matthew H. Turnbull,[5] Victor Engel,[3,*] Kim J. Brown[3,*] and William S.F. Schuster[6]

[1]Landcare Research, PO Box 69, Lincoln 8152, New Zealand; [2]Landcare Research, Private Bag 11052, Palmerston North, New Zealand; [3]Lamont–Doherty Earth Observatory, Columbia University, Palisades, NY 10964-8000, USA; [4]Department of Biological Sciences, Texas Tech University, Lubbock, TX 79409-3131, USA; [5]Department of Plant and Microbial Sciences, University of Canterbury, Private Bag 4800, Christchurch, New Zealand; [6]Black Rock Forest Consortium, 129 Continental Road, Cornwall, NY 12518, USA

Abstract

A multilayer, one-dimensional model is used to explore the processes regulating photosynthesis in two forest canopies with contrasting structural and physiological properties. An evergreen coniferous forest dominated by *Dacrydium*, growing at a site with very low nutrient availability in New Zealand, is compared with a deciduous mixed *Quercus* forest, growing at a site with high nitrogen inputs from atmospheric deposition at almost the same latitude in the north-eastern USA. The canopy model integrates radiative transfer, energy balance, evaporation and photosynthesis, coupled with stomatal conductance, in 20 layers through the canopy and is combined with a simple water-balance model to introduce the effects of seasonal root-zone water balance and allow the estimation of net annual carbon uptake. Estimates of the vertical distribution of foliage area and measurements of physiological parameters for the two sites were made for the model.

For *Dacrydium*, 54% of the foliage area occurred in the upper 11 layers of the canopy. In contrast, 52% of the foliage area occurred in the upper five layers of the *Quercus* canopy. At the top of the canopy, the maximum rates of carboxylation activity, V_{cmax}, for the *Dacrydium* and *Quercus* canopies (half-surface area basis) were 11.9 and 50.2 μmol/m^2/s, respectively. In midsummer, this resulted in 50% of canopy photosynthesis taking place in layers 5 to 10 in the *Dacrydium* canopy and 97% of photosynthe-

*See p. vii for present addresses.

(eds M. Mencuccini, J. Grace, J. Moncrieff and K.G. McNaughton)

sis in layers 1 to 5 in the *Quercus* canopy. The annual net carbon uptake for the *Quercus* canopy (1.22 kg C/m^2) was 14% greater than that for the *Dacrydium* canopy (1.07 kg C/m^2), but the difference in the ratios of net annual carbon uptake to intercepted radiation was much larger (75%) for the *Quercus* canopy (0.96 g C/MJ (400–700 nm)) than that for the *Dacrydium* canopy (0.55 g C/MJ).

Analysis of the component processes revealed that, at midday in typical midsummer conditions, photosynthesis for sunlit foliage throughout both canopies was limited by carboxylation activity. Shaded foliage in the *Quercus* canopy was strongly limited by the rate of electron transport, but, in contrast, photosynthesis in shaded foliage in the upper layers of the *Dacrydium* canopy was limited by carboxylation activity. The model is used to show that the consequence of the contrasting values of V_{cmax} and canopy structure is that carbon uptake in the *Quercus* canopy is much more sensitive to changes in irradiance than that in the *Dacrydium* canopy. Furthermore, at high incident irradiance, carbon uptake by the *Quercus* canopy is much more sensitive to increasing diffuse fraction of radiation than that by the *Dacrydium* canopy.

Introduction

Carbon uptake by forest canopies is dependent on the distribution of foliage area, the physiological properties of foliage and environmental variables. These biophysical processes are well understood, but, in order to estimate annual rates of carbon uptake, it is necessary to scale the processes in canopies both spatially and temporally, using models (Baldocchi and Meyers, 1998). There is considerable interest in applying such models to forests to determine environmental limits to productivity, and also in integrating carbon exchange across ecosystems to provide regional (Lloyd *et al.*, 1995) and global estimates of carbon exchange by interfacing models with data obtained from remote sensing (Sellers *et al.*, 1997). Other models used to estimate productivity at regional (Haxeltine and Prentice, 1996; Landsberg and Waring, 1997) and global (Potter *et al.*, 1993) scales require estimates of light-use efficiency. There is some evidence that light-use efficiency at the landscape scale is conservative within plant functional types (Gower *et al.*, 1999), or more widely when expressed in terms of gross productivity (Goetz and Prince, 1999). However, the application of models at smaller scales to explain differences in productivity between sites (Landsberg *et al.*, 2001) depends on knowing a value for light-use efficiency and the constraints imposed on it by limiting environmental and site variables. Williams *et al.* (1997) and Whitehead *et al.* (2002) have demonstrated success in using fine-scale models based on detailed measurements within canopies to derive aggregated parameters for use in models at a coarser scale over much longer time periods.

Models using complex spatial arrangement of foliage in discontinuous canopies to integrate processes of radiative transfer (Norman and Jarvis, 1974; Cescatti, 1997) and photosynthesis (Wang and Jarvis, 1990) for different cohorts in respect of age and position provide a detailed description of carbon uptake for canopies. However, such models are complex and require values for a large number of parameters and driving variables and considerable knowledge of their distributions within canopies (Sellers *et al.*, 1992; Wang *et al.*, 1992). An alternative treatment is the 'big-leaf' approach, where the canopy is considered as one large entity. Simpler and less demanding in terms of parameters, this approach has been used successfully to model net carbon exchange in canopies (Amthor, 1994; Lloyd *et al.*, 1995). However, difficulties in dealing with non-linear responses of stomatal conductance and photosynthesis to environmental variables and the inability to measure parameters directly can lead to problems with scaling (Raupach and Finnigan, 1988), resulting in overestimation of the rates of photosynthesis (Leuning *et al.*, 1995; de Pury and Farquhar, 1997).

Baldocchi and Meyers (1998) argued that the employment of multilayered models overcomes the limitations of the 'big-leaf' approach. The canopy is divided into a number of layers, with values for parameters, obtained directly from measurements, assigned to each level. Incorporation of coupling between stomatal conductance and photosynthesis at the leaf level (Wong *et al.*, 1979; Collatz *et al.*, 1991; Dang *et al.*, 1998) with calculations of radiative transfer and energy balance for sun and shaded foliage (de Pury and Farquhar, 1997) allows estimates of transpiration and photosynthesis to be made for each layer. These are then integrated to give canopy values. Such an approach embraces correct scaling procedures (Baldocchi and Harley, 1995; Leuning *et al.*, 1995; Williams *et al.*, 1996).

For annual estimates of carbon exchange, it is necessary to include seasonal changes both in foliage development (Law *et al.*, 2000; Wilson *et al.*, 2001) and in values for the parameters describing photosynthesis associated with leaf age (Hollinger *et al.*, 1999; Wilson *et al.*, 2001) and position in the canopy (Carswell *et al.*, 2000; Wilson *et al.*, 2000). It is also necessary to consider the effects of 'clumping' of foliage on radiative transfer processes in deciduous (Baldocchi and Meyers, 1998) and open coniferous (Law *et al.*, 2001) stands.

Multilayered models synthesize current knowledge of biophysical processes in canopies (Leuning *et al.*, 1995) and allow interpretation of the processes regulating photosynthesis and of the interactions between physiology and microclimate along non-linear gradients of biological and environmental conditions in canopies (Williams *et al.*, 1996). This can be useful for interpreting the contributions from different cohorts of foliage to total fluxes and how these change both diurnally and seasonally.

Models using the big-leaf approach to simulate net carbon uptake for canopies (Amthor, 1994; Lloyd *et al.*, 1995; Waring *et al.*, 1995) have been tested by comparing outputs with independent measurements, using eddy covariance. More recently, multilayered models have been tested more rigorously against measurements of flux over longer time periods (Dang *et al.*, 1998; Law *et al.*, 2000; Wilson *et al.*, 2001). Such comparisons are likely to continue with the proliferation of sites where there are continuous measurements of fluxes of CO_2 and water vapour (Running *et al.*, 1999; Valentini *et al.*, 2000). Our own experience in validating the model described in this chapter showed very good agreement between modelled seasonal estimates of intercellular CO_2 partial pressure integrated through the canopy of a young *Pinus radiata* D. Don canopy and independent measurements made across annual rings of wood, using carbon-isotope techniques (Walcroft *et al.*, 1997, 2002).

In this chapter we apply a multilayered modelling approach to resolving the processes regulating net carbon uptake in forest canopies. We selected two canopies with contrasting phenology of foliage production and loss, spatial distribution of foliage and leaf-scale physiological properties as case-studies for comparison: an evergreen, coniferous canopy consisting of slow-growing *Dacrydium cupressinum* Sol. ex Lamb. (rimu) trees growing with low nutrient availability in New Zealand is contrasted with a canopy in a mixed *Quercus* (oak) forest with high nutrient availability in New York, USA. Collaborative research at these sites by the authors is designed to identify the interactions of site, environmental and physiological variables on the mechanisms regulating carbon uptake in the contrasting ecosystems. The two sites are at similar latitudes with similar annual incident radiation and the effects of seasonal root-zone water deficits were minimal during the years selected for this study. We focus our analysis on the response of net carbon uptake by the canopies to changes in incident radiation and diffuse fraction in relation to calculating values of light-use efficiency.

Description of the Model

A one-dimensional, multilayer model of radiative transfer, energy balance, evaporation and photosynthesis (Leuning *et al.*, 1995) was combined with a simple water-balance model (Whitehead *et al.*, 2001) to incorporate the effects of root-zone water deficit on canopy carbon uptake and transpiration. Leaf-area index, L, and a clumping factor, Ω (Baldocchi and Myers, 1998), to represent the spatial distribution of foliage, are required and the canopy is divided into horizontally homogeneous layers, each with an assigned value of foliage-area density. A radiation-transfer model, incorporating direct and diffuse components of radiation in the photosynthetically active, near-infrared and thermal wavebands, is used to calculate the energy balance for sunlit and shaded foliage in each layer. A coupled photosynthesis–stomatal conductance model is used to calculate photosynthesis for sunlit and shaded foliage in each layer and these values are summed to give totals for the canopy.

The coupled photosynthesis–stomatal conductance model for individual leaves is taken from Leuning (1995), where the rate of photosynthesis, A, following Farquhar *et al.* (1980) and von Caemmerer and Farquhar (1981), is given as:

$$A = \min\{A_c, A_q\} - R_d \tag{1}$$

where A_c is the rate of photosynthesis limited by the ribulose 1,5-bisphosphate (RuBP) carboxylation activity of the enzyme ribulose 1,5-bisphosphate carboxylase-oxygenase (rubisco), A_q is the rate limited by RuBP regeneration by the electron-transport system and min{} refers to the minimum of the two rates. R_d is the rate of respiration in the daytime resulting from processes other than photorespiration.

The rate of photosynthesis limited by rubisco activity, A_c, is described by:

$$A_c = V_{cmax} \frac{c_i - \Gamma^*}{c_i + K_c[1 + (o_i / K_o)]} \tag{2}$$

where V_{cmax} is the maximum rate of carboxylation activity by rubisco when the substrates RuBP and CO_2 are saturating, c_i and o_i are intercellular partial pressures of CO_2 and O_2, respectively, Γ^* is the CO_2 partial pressure in the absence of day respiration and K_c and K_o are the Michaelis constants for CO_2 and O_2, respectively. The rate of photosynthesis limited by the regeneration of RuBP, A_q, is described by:

$$A_q = \frac{J(c_i - \Gamma^*)}{4(c_i + 2\Gamma^*)} \tag{3}$$

where J is the electron-transport rate at a given irradiance, Q, and the response of J to Q is described, using a non-rectangular hyperbola (Farquhar and Wong, 1984), as:

$$\beta J^2 - (\alpha Q + J_{max})J + \alpha Q J_{max} = 0 \tag{4}$$

where J_{max} is the maximum rate of electron transport at saturating irradiance, β defines the convexity of the hyperbola and α is the quantum yield of electron transport. Values for V_{cmax}, J_{max}, Γ^* and R_d are all dependent on leaf temperature. The forms of functions used to determine the responses to temperature were taken from Leuning (1995), using values for parameters given by Walcroft *et al.* (1997).

The intercellular partial pressure of CO_2, c_i, is calculated from the partial pressure at the leaf surface, c_s, and the stomatal conductance to CO_2 transfer, g_{sc}, where:

$$c_i = c_s - (A/g_{sc}) \tag{5}$$

The coupling of photosynthesis and stomatal conductance and the relationship with c_s and air-saturation deficit at the leaf surface, D_s, is modified from Leuning (1995) as:

$$g_{sc} = g_{sc0} + \frac{\tau a A}{(c_s - \Gamma)(1 + [D_s - D_{smin}] / D_{s0})} \tag{6}$$

where g_{sc0} is the residual conductance at the light compensation point, D_{s0} describes the sensitivity of g_{sc} to D_s, D_{smin} is the minimum value, Γ is the CO_2 partial pressure at compensation in the presence of photorespiration of air-saturation deficit below which g_{sc} remains at its maximum value and a is related to the intercellular CO_2 partial pressure. The coefficient τ decreases photosynthesis when daily root-zone water storage falls below a minimum value (see below).

Using a simple daily water-balance model (Whitehead *et al.*, 2001), root-zone water storage on the first day, W_i, is calculated from:

$$W_i = W_{i-1} + P_i - E_{ti} - E_{wi} - E_{ui} - F_i \tag{7}$$

where P_i is the daily rainfall, E_{ti} is transpiration from the dry tree canopy, E_{wi} is evaporation from the wet canopy, E_{ui} is evaporation from the understorey and soil and F_i is drainage from the root zone. E_{wi} is assumed to be a constant fraction of rainfall ($= 0.2P_i$) (Kelliher *et al.*, 1992).

Transpiration from individual leaves, E_t, in each layer in the canopy is calculated using the Penman–Monteith equation following Leuning *et al.* (1995).

This requires calculation of the net radiation absorbed by leaves and boundary-layer conductances for water vapour and heat transfer. Net radiation is calculated from leaf energy-balance considerations and the conductances are calculated from leaf width, *d*, and wind speed (assumed to decrease exponentially with depth in the canopy).

Evaporation from the understorey and soil, E_u, is calculated from the available energy-flux density reaching the forest floor, G_u, as:

$$E_u = \tau\omega s G_u / [\lambda(s + \gamma)] \quad (8)$$

where the coefficient ω (= 1.4) enhances evaporation due to coupling between the understorey and the air above the forest canopy (Kelliher *et al.*, 1990), *s* is the slope of the relationship between saturated vapour pressure and temperature, λ is the latent heat of vaporization of water and γ is the psychrometric constant. Using Beer's law, G_u is calculated from ($e^{-kL'}G_a$) where G_a is the available energy-flux density above the canopy (assumed to be equal to 0.7 × short-wave radiation), L' is the the leaf-area index on a projected basis and *k* is the canopy light extinction (assumed to be 0.5 for a spherical leaf-angle distribution).

If W_{max} and W_{min} are the maximum and minimum root-zone water storage, the value for τ in Equations (6) and (8) is set to 1 while $W = W_{max}$, and decreases linearly to zero when $W = W_{min}$ (McMurtrie *et al.*, 1990; Wang and Leuning, 1998). Drainage from the root zone occurs after rainfall when $W_i > W_{max}$.

The model requires daily values of minimum and maximum air temperature, short-wave radiation and rainfall. Hourly values of temperature, radiation and air-saturation deficit, *D*, are calculated following the procedures described by Goudriaan and van Laar (1994). Canopy fluxes are calculated five times each day from hourly data using weighted values according to a Gaussian distribution, following Leuning *et al.* (1995).

The model requires values for *L*, 11 parameters associated with foliage properties, V_{cmax}, J_{max}, R_d, α, β, a, g_{s0}, D_{s0}, D_{smin}, *d* and Ω, and two parameters describing root-zone water storage, W_{max} and W_{min}. For the analysis used in this chapter, the canopies were divided into 20 layers with values of foliage-area density and weighting factors for physiological parameters assigned to each layer.

Site Descriptions

Okarito Forest

The site is a temperate mixed conifer–broad-leaved forest situated on extensive lowland terraces at Okarito Forest, South Westland, New Zealand (lat. 43.2°S, long. 170.3°E, elevation above sea level 50 m). The study area is dominated by 100- to 400-year-old *Dacrydium* trees, with an average tree number of 560 stems/ha and a basal area of 32 m^2/ha (68% of the total basal area at the study site (I.L. James, 2000; personal communication), with a number of other conifers and understorey broad-leaved species present (James and Norton, 2001). Canopy height ranges from 20 to 25 m, with a canopy depth of approximately 10 m. The forest comprises distinct even-aged cohorts of trees, resulting from infrequent, but catastrophic disturbance from earthquakes and smaller-scale disturbances from windfalls.

Annual rainfall at the site is high (approximately 3400 mm) and evenly distributed throughout the year, attributable to prevailing, onshore westerly winds and the rapid increase in altitude associated with the southern Alps east of the forest. Mean annual temperature is 11.3°C, with a small range between winter and summer of 8.6°C. Frosts rarely occur at the site.

The landform is glacial in origin and the forest is established on terrace outwashes from moraines, formed approximately 20,000 years ago. The soil taxonomy is described as Entisols that have evolved to Inceptisols or Spodosols (Soil Survey Staff, 1994). The loess is poorly preserved because of erosion and acid dissolution from extreme leaching resulting from high rainfall (Almond, 1996). The acid humic organic soils are characterized by a layer more than 400 mm deep of very fluid, highly decomposed peat over a very fluid peaty silt loam, below which is a massive blue-grey silt loam formed in loess. The soils have a high organic-matter content (approximately 30%; J.A. Adams and D.A. Norton, personal communication), very low permeability and low porosity and are frequently waterlogged. Maximum available water ($W_{max} - W_{min}$) was estimated to be low (43 mm, Table 15.1) because of the shallow rooting-zone depth (Jackson, 1984). The soils are extremely acid (pH 3.8–4.4 to a depth of 500 mm) with low availability of nitrogen and phosphorus (Mew and Lee, 1981). The input of nitrogen from the atmosphere is low because of the remote location distant from anthropogenic sources of pollutants. The average wet deposition rate is likely to be about 0.7 kg N/ha/year (Nichol *et al.*, 1997). A more detailed description of the site can be found in Whitehead *et al.* (2002).

Black Rock Forest

This site comprises an area at low elevation in the Cascade Brook catchment, set within a scientific reserve established in 1927 in the Highlands Province of New York, USA (lat. 41.4°N, long. 74.0°E, elevation above sea level 270 m). The study area is representative of temperate mixed deciduous forest, dominated by *Quercus* species, distributed widely across eastern North America. Average tree number at the site is 650 stems/ha, dominated by red oak (*Quercus rubra* L., 41% of the total basal area; W.S.F. Schuster, 2001, personal communication), red maple (*Acer rubrum* L., 32%) and chestnut oak (*Quercus prinus* L. syn. *Quercus montana* Willd., 6%). Canopy height ranged from 20 to 30 m, with a depth of approximately 12 m.

Mean annual rainfall at the site is 1190 mm and air temperature is strongly seasonal, with a sharp contrast between summer (mean July temperature 23.4°C) and winter (mean January temperature −2.7°C). The soils are shallow, poorly drained and classified as brown forest type of medium texture (Charlton/Paxton soil) to a depth of 1 m, underlain with granite-gneiss bedrock or glacial-till material (Olsson, 1981). Maximum available water was estimated to be 75 mm (Table 15.1; V. Engel, unpublished data). The annual input of nitrogen to the site from the atmosphere is high, with an average wet deposition rate of 4.7 kg N/ha/year; National Atmospheric Deposition Programme (NADP), 2001, personal communication). Further details of the site are given by Turnbull *et al.* (2001).

Table 15.1. Values for parameters used in the model for the *Dacrydium* and *Quercus* canopies. All values are expressed on a half-surface area basis. Descriptions of the symbols are given in Equations (1)–(6) and (9) in the text. The values for L are maximum seasonal values and values for V_{cmax} are from measurements at a base temperature of 20°C for foliage at the top of the canopy.

Symbol	Units	*Dacrydium*	*Quercus*	Source
L	m^2/m^2	5.4	5.8	K.J. Brown (unpublished data)
Ω		0.83	0.84	K.J. Brown (unpublished data), Baldocchi and Meyers (1998)
d	m	0.001	0.1	
V_{cmax}	$\mu mol/m^2/s$	11.9	50.2	D.T. Tissue (unpublished data), Turnbull *et al.* (2002)
J_{max}	$\mu mol/m^2/s$	$2.7 \times V_{cmax}$	$2.4 \times V_{cmax}$	D.T. Tissue (unpublished data), Turnbull *et al.* (2002)
R_d	$\mu mol/m^2/s$	$0.06 \times V_{cmax}$	$0.02 \times V_{cmax}$	K.L. Griffin (unpublished data)
α	mol E/mol quanta	0.22	0.22	D.T. Tissue (unpublished data), Turnbull *et al.* (2002)
β		0.66	0.66	D.T. Tissue (unpublished data), Turnbull *et al.* (2002)
a		4	4	Whitehead *et al.* (1996)
g_{s0}	$mmol/m^2/s$	10	10	Whitehead *et al.* (1996)
D_{s0}	Pa	885	1001	D.T. Tissue and V. Engel (unpublished data)
D_{smin}	Pa	498	480	D.T. Tissue and V. Engel (unpublished data)
W_{max}	mm	57	125	Whitehead *et al.* (2002), V. Engel (unpublished data)
W_{min}	mm	14	50	Whitehead *et al.* (2002), V. Engel (unpublished data)

Foliage Area and Foliage Distribution

For the purposes of comparison between *Dacrydium* and *Quercus*, values of L and parameters in the photosynthesis–stomatal-conductance model are expressed on the basis of half the total surface area (Chen and Black, 1992). For *Quercus*, this is equivalent to a one-sided basis. For the complex shape of the foliage in *Dacrydium*, to convert from half-surface area to projected area, values should be multiplied by 0.54 (Whitehead *et al.*, 2002).

Measurements of leaf-area index, L, for the sites were made using a plant-canopy analyser (Model LAI-2000, Li-Cor Inc., Lincoln, Nebraska, USA). Values of L in midsummer were close for the two sites (Table 15.1). The value for Ω for the *Quercus* canopy was taken from Baldocchi and Meyers (1998) and estimated from measurements of shoot structure and analysis of gap frequency for the *Dacrydium* canopy. The degree of clumping in both canopies was very similar (Table 15.1). Measurements of phenology of foliage growth and loss showed that bud burst in *Quercus* started in May (spring), with maximum foliage area occurring from early June to late September (summer) and leaf fall being complete by the end of October (autumn). Measurements in the *Dacrydium* canopy showed that new foliage growth started in November (spring) but the contribution of new foliage was estimated to be less than 10% of existing foliage area.

In the absence of measurements, the vertical distribution of foliage-area density for the *Quercus* canopy was assumed to be the same as that for mixed *Quercus* forest with a similar leaf-area index (Hutchison *et al.*, 1986). For the *Dacrydium* canopy, the vertical distribution of foliage-area density was inferred from analysing photographic images of tree crowns. This resulted in markedly different distributions of foliage area (Fig. 15.1). For the *Dacrydium* canopy, foliage-area density was low in the top layers, increasing gradually with depth to reach a maximum in the mid-canopy. In cumulative terms, the upper 11 layers of the canopy contained 54% of the total foliage area and 90% of the foliage area occurred in the upper 16 layers. In contrast, foliage in the *Quercus* canopy was much more concentrated in the upper layers, with 52% of the foliage area in the upper five layers and 90% in the upper 13 layers.

Estimation of Parameters at the Leaf Scale for the Two Sites

Values for the parameters associated with photosynthesis were calculated from measurements of the response of photosynthesis to intercellular CO_2 partial pressure and irradiance for leaves (*Quercus*) or shoots (*Dacrydium*) in the top, middle and lowest canopy layers and linear decreases with depth were assumed. For the *Dacrydium* canopy, it was assumed that there were no seasonal changes in values for the photosynthetic parameters. For the *Quercus* canopy, V_{cmax} was assumed to be low when the leaves emerged in spring, increasing to reach a maximum value in early summer and then declining gradually during late summer and autumn, following Wilson *et al.* (2001). The value of V_{cmax} at the top of the *Dacrydium* canopy was much lower than that

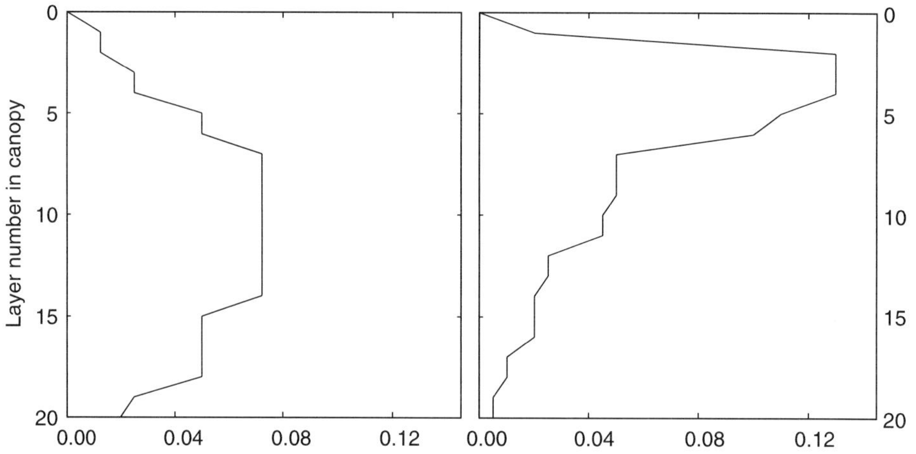

Fig. 15.1. The distribution of foliage-area density within each of the 20 layers in the canopies for *Dacrydium* and *Quercus*. Foliage-area density is normalized in respect of leaf-area index, shown in Table 15.1, and layer numbers are equidistant intervals through the canopies, with layer 1 at the top.

for *Quercus*, consistent with the general difference between temperate broad-leaved trees and conifers (Wullschleger, 1993). Analysis of the response of photosynthesis to irradiance showed that the quantum yield and the convexity of the curve were the same for both species. Measurements showed that stomatal conductance at low values of air-saturation deficit was much higher for *Quercus* than for *Dacrydium* and the sensitivity of the decrease in conductance with increasing saturation deficit, D_{s0}, was greater for *Quercus* than for *Dacrydium* (Table 15.1).

Canopy Annual Net Carbon Uptake

The climate at the two sites during the periods when the trees were growing was very similar. Annual daily minimum and maximum air temperatures at the *Dacrydium* site were −1.9 and 25.9°C, respectively, and the equivalent temperatures for the *Quercus* site during the period when leaves were present on the trees were −2.7 and 33.1°C. The average air-saturation deficit for these same periods was 598 and 655 Pa for the *Dacrydium* and *Quercus* sites, respectively.

Annual incident radiation for the two forests was similar, being only 10% higher for the *Dacrydium* site than for the *Quercus* site (Table 15.2). However, the incident radiation during the period when leaves were present at the *Quercus* site was much less, at 70% of the annual total. This resulted in 53% more radiation being intercepted by the *Dacrydium* canopy than by the *Quercus* canopy when leaves were present.

Table 15.2. Annual estimates of intercepted radiation, diffuse fraction of radiation in the summer months, carbon uptake and transpiration for the *Dacrydium* and *Quercus* canopies.

Quantity	Units	*Dacrydium*	*Quercus*
Incident radiation (400–700 nm)	MJ/m^2	2493	2261 1578 (when leaves present)
Average summer diffuse fraction, φ		0.54	0.62
Intercepted radiation (400–700 nm), Q_i	MJ/m^2	1942	1267
Rainfall	mm	3210	1108
Net carbon uptake, A_{can}	kg C/m^2	1.07	1.22
Tree transpiration	mm	217	293
Wet canopy and understorey evaporation	mm	875	590
Light-use efficiency A_{can}/Q_i	g C/MJ	0.55	0.96

Rates of daily carbon uptake were much lower for the *Dacrydium* canopy than for the *Quercus* canopy, but carbon uptake continued throughout the year for *Dacrydium*. Daily maximum carbon uptake reached 0.47 mol/m^2 ground area/day in summer for the *Dacrydium* canopy, but rates for *Quercus* were more than double this, reaching a maximum value of 1.04 $mol/m^2/day$ (Fig. 15.2). This difference was also reflected in canopy conductance, which was reasonably constant throughout the year for *Dacrydium*, with a maximum average daily value of 0.11 $mol/m^2/s$, and much higher for *Quercus*, reaching a maximum average daily value of 0.46 $mol/m^2/s$ during conditions of low air-saturation deficit in spring (Fig. 15.2).

Annual rainfall was much higher at the *Dacrydium* site than at the *Quercus* site during the growing season (Table 15.2). For *Dacrydium*, there were two periods in summer when intercellular CO_2 partial pressure for the canopy decreased in response to root-zone water deficit as root-zone water content fell to 16% of its maximum value (Fig. 15.2). Due to the larger root-zone water storage at the *Quercus* site, root-zone water content fell to a minimum of 25% of the maximum storage. This did not lead to marked decreases in intercellular CO_2 partial pressure during the summer. Generally, values of intercellular CO_2 partial pressure were lower for the *Quercus* canopy than for the *Dacrydium* canopy.

In cumulative terms, the rate of radiation interception was slightly greater for *Quercus* than for *Dacrydium* during the spring and early summer. However, the rate of cumulative carbon uptake was much higher for the *Quercus* canopy throughout spring and summer (Fig. 15.3). This resulted in a larger annual carbon uptake for *Quercus* and a much higher value for light-use efficiency compared with the values for *Dacrydium* (Table 15.2).

Distribution of Carbon Uptake within the Canopies

Analysis of radiation interception and carbon uptake with depth in the canopy illustrates the effects of the differences in the distribution of foliage-area density and the values for the photosynthetic parameters on canopy car-

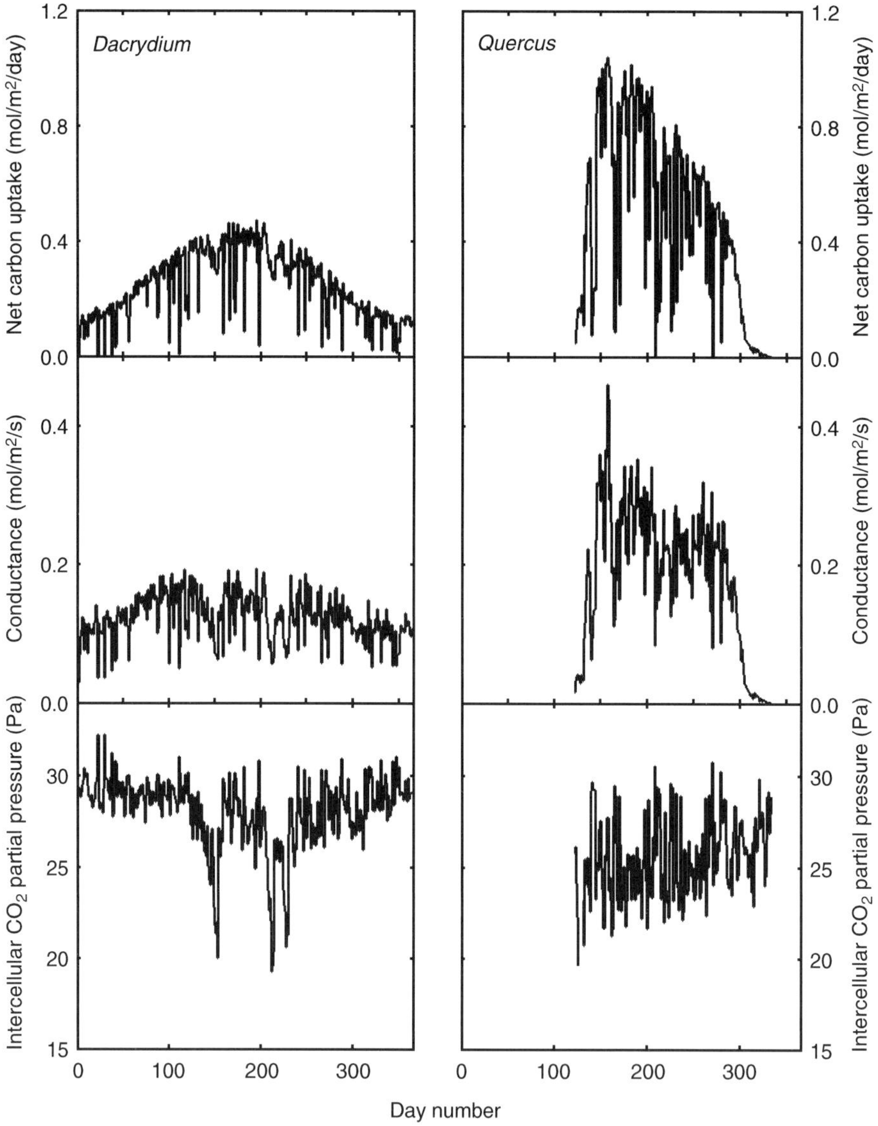

Fig. 15.2. Daily net carbon uptake, average canopy conductance to water vapour and intercellular CO_2 partial pressure during 1 year for the *Dacrydium* and *Quercus* canopies.

bon uptake. Of the total radiation intercepted during the year, 0.25 was intercepted after the first 3 months, 0.65 after 6 months and 0.85 after 9 months for *Dacrydium*. For *Quercus*, only 0.3 was intercepted during the first 3 months as the foliage developed, but the proportion had increased to 0.59 after 6 months and 0.89 after 9 months (Fig. 15.4). Only a small proportion of radiation (0.26) was intercepted in the upper five layers of the *Dacrydium* canopy. In contrast, a very large proportion (0.81) of the radiation was intercepted in these layers in the *Quercus* canopy. Interception of 0.9 of the radia-

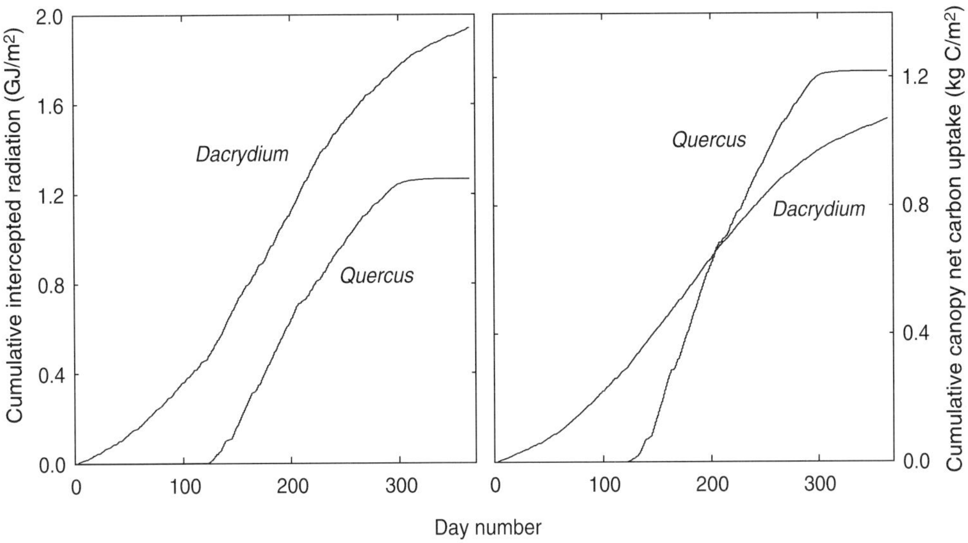

Fig. 15.3. Cumulative daily intercepted radiation and net carbon uptake for 1 year for the *Dacrydium* and *Quercus* canopies. The day numbers start in winter from 1 July for *Dacrydium* and 1 January for *Quercus*.

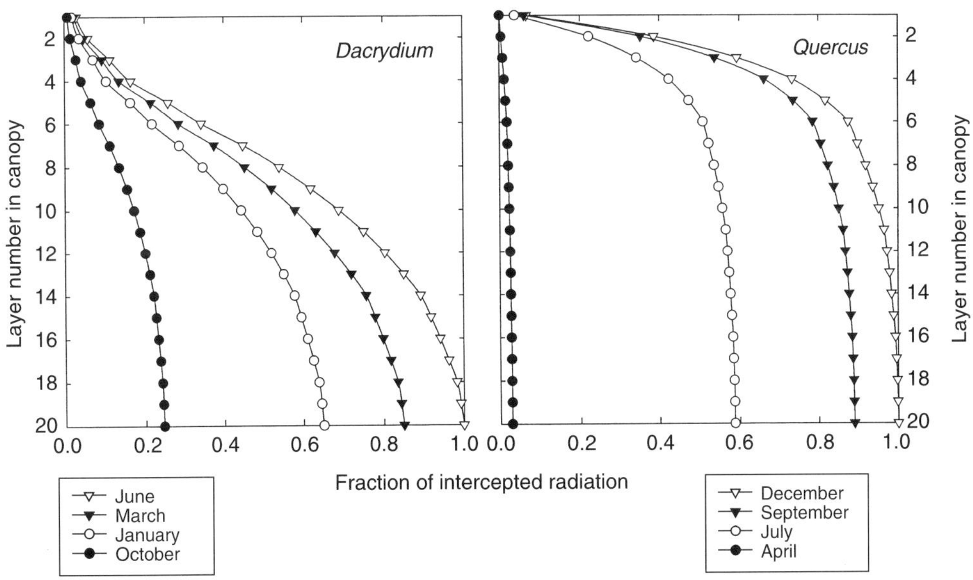

Fig. 15.4. Cumulative proportional interceptance of radiation for each layer in the *Dacrydium* and *Quercus* canopies. The different symbols refer to cumulative values at 3-monthly intervals throughout the year. Note that the months refer to the southern hemisphere for *Dacrydium* and the northern hemisphere for *Quercus*.

tion was not reached until layer 14 in the *Dacrydium* canopy but this occurred by layer 5 in the *Quercus* canopy. This leads to contrasting proportional contributions to carbon uptake by layers for the two canopies. During the 2 months in midsummer, the largest contribution to carbon uptake occurred in layers 5–10 (50%) in the *Dacrydium* canopy and in layers 1–5 (97%) in the *Quercus* canopy (Fig. 15.5).

From Equation (1), the rate of photosynthesis for a leaf is given by the minimum of the rate limited by carboxylation activity, A_c, and the rate limited by electron transport, A_q. Analysis of the rate-limiting process for each layer in the canopy reveals contrasting properties for the two canopies. For typical conditions at midday in summer, the proportion of sunlit foliage decreased with depth in the canopy more rapidly in the *Quercus* canopy than for the *Dacrydium* canopy (Fig. 15.6). Leaf temperature was close to air temperature for foliage in all layers in the *Dacrydium* canopy, but increased with depth in the *Quercus* canopy due to energy-balance considerations associated with the larger leaves. Sunlit foliage in all layers in both canopies was limited by carboxylation activity, and rates of photosynthesis were lower for *Dacrydium* than for *Quercus* because of the difference in V_{cmax}. Shaded foliage in the upper seven layers of the *Dacrydium* canopy was limited by carboxylation activity, but in the lower layers the rate of electron transport was limiting. In contrast, shaded foliage in the *Quercus* canopy was strongly limited by the rate of electron transport throughout the canopy, especially in the layers below the upper five.

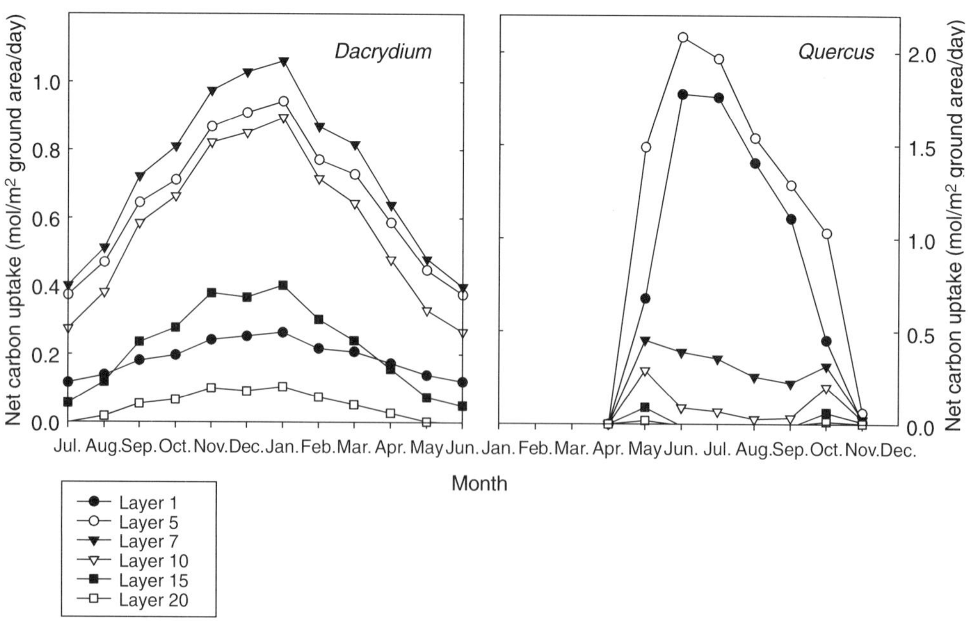

Fig. 15.5. Seasonal net carbon uptake per unit ground-area for different layers in the *Dacrydium* and *Quercus* canopies. Note that the months refer to the southern hemisphere for *Dacrydium* and the northern hemisphere for *Quercus*.

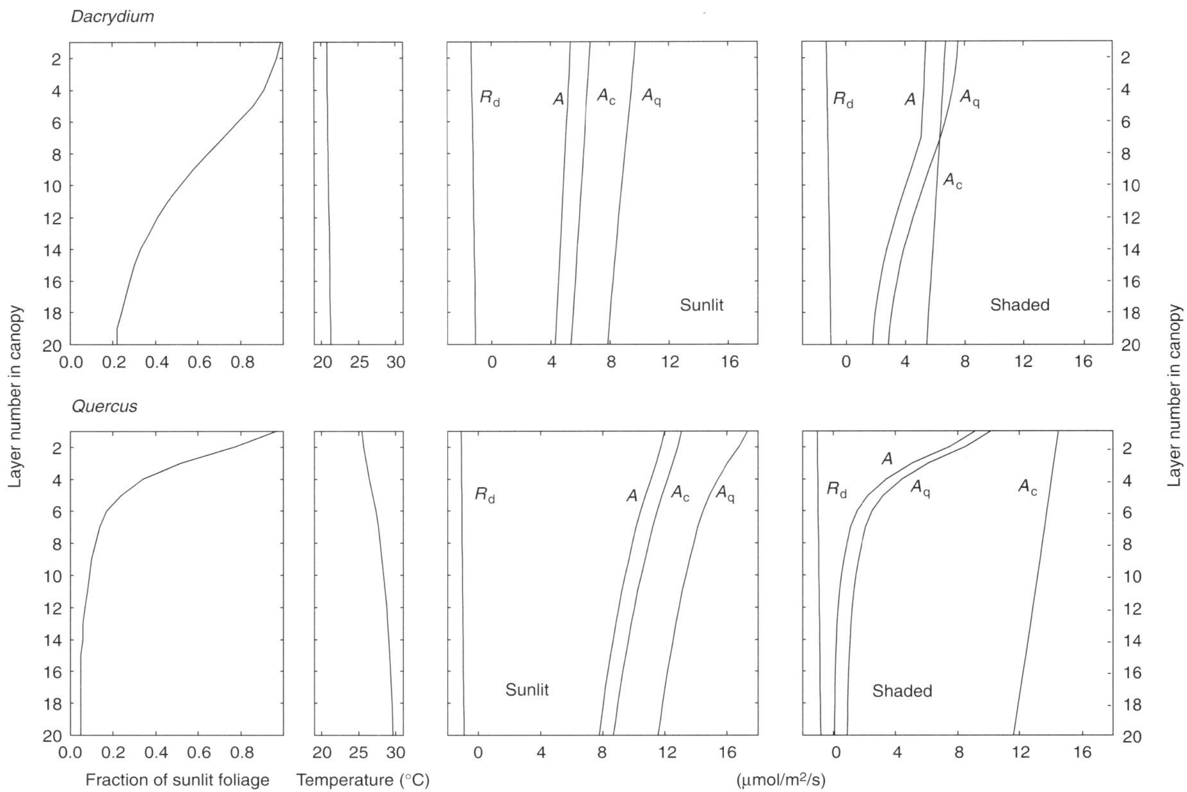

Fig. 15.6. Vertical profiles of the proportion of sunlit foliage, temperature of sunlit foliage and the components of photosynthesis through the *Dacrydium* and *Quercus* canopies for typical conditions on a bright, clear day at midday in midsummer with incident irradiance (400–700 nm) 500 W/m^2, diffuse fraction 0.2, solar elevation angle 70°, air temperature 20°C, air-saturation deficit 500 Pa and a wind speed of 2.5 m/s. The components of photosynthesis are shown on a unit foliage-area basis for sunlit and shaded foliage and are defined in Equations (1)–(3), where the rate of photosynthesis, A, is given as the minimum of the rate limited by carboxylation, A_c, and the rate limited by electron transport, A_q, minus the rate of daytime respiration, R_d.

Response to Irradiance and Diffuse Fraction

Since shaded foliage in the upper layers of the *Quercus* canopy was strongly limited by the rate of electron transport at high irradiance, increases in the diffuse fraction of incident radiation resulted in increased penetration of radiation to lower depths, enhancing carbon uptake (Fig. 15.7). For the *Dacrydium* canopy an increase in the diffuse fraction of radiation did not result in increased carbon uptake in shaded foliage in the top layers because photosynthesis in these layers was limited by carboxylation activity, except on very clear days. Combining the effects of irradiance and the diffuse fraction of radiation on carbon uptake at the canopy scale reveals differences in the sensitivity of the canopies to the light environment (Fig. 15.8). In overcast conditions, with low irradiance, the proportional increase in carbon uptake with increasing diffuse fraction was small for both canopies. However, the response for the *Dacrydium* canopy was slightly greater than that for the *Quercus* canopy as more shaded leaves in the lower layers of the canopy became limited by carboxylation activity (Fig. 15.7). However, at high irradiance, carbon uptake in the *Quercus* canopy was very sensitive to increasing diffuse fraction as rates of electron transport increased in the top layers and at increasing depths in the canopy. The response in the *Dacrydium* canopy to increasing diffuse fraction at high irradiance was much less than that for the *Quercus* canopy because shaded foliage in the upper layers remained limited by carboxylation activity.

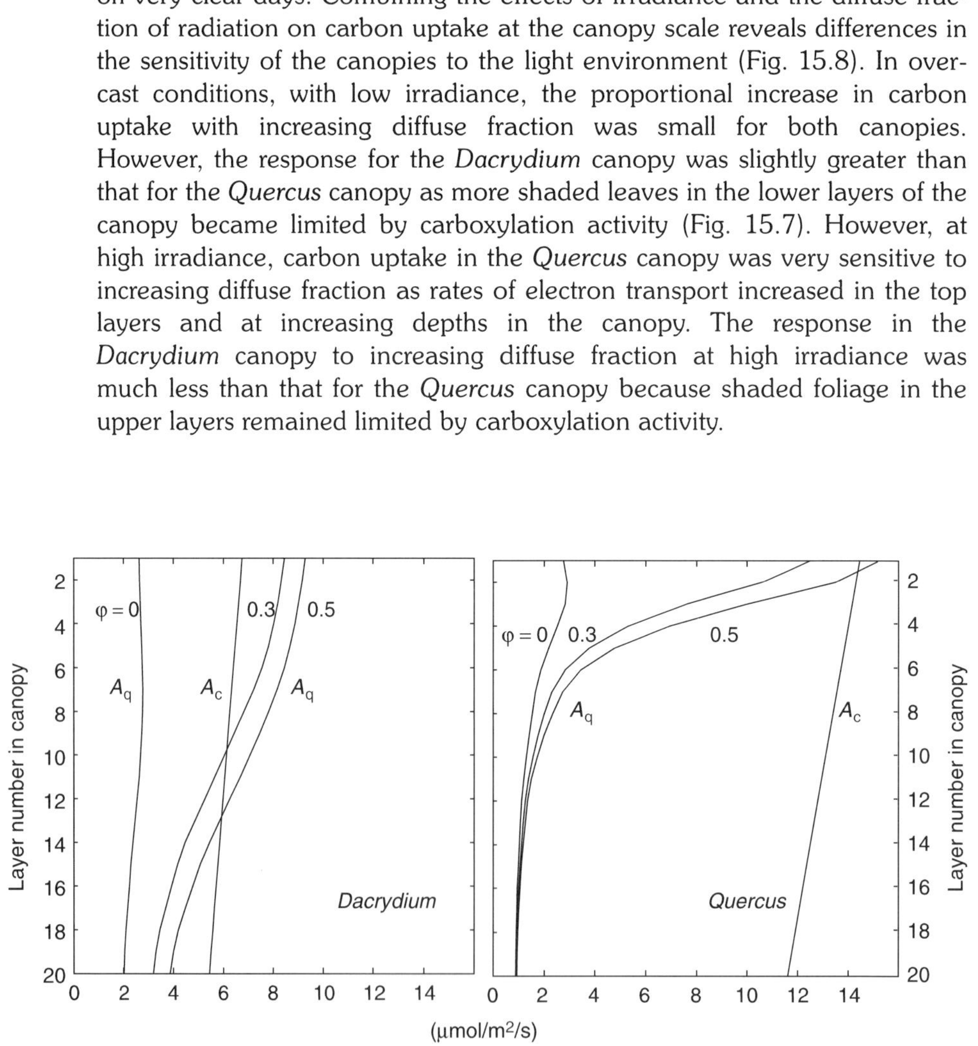

Fig. 15.7. Vertical profiles of the components of photosynthesis A_c and A_q for shaded foliage through the *Dacrydium* and *Quercus* canopies for the same typical conditions at midday in midsummer as those shown in Fig. 15.6, except that the diffuse fraction of radiation, φ, is set at 0, 0.3 and 0.5. The profiles of A_c remain the same as φ is increased.

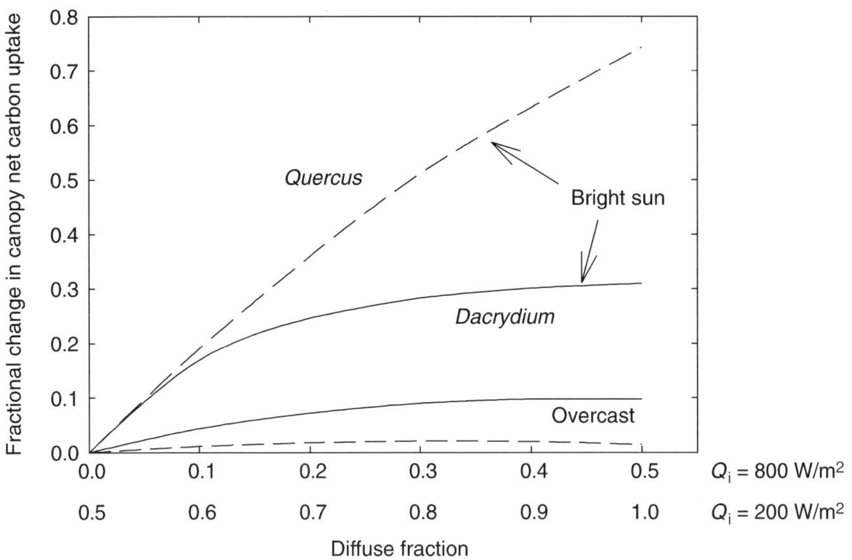

Fig. 15.8. Proportional change in net carbon uptake for the *Dacrydium* (solid line) and *Quercus* (dashed line) canopies with increasing diffuse fraction of radiation at midday for a bright sunny day and an overcast day. Apart from the irradiance and diffuse fraction of radiation, the other conditions are the same as those in Fig. 15.6. For the sunny (incident irradiance (400–700 nm), Q_i = 800 W/m^2) and overcast (Q_i = 200 W/m^2) days, the diffuse fraction is increased from 0 to 0.5 and 0.5 to 1.0, respectively.

Implications for Carbon Uptake

In summer, rates of photosynthesis for the *Quercus* canopy were much higher than those for the *Dacrydium* canopy. Despite the much longer period for intercepting radiation in the *Dacrydium* canopy, annual carbon uptake was 14% greater in the *Quercus* canopy. Light-use efficiency for *Quercus* was 75% greater than that for *Dacrydium*, emphasizing the much higher interceptance in the *Quercus* canopy. Wilson *et al.* (2001) emphasized the sensitivity of modelled annual carbon uptake in *Quercus* to seasonal changes in photosynthetic parameters. Similarly, carbon uptake in *Quercus* is sensitive to the timing of foliage production and loss, where delays of several days at the beginning or end of the growing season can result in marked changes in annual carbon uptake.

Our analysis of the vertical distribution of the processes limiting photosynthesis for the *Quercus* canopy confirms the interpretation described for a similar canopy by Baldocchi and Harley (1995), while no other data are available to support our results for the *Dacrydium* canopy. A consequence of the vertical distribution of sunlit and shaded foliage area and the associated photosynthetic parameters is that the response of photosynthesis to irradiance is non-linear for sunlit foliage but linear for shaded foliage. Values for V_{cmax} in the *Dacrydium* canopy were sufficiently low for photosynthesis to be limited by carboxylation activity, rather than electron transport, in shaded foliage in the upper canopy layers (Fig. 15.6). The inability of shaded foliage in the upper layers of the *Dacrydium* canopy to benefit from increased irradiance explains the weak response of canopy carbon uptake to increased irradiance in the *Dacrydium* canopy (Fig. 15.8). In their early analysis, von Caemmerer and Farquhar (1981) suggested that, in crop plants, photosynthesis in most leaves is likely to be limited by the rate of electron transport, so canopy photosynthesis was limited by the values for J_{max} (and thus α) and R_d and regulated by irradiance, temperature and CO_2 partial pressure. Theoretical predictions (Leuning *et al.*, 1995), results for a wheat crop (de Pury and Farquhar, 1997) and our analysis of the behaviour in the *Quercus* canopy (Fig. 15.6) are consistent with photosynthesis being limited by the rate of electron transport. Available data for photosynthetic parameters show a strong correlation between V_{cmax} and J_{max} for a wide range of species (Wullschleger, 1993). The combination of low values for V_{cmax} and low interceptance of radiation (resulting from the low ratio of projected to half-surface area for foliage) in the *Dacrydium* canopy is much more likely to occur in natural ecosystems with low nutrient availability than in systems managed for maximizing productivity. This can be attributed to the low availability of nitrogen at the *Dacrydium* site due to pronounced leaching in the highly organic, acid soils (Mew and Lee, 1981). A strong correlation exists between photosynthetic capacity, or V_{cmax}, and foliage nitrogen concentration because a large proportion of nitrogen in foliage is stored in the carboxylating enzyme rubisco (Field, 1983). There are also many examples showing strong relationships between nitrogen availability and foliage production in trees (Raison and Myers, 1992). Thus, increased nitrogen availability at the *Dacrydium* site, for example by the application of fertilizer, would be expected to increase V_{cmax} and foliage area and thus canopy carbon uptake, resulting from increased photosynthesis in shaded foliage in the upper canopy. In contrast, the increase in canopy photosynthesis in response to the addition of nitrogen fertilizer in the *Quercus* stand would be expected to be much less because of the high fraction of shaded leaves in the upper canopy limited by A_q. It should be borne in mind that this analysis is based on conditions at midday and provides a general, comparative explanation for the behaviour of the two canopies. In reality, the processes limiting photosynthesis in different layers of the canopies are dynamic (de Pury and Farquhar, 1997) and depend on solar elevation, irradiance, temperature and the effect of air-saturation deficit on stomatal conductance. In a more detailed analysis in a tropical rain forest using a 'big-leaf' approach, Lloyd *et al.* (1995) showed that the canopy was limited by the rate of electron transport throughout the day, except for a brief period around midday when photosynthesis was limited by carboxylation efficiency.

It is well established from models of radiative transfer that the penetration of radiation to lower parts of canopies increases with increasing diffuse fraction of incident radiation (Norman and Jarvis, 1974; Wang and Jarvis, 1990) and that this results in enhanced photosynthesis (Wang *et al.*, 1992). It follows that light-use efficiency is higher in cloudy conditions than it is on clear days because of a more even distribution of irradiance across sunlit and shaded foliage through the canopy (Leuning *et al.*, 1995; Gower *et al.*, 1999). This effect was demonstrated in broad-leaved (Hollinger *et al.*, 1994) and conifer (Hollinger *et al.*, 1999) forests from direct measurements, using eddy covariance. In a more detailed analysis of this issue, Roderick *et al.* (2001) demonstrated that an increase in the diffuse fraction is expected to increase net carbon uptake by terrestrial ecosystems at continental and global scales. From our analysis, carbon uptake in the *Quercus* canopy is much more sensitive to increasing diffuse fraction than the *Dacrydium* canopy. The effect is likely to be enhanced at the site in New York because of the observed gradual increase in haziness in northern-hemisphere temperate zones (Hollinger *et al.*, 1994).

While multilayered models are useful for synthesizing the processes operating in canopies and provide detail of the vertical distributions of the contribution to canopy carbon uptake in relation to driving variables, their complexity limits their incorporation into regional or global climate models (Sellers *et al.*, 1992, 1997). While these models are relevant, and need to be applied and tested, in complex, mixed-species canopies, a new generation of models is emerging, which are simpler but retain the rigour of correct scaling procedures. By considering sunlit and shaded fraction of the canopy separately, de Pury and Farquhar (1997) and Wang and Leuning (1998) show that the foliage in each class can be treated as a single entity for the purposes of the calculations, thus reducing the number of parameters required and the computing time. Good agreement was obtained when these models were tested against a multilayer approach and measurements made in relatively simple wheat canopies (Leuning *et al.*, 1998).

Conclusions

This analysis illustrates the response of the processes of photosynthesis in two contrasting canopies with an apparent difference in light-use efficiency. The high proportion of foliage area in the upper layers and high values for V_{cmax} and J_{max} in the *Quercus* canopy resulted in high light-use efficiency and high rates of carbon uptake, which were highly sensitive to irradiance and diffuse fraction. Conversely, these conditions were all low in the *Dacrydium* canopy. More generally, the *Dacrydium* canopy represents sites with very low nitrogen (or other nutrient in some circumstances) availability, while the *Quercus* canopy is representative of sites where the nitrogen availability is high, which would include most ecosystems managed for high productivity. The availability of nitrogen determines the leaf-area index, the vertical distribution of foliage area and the capacity for photosynthesis, which is regulated by the relative magnitude of A_c and A_q, the distribution of irradiance and the effect of air-saturation deficit on stomatal conductance. The consequence of improved availability of nitrogen would be that the behaviour of the *Dacrydium* canopy would be expected to move gradually towards that of the *Quercus* canopy. The analysis could be extended to test this hypothesis for forests growing along a nutritional gradient and would be useful for predicting the consequences of changing climate, light environment or rates of atmospheric nitrogen deposition for ecosystem carbon uptake.

Acknowledgements

We gratefully acknowledge funding for this work from the Foundation for Research, Science and Technology (New Zealand) and the Andrew W. Mellon Foundation (USA). We thank Timberlands West Coast Ltd and the Black Rock Forest Consortium for allowing us access to the field sites. Ray Leuning kindly assisted us with the use of his model and Graeme Rogers, Tony McSeveny, John Byers, Nicole Lauren and Ardis Thompson provided us with excellent technical support at the field sites and with laboratory analyses.

References

Almond, P.C. (1996) Loess, soil stratigraphy and Aokautere ash on late Pleistocene surfaces in south Westland, New Zealand: interpretation and correlation with the glacial stratigraphy. *Quaternary International* 34–36, 163–176.

Amthor, J.S. (1994) Scaling CO_2–photosynthesis relationships from the leaf to the canopy. *Photosynthesis Research* 39, 321–350.

Baldocchi, D. and Harley, P.C. (1995) Scaling carbon dioxide and water vapour exchange from leaf to canopy in a deciduous forest. II. Model testing and application. *Plant, Cell and Environment* 18, 1157–1173.

Baldocchi, D. and Meyers, T. (1998) On using eco-physiological, micrometeorological and biogeochemical theory to evaluate carbon dioxide, water vapour and trace gas fluxes over vegetation: a perspective. *Agricultural and Forest Meteorology* 90, 1–25.

Carswell, F.E., Meir, P., Wandelli, E.V., Bonates, L.C.M., Kruijt, B., Barbosa, E.M., Nobre, A.D., Grace, J. and Jarvis, P.G. (2000) Photosynthetic capacity in a central Amazonian rain forest. *Tree Physiology* 20, 179–186.

Cescatti, A. (1997) Modelling the radiative transfer in discontinuous canopies of asymmetric crowns. I. Model structure and algorithms. *Ecological Modelling* 101, 263–274.

Chen, J.M. and Black, T.A. (1992) Determining leaf area index for non-flat leaves. *Plant, Cell and Environment* 15, 421–429.

Collatz, G.J., Ball, J.T., Grivet, C. and Berry, J.A. (1991) Physiological and environmental regulation of stomatal conductance, photosynthesis and transpiration: a model that includes a laminar boundary layer. *Agricultural and Forest Meteorology* 54, 107–136.

Dang, Q.-L., Margolis, H.A. and Collatz, G.J. (1998) Parameterization and testing of a coupled photosynthesis–stomatal conductance model for boreal trees. *Tree Physiology* 18, 141–153.

de Pury, D.G.G. and Farquhar, G.D. (1997) Simple scaling of photosynthesis from leaves to canopies without the error of big-leaf models. *Plant, Cell and Environment* 20, 537–557.

Farquhar, G.D. and Wong, S.C. (1984) An empirical model of stomatal conductance. *Australian Journal of Plant Physiology* 11, 191–210.

Farquhar, G.D., von Caemmerer, S. and Berry, J.A. (1980) A biochemical model of photosynthetic CO_2 assimilation in leaves of C_3 species. *Planta* 149, 78–90.

Field, C. (1983) Allocating leaf nitrogen for the maximisation of carbon gain: leaf age as a control on the allocation program. *Oecologia* 56, 341–347.

Goetz, S.J. and Prince, S.D. (1999) Modelling terrestrial carbon exchange and storage: evidence and implications of functional convergence in light-use efficiency. *Advances in Ecological Research* 28, 57–91.

Goudriaan, J. and van Laar, H.H. (1994) *Modelling Potential Crop Growth Processes*. Kluwer Academic Publishers, Dordrecht, 238 pp.

Gower, S.T., Kucharik, C.J. and Norman, J.M. (1999) Direct and indirect estimation of leaf area index, f_{APAR}, and net primary production of terrestrial ecosystems. *Remote Sensing of Environment* 70, 29–51.

Haxeltine, A. and Prentice, I.C. (1996) A general model for the light-use efficiency of primary production. *Functional Ecology* 10, 551–561.

Hollinger, D.Y., Kelliher, F.M., Byers, J.N., Hunt, J.E., McSeveny, T.M. and Weir, P.L. (1994) Carbon dioxide exchange between an undisturbed old-growth temperate forest and the atmosphere. *Ecology* 75, 134–150.

Hollinger, D.Y., Goltz, S.M., Davidson, E.A., Lee, J.T., Tu, K. and Valentine, H.T. (1999) Seasonal patterns and environmental control of carbon dioxide and water vapour exchange in an ecotonal boreal forest. *Global Change Biology* 5, 891–902.

Hutchison, B.A., Matt, D.R., McMillen, R.T., Gross, L.J., Tajchman, S.J. and Norman, J.M. (1986) The architecture of a deciduous forest canopy in eastern Tennessee, USA. *Journal of Ecology* 74, 635–676.

Jackson, R.J. (1984) Investigation of the properties and genesis of West Coast wet land soils, South Island, New Zealand. 5. Physical characteristics of soil profiles and soil water regimes. *New Zealand Journal of Science* 27, 155–174.

James, I.L. and Norton, D.A. (2001) Helicopter based natural forest management for New Zealand's rimu (*Dacrydium cupressinum*, Podocarpaceae) forests. *Forest Ecology and Management* 155, 337–346.

Kelliher, F.M., Whitehead, D., McAneney, K.J. and Judd, M.J. (1990) Partitioning evapotranspiration into tree and understorey components in two young *Pinus radiata* D. Don stands. *Agricultural and Forest Meteorology* 50, 211–227.

Kelliher, F.M., Whitehead, D. and Pollock, D.S. (1992) Rainfall interception by trees and slash in a young *Pinus radiata* D. Don stand. *Journal of Hydrology* 131, 187–204.

Landsberg, J.J. and Waring, R.H. (1997) A generalised model of forest productivity using simplified concepts of radiation-use efficiency, carbon balance and partitioning. *Forest Ecology and Management* 95, 209–228.

Landsberg, J.J., Johnson, K.H., Albaugh, T.J., Allen, H.L. and Keand, S.E. (2001) Applying 3-PG, a simple process-based model designed to produce practical results, to data from loblolly pine experiments. *Forest Science* 47, 43–51.

Law, B.E., Williams, M., Anthoni, P.M., Baldocchi, D.D. and Unsworth, M.H. (2000) Measuring and modelling seasonal variation of carbon dioxide and water vapour exchange of a *Pinus ponderosa* forest subject to soil water deficit. *Global Change Biology* 6, 613–630.

Law, B.E., Van Tuyl, S., Cescatti, A. and Baldocchi, D.D. (2001) Estimation of leaf area index in open-canopy ponderosa pine forests at different successional stages and management regimes in Oregon. *Agricultural and Forest Meteorology* 108, 1–14.

Leuning, R. (1995) A critical appraisal of a combined stomatal–photosynthesis model for C_3 plants. *Plant, Cell and Environment* 18, 339–357.

Leuning, R., Kelliher, F.M., de Pury, D.G.G. and Schulze, E.-D. (1995) Leaf nitrogen, photosynthesis, conductance and transpiration: scaling from leaves to canopies. *Plant, Cell and Environment* 18, 1183–1200.

Leuning, R., Dunin, F.X. and Wang, Y.-P. (1998) A two-leaf model for canopy conductance, photosynthesis and partitioning of available energy. II. Comparison with measurements. *Agricultural and Forest Meteorology* 91, 113–125.

Lloyd, J., Grace, J., Miranda, A.C., Meir, P., Wong, S.C., Miranda, H.S., Wright, I.R., Gash, J.H.C. and McIntyre, J. (1995) A simple calibrated model of Amazon rainforest productivity based on leaf biochemical properties. *Plant, Cell and Environment* 18, 1129–1145.

McMurtrie, R.E., Rook, D.A. and Kelliher, F.M. (1990) Modelling the yield of *Pinus radiata* on a site limited by water and nitrogen. *Forest Ecology and Management* 30, 381–413.

Mew, G. and Lee, R (1981) Investigation of the properties and genesis of West Coast wet land soils, South Island, New Zealand. 1. Type localities, profile morphology, and soil chemistry. *New Zealand Journal of Science* 24, 1–24.

Nichol, S.E., Harvey, M.J. and Boyd, I.S. (1997) Ten years of rainfall chemistry in New Zealand. *Clean Air* 31, 30–37.

Norman, J.M. and Jarvis, P.G. (1974) Photosynthesis in Sitka spruce (*Picea sitchensis* (Bong.) Carr). III. Measurements of canopy structure and interception of radiation. *Journal of Applied Ecology* 11, 375–398.

Olsson, K.S. (1981) *Soil Survey of Orange County, New York*. USDA Soil Conservation Service, US Government Printing Office, Washington, DC, 192 pp.

Potter, C.S., Randerson, J.T., Field, C.B., Matson, P.A., Vitousek, P.M., Mooney, H.A. and Klooster, S.A. (1993) Terrestrial ecosystem production: a process model based on global satellite and surface data. *Global Biogeochemical Cycles* 7, 811–841.

Raison, R.J. and Myers, B.J. (1992) The biology of forest growth experiment: linking water and nitrogen availability to the growth of *Pinus radiata*. *Forest Ecology and Management* 52, 279–308.

Raupach, M.R. and Finnigan, J.J. (1988) Single-layer models of evaporation from plant canopies are incorrect but useful, whereas, multiplayer models are correct but useless: discuss. *Australian Journal of Plant Physiology* 15, 705–716.

Roderick, M.L., Farquhar, G.D., Berry, S.L. and Noble, I.R. (2001) On the direct effect of clouds and atmospheric particles on the productivity and structure of vegetation. *Oecologia* 129, 21–30.

Running, S.W., Baldocchi, D.D., Turner, D.P., Gower, S.T., Bakwin, P.S. and Hibbard, K.A. (1999) A global terrestrial monitoring network integrating tower fluxes, flask sampling, ecosystem modelling and EOS satellite data. *Remote Sensing of Environment* 70, 108–127.

Sellers, P.J., Berry, J.A., Collatz, G.J., Field, C.B. and Hall, F.G. (1992) Canopy reflectance, photosynthesis, photosynthesis and transpiration. III. A reanalysis using improved leaf models and a new canopy integration scheme. *Remote Sensing of Environment* 42, 187–216.

Sellers, P.J., Dickinson, R.E., Randall, D.A., Betts, A.K., Hall, F.G., Berry, J.A., Collatz, G.J., Dennings, A.S., Mooney, H.A., Nobre, C.A., Sato, N., Field, C.B. and Henderson-Sellers, A. (1997) Modeling the exchange of energy, water, and carbon between continents and the atmosphere. *Science* 275, 502–509.

Soil Survey Staff (1994) *Keys to Soil Taxonomy*, 6th edn. Soil Conservation Service, US Department of Agriculture, Washington, DC.

Turnbull, M.H., Whitehead, D., Tissue, D.T.T., Schuster, W.S.F., Brown, K.J. and Griffin, K.L. (2001) The response of leaf respiration to temperature and leaf characteristics in three deciduous tree species differs at sites with contrasting water availability. *Tree Physiology* 21, 571–578.

Turnbull, M.H., Whitehead, D., Tissue, D.T., Schuster, S.F., Brown, K.J., Engel, V.C. and Griffin, K.L. (2002) Photosynthetic characteristics in canopies of *Quercus rubra*, *Quercus prinus* and *Acer rubrum* differ in response to soil water availability. *Oecologia* 130, 515–524.

Valentini, R., Matteucci, C., Dolman, A.J., Schulze, E.-D., Rebmann, C., Moors, E.J., Granier, A., Gross, P., Jensen, N.O., Pilegaard, K., Lindroth, A., Grelle, A., Bernhofer, C., Grünwald, T., Aubinet, M., Ceulemans, R., Kowalski, A.S., Vesala, T., Rannik, Ü., Berbigier, P., Loustau, D., Gumundsson, J., Thorgeirsson, H., Ibrom,

A., Morgenstem, K., Clement, R., Moncrieff, J., Montagnani, L., Minerbi, S. and Jarvis, P.G. (2000) Respiration as the main determinant of carbon balance in European forests. *Nature* 404, 861–865.

von Caemmerer, S. and Farquhar, G.D. (1981) Some relationships between the biochemistry of photosynthesis and the gas exchange of leaves. *Planta* 153, 376–387.

Walcroft, A.S., Whitehead, D., Silvester, W.B. and Kelliher, F.M. (1997) The response of photosynthetic model parameters to temperature and nitrogen concentration in *Pinus radiata* D. Don. *Plant, Cell and Environment* 20, 1338–1348.

Walcroft, A.S., Whitehead, D., Kelliher, F.M., Arneth, A. and Silvester, W.B. (2002) The effects of long-term, partial shading on growth and photosynthesis of young *Pinus radiata* D. Don. trees. *Forest Ecology and Management* 163, 151–163.

Wang, Y.-P. and Jarvis, P.G. (1990) Description and validation of an array model – MAESTRO. *Agricultural and Forest Meteorology* 51, 257–280.

Wang, Y.-P. and Leuning, R. (1998) A two-leaf model for canopy conductance, photosynthesis and partitioning of available energy I: model description and comparison with a multi-layered model. *Agricultural and Forest Meteorology* 91, 89–111.

Wang, Y.-P., McMurtrie, R.E. and Landsberg, J.J. (1992) Modelling canopy photosynthesis and productivity. In: Baker, N.R. and Thomas, H. (eds) *Crop Photosynthesis: Spatial and Temporal Determination*. Elsevier, Amsterdam, pp. 43–67.

Waring, R.H., Law, B.E., Goulden, M.L., Bassow, S.L., McCreight, R.W., Wofsy, S.C. and Bazzaz, F.A. (1995) Scaling gross ecosystem production at Harvard Forest with remote sensing: a comparison of estimates from a constrained quantum-use efficiency model and eddy correlation. *Plant, Cell and Environment* 18, 1201–1213.

Whitehead, D., Livingston, N.J., Kelliher, F.M., Hogan, K.P., Pepin, S., McSeveny, T.M. and Byers, J.N. (1996) Response of transpiration and photosynthesis to a transient change in illuminated foliage area for a *Pinus radiata* D. Don tree. *Plant, Cell and Environment* 19, 949–957.

Whitehead, D., Leathwick, J.R. and Walcroft, A.S. (2001) Modeling annual carbon uptake for the indigenous forests of New Zealand. *Forest Science* 47, 9–19.

Whitehead, D., Hall, G.M.J., Walcroft, A.S., Brown, K.J., Landsberg, J.J., Tissue, D.T., Turnbull, M.H., Griffin, K.L., Schuster, W.S.F., Carswell, F.E., Trotter, C.M., James, I.L. and Norton, D.A. (2002) Analysis of the growth of rimu (*Dacrydium cupressinum*) in south Westland, New Zealand, using process-based simulation models. *International Journal of Biometeorology* 46, 66–75.

Williams, M., Rastetter, E.B., Fernandes, D.N., Goulden, M.L., Wofsy, S.C., Shaver, G.R., Melillo, J.M., Munger, J.W., Fan, S.-M. and Nadelhoffer, K.J. (1996) Modelling the soil–plant–atmosphere continuum in a *Quercus–Acer* stand at Harvard Forest: the regulation of stomatal conductance by light, nitrogen and soil/plant hydraulic properties. *Plant, Cell and Environment* 19, 911–927.

Williams, M., Rastetter, E.B., Fernandes, D.N., Goulden, M.L., Shaver, G.R. and Johnson, L.C. (1997) Predicting gross primary productivity in terrestrial ecosystems. *Ecological Applications* 7, 882–894.

Wilson, K.B., Baldocchi, D.D. and Hanson, P.J. (2000) Spatial and seasonal variability of photosynthetic parameters and their relationship to leaf nitrogen in a deciduous forest. *Tree Physiology* 20, 565–578.

Wilson, K.B., Baldocchi, D.D. and Hanson, P.J. (2001) Leaf age affects the seasonal pattern of photosynthetic capacity and net ecosystem exchange of carbon in a deciduous forest. *Plant, Cell and Environment* 24, 571–583.

Wong, S.C., Cowan, I.R. and Farquhar, G.D. (1979) Stomatal conductance correlates with photosynthetic capacity. *Nature* 282, 424–426.

Wullschleger, S.D. (1993) Biochemical limitations to carbon assimilation in C3 plants – a retrospective analysis of the A/C_i curves from 109 species. *Journal of Experimental Botany* 44, 907–920.

16

Links between Science and Forest Management, as Illustrated by a Model of Branch Development

J.C. Grace

New Zealand Forest Research Institute, Private Bag 3020, Rotorua, New Zealand

Abstract

Branches play a pivotal role in a forest. They support the foliage required for growth but also have a detrimental influence on the quality of the timber. New Zealand forest companies have supported the development of a branch model, Branch Location, Orientation and Size Simulator (BLOSSIM), that is based on knowledge of crown morphology. Detailed branching data from 66 trees have been used for model development. Independent data are being used to determine how the model performs for a wider range of sites throughout New Zealand. For science to provide an effective contribution to forest management, a thorough understanding of tree development is required and the results need to be synthesized to be applicable to forest managers.

Introduction

The ability to accurately predict tree growth and the development of stem quality in response to differences in tree breed, site and silviculture is necessary for the effective management of a production forest. The use of computer-simulation models for this purpose goes back to the early 1960s. Munro (1974) classified forest-growth models into three categories according to the underlying structure of the model. Whole-stand distant-independent models were formulated using stand-level data and predicted stand-level variables relevant to forest managers, such as mean top height, basal area, volume and stocking. Single-tree distant-independent models were formulated using individual tree data. Predicted individual tree variables, such as tree diameter at breast height and tree height, were summed/averaged to obtain stand-level values. In single-tree distant-dependent models, the locations of the individual trees were used to derive competition indices. These indices were incorporated into equations

 Forests at the Land–Atmosphere Interface (eds M. Mencuccini, J. Grace, J. Moncrieff and K.G. McNaughton)

predicting individual tree growth. Again individual tree predictions were summed/averaged to give stand-level values. Such models, which did not consider the underlying biological processes influencing tree growth, were later termed empirical models.

During the 1960s and 1970s computer-simulation models became common in many aspects of science. The idea of using process-based models, models that simulate the biological processes influencing tree growth, for forest management appears to have originated from the premise that 'increased knowledge of the production processes is the best and only way to an increased and sustained forest production' (Linder, 1981). The idea for this approach had its origins in crop-physiology research and in particular the work of de Wit (1970) (J. Landsberg, personal communication). To promote/discuss the above philosophy, a workshop on 'Understanding and Predicting Tree Growth' was organized in 1979 by the research groups of P.G. Jarvis, E.D. Ford and S. Linder. The papers that were presented are included in a publication edited by S. Linder (Linder, 1981).

During the 1980s the New Zealand Forest Research Institute had separate research groups that were developing stand-level and process-based models for the management of the radiata pine resource. The research group developing process-based models was influenced by and collaborated with P.G. Jarvis, J.M. Norman and the team of J.J. Landsberg and S. Linder at CSIRO in Canberra, Australia. While a number of research papers were published (see Goulding, 1994; Medlyn, Chapter 8, this volume), no model reached the stage of being applied by forest managers. One reason was that the resource had been intensively monitored, which enabled stand-level models to be developed and incorporated into a software package for forest managers in a short time frame. In retrospect, the aim of the process-based modelling should have been to provide scientific knowledge that would contribute to the development of simpler models, more suited to forest management. This approach of simplifying process-based models has been suggested by Penning de Vries (1983) and Battaglia and Sands (1998) and was being investigated (Grace, 1990; Ludlow *et al.*, 1990).

Mohren and Burkhart (1994) considered that process-based models were more suited to research purposes than to forest management. However, they did consider that process-based models, i.e. models that capture the essential dynamics of the system in relation to site conditions, would be of increasing importance in guiding forest-policy evaluation in respect of changing environmental conditions. This was confirmed by Battaglia and Sands (1998), who found that the main uses of 14 process-based models, reported in the literature between 1990 and 1996, were to answer long-term questions related to climate change, pollution effects and ecological succession. More recently, Aracruz Celulose, a large forest company in Brazil, has started to use the process-based model 3-PG (Landsberg and Waring, 1997) for the management of its eucalyptus plantations. This model was developed in a deliberate attempt to bridge the gap between empirical growth and yield models and process-based carbon-balance models (Landsberg, 2003).

Many papers have emphasized the point that the end-use of the model is an important consideration in model design (e.g. Bruce and Wensel, 1988;

Bruce, 1990; Battaglia and Sands, 1998; Grace *et al.*, 1999; Robinson and Ek, 2000). Penning de Vries (1983) and Battaglia and Sands (1998) state that the process of defining an appropriate model structure starts with the end-user, whose needs determine the questions to be addressed.

Currently New Zealand forest managers are interested in the growth of branches, as well as the growth of the stem. Grading rules, in New Zealand and overseas, limit the number and size of bark-encased or partially bark-encased knots allowed in timber sold for appearance. They also limit the size of knots allowed in structural timber as the deviation of stem-wood fibres around branches reduces the strength of a piece of timber. Branches therefore play a pivotal role in the forest. They support the foliage required for photosynthesis and hence growth, but result in knots within the timber with a detrimental effect on timber quality.

The objectives of the remainder of this chapter are to:

- outline the practice of plantation forestry in New Zealand – this has a bearing on model design;
- describe a branch model, Branch Location, Orientation and Size Simulator (BLOSSIM), for radiata pine (*Pinus radiata* D. Don), which was designed to be used by forest managers and is being developed with funding from a partnership between Forest Research and the forest companies belonging to the Stand Growth Modelling Cooperative;
- discuss some ideas that have arisen from this project, about the interaction of science with forest management.

Plantation Forestry in New Zealand

Radiata pine accounts for 90% of the estimated 1.77 million ha of planted production forests in New Zealand (Ministry of Agriculture and Forestry, 2001). Most radiata pine plantations are intensively managed for profit from the export of wood and fibre products. To maximize clearwood production, 51% of the plantations receive green-crown pruning with early thinning to waste and no production thinning. James (1990) discussed the evolution of this regime. The remaining 49% are managed in one of three other alternative ways: green-crown pruning with production thinning (19%), or no pruning with (4%) and without (26%) production thinning. The average (area-weighted) rotation age is 26.8 years (Ministry of Agriculture and Forestry, 2001).

There has been an active tree-improvement programme for radiata pine since the 1950s (Shelbourne *et al.*, 1986). Selections, established as parent trees in clonal seed orchards have resulted in trees with improved diameter growth, a multinodal branching habit, less sweep and less malformation. Separate breeds have also been developed for disease resistance, long-internode branching habit and stiffer wood. Currently seed lots can be ranked according to stem growth, stem straightness, branching characteristics, wood density, spiral grain and resistance to *Dothistroma septospora* (Dorog.) Morelet (also known as *D. pini*). (Vincent, 1998).

In 1998, one large forest company announced a regime of planting 555 stems/ha and not thinning or pruning for most of its sites. Reasons given for this regime include a perceived lower premium for clearwood in 20 years due to increased availability of pruned trees, better techniques for manufacturing clearwood from unpruned logs and improved plant quality with exceptional survival rates (Dyck and Thomson, 1999).

Knowledge of the planted resource is relatively high. Forest companies routinely use inventory methods to assess the quality of the resource. One commonly used technique is the method for assessment of recoverable volume by log types (MARVL) (e.g. Deadman and Goulding, 1979). Tree diameters and heights are measured, while branch diameters and stem shape are assessed visually. This information, together with a list of product constraints, is input into a computer model to estimate the amount of each product that can be obtained from the forest.

In general, inventories should be accurate (unbiased) and precise (narrow confidence interval around the estimates). Foresters try to estimate stand wood volume to within 10% at a 95% probability level. The number of trees measured in a stand will depend on its variability but would be of the order of 15–20 plots, each containing 15–20 trees (A. Gordon, Forest Research, personal communication).

When MARVL inventories are conducted several years prior to harvest, the tree height and diameter at the end of the rotation can be estimated using an empirical growth model, GROWMARVL (Gordon *et al.*, 1995). Industry planners realized that the estimates of products available from the forest could be improved if they had models to predict the changes in branch diameter between the inventory and harvest. As a consequence, many New Zealand forest companies (through the Stand Growth Modelling Cooperative) have been contributing to the development of both an individual tree-growth model and a branch model for radiata pine that would link with the current inventory techniques.

Model Design – BLOSSIM

Crown characteristics and stem shapes of radiata pine vary with location. For example, trees with large branches and poor form occur when grown at wide spacing on ex-pasture sites, while trees with fine branching habit are found on former sand-dune sites.

The branching pattern is also highly variable due to genetics. Tree breeders have taken advantage of this variability and developed both a long-internode breed with fewer branch clusters in an annual shoot and a multinodal breed with more clusters but smaller branch diameters.

Many different modelling approaches could be used to develop the model required by the forest industry. The model could predict growth at stand level or at tree level. At the tree level, the model could either consider or ignore the spatial location of trees (i.e. be either distant-dependent or distant-independent). At the branch level, the model could either consider or ignore the spatial location of the branches. The model could predict growth from the underlying growth processes (light interception, photosynthesis, respiration and allocation) (i.e. be process-based) or predict growth directly from field measurements (i.e. be empirical).

Choosing the appropriate strategy will depend on many factors. A scientist is motivated by discovering/developing new knowledge/technology. A forest manager is motivated to fund research to obtain a management tool. Satisfying both needs at the same time is likely to be difficult.

As inventory data are collected on individual trees, an individual-tree rather than stand-level model was considered appropriate. A distant-independent model was considered appropriate in order to link with the individual-tree distant-independent growth model (ITGM) being developed for the Stand Growth Modelling Cooperative (A.D. Gordon and R.G. Shula, personal communication). Spatially explicit branching was chosen in order to provide sufficient detail to link with AUTOSAW – a sawing simulator (Todoroki, 1991, 1997). These decisions are clearly in the interest of the forest manager in that the model will link with other available software, rather than being a completely independent system.

A model structure was developed that incorporated our understanding of the underlying mechanisms of branch development. The model functions were developed using field data obtained by destructive sampling of near-rotation-age trees. This was a deliberate attempt to investigate whether there was a compromise between 'best-fit' models and very detailed process-based models and was in the interest of the scientist.

At the outset, it was realized that several components were required to develop a system acceptable to forest managers:

- A thorough understanding of the branching pattern of a radiata pine was required to provide a realistic description of the location and diameter of branches on individual trees at any age.
- A method to predict branch-diameter increment, particularly in response to management practices, was required in order to be able to determine how branch diameters would change through time.
- Methods were required to be able to convert the inventory description of branching into a more detailed description that could be input into BLOSSIM so that future branch growth could be simulated.

Realistic description of branching patterns

The resulting branch model, BLOSSIM (Grace *et al.*, 1998, 1999) is hierarchical in structure and considers the development of branching at the annual shoot level, the cluster (or whorl) level and the branch level. Destructive sampling of near-rotation-age trees is being used to quantify the effects of site, silviculture and genetics. To date, 66 trees from nine treatments have been sampled but this number is considered to be far too small to realistically predict branch development for any seed lot on any site in New Zealand. Each data set has been analysed individually to give model coefficients for specific regions and seed lots. The most appropriate set of coefficients is used for regions/seed lots where no data are available.

There are two functions pertaining to the annual shoot. These predict the number of branch clusters in an annual shoot and their relative position within the shoot. Data from four genetically improved families at three sites support the observation that the number of clusters in an annual shoot is influenced by temperature (e.g. Bollmann and Sweet, 1976). Bannister (1962) suggested that the relative positions of clusters within an annual shoot with a given number of clusters might also vary with environmental conditions. The data from the above 66 trees have not shown any major differences between sites. Simulations (Pont *et al.*, 1999) have indicated that the relative position of clusters had little impact on the value of visually graded timber. Hence the positions have been assumed to conform to constant ratios for all sites.

The branch clusters within an annual shoot are not identical and it is proving difficult to determine the underlying patterns for the diameter of the largest branch in each cluster. One reason is that annual shoot lengths have been determined from ring counts. This gives midwinter to midwinter annual shoots (the mensurational definition), whereas Burdon (1994) suggests that morphologically an annual shoot should be measured from midsummer to midsummer as the resting bud forms in midsummer and autumn elongation has essentially a 'head-start' on spring elongation. For the more recently sampled trees, annual shoots have been determined using both ring counts and morphological features. Visually, there appear to be more consistent patterns when trees are aged using morphological features compared with stem ring counts, but more detailed analyses are required to confirm this. The mensurational definition of an annual shoot is being used within BLOSSIM since switching to the morphological definition would require major modifications to mensurational height-growth models.

Within a cluster, the spatial location of branches and their diameter are associated with the underlying phyllotaxis (Pont, 2001). The majority of branch clusters have the Fibonacci phyllotactic pattern (i.e. 137.5° between adjacent branches in the spiral). In general, the higher the position of a branch in the phyllotactic spiral, the larger its diameter. In BLOSSIM, branches are arranged according to the Fibonacci phyllotactic pattern and their diameter is a function of the position in the spiral, with the largest branch at the top of the spiral.

Currently the azimuth angle of the largest branch is assumed to be random, since simulations (Pont *et al.*, 1999) indicated that different assumptions had little influence on value when graded for visual products. The random assumption is not strictly correct. For the 66 trees sampled (all from within plantations), the compass direction of the largest branch in each cluster was found to be highly variable but tended to avoid south. At stand boundaries, branches are much larger towards the boundary.

Once the tree has reached reproductive maturity (usually 3–6 years from seed), many branch clusters contain stem cones. Using the mensurational definition of an annual shoot, clusters containing cones appear to be randomly distributed, but there are definite patterns associated with the morphological definition of an annual shoot. In BLOSSIM, the age of reproductive maturity is predicted using a logistic function and clusters containing cones are assumed to be randomly distributed within the annual shoot.

Apart from branch-cluster position and branch-diameter growth, there are three other characteristics associated with a branch that affect its suitability for particular end uses. These are the angle of the branch, whether it is live (intergrown) or dead (surrounded by bark) and whether there is a layer of bark trapped above the upper surface of the branch of a live branch.

Branch growth and response to thinning

Initial attempts to develop a function to predict branch-diameter growth from branch diameter and branch age pointed to the idea that the ability of a branch to grow in diameter was to some extent predetermined. The hypothesis was that each branch had a 'potential' that influenced its growth. This is a multiplicative function of individual potentials related to:

- position of the branch in the phyllotactic spiral within its cluster;
- the competitiveness (potential for growth) of the cluster compared with other clusters in the tree;
- the competitiveness (potential for growth) of the tree within the stand;
- stand average stocking;
- site characteristics;
- genetic characteristics.

Measurements of branch diameter over time for selected trees in a thinning trial indicated that the younger and larger branches on the larger trees would show increased diameter growth following a thinning. In the case of a thinning, the 'potential' is therefore multiplied by a further term, 'a response potential', which influences the branch growth rate for a number of years until the branch is growing at a rate appropriate to the wider spacing.

Branch-diameter growth at a given age is calculated as the difference between predicted branch diameter for the 2 consecutive years. Branch diameter itself is predicted from branch age and branch potential.

Validation

Clearly, the branching pattern of radiata pine is exceedingly complex. It is considered that the modular structure of BLOSSIM is sound. However, a model derived from detailed measurement of a limited number of trees on a few sites is unlikely to give sufficiently accurate predictions for the forest manager on all sites. Methods are needed to determine the critical components within the model and to determine the accuracy of the model across a wider range of forest stands. The results of such analyses will be used to determine future data-collection strategies and model development.

One method has been used to determine the critical model functions. BLOSSIM was linked with AUTOSAW and simulations were run to determine the relative importance of the different functions for timber visually graded for premium products, such as clear and cuttings grades (Pont *et al.*, 1999). The most important function was the number of branch clusters in an annual shoot. Number and diameter of branches and the occurrence of defects were also important. The relative importance of the functions may change for structural timber. With the move to machine stress-grading of structural timber, a model simulating the machine stress-grading of boards is required to determine the relative importance of the different branching functions for structural timber.

Based on the above simulations, it has been suggested that the following variables should be recorded in a routine inventory:

- number of branches in the lowest cluster;
- diameter of the largest branch in the lowest cluster;
- cluster count for a given stem length.

Recent tests have confirmed that it is practical to collect these data during an inventory. Measuring these variables will enable the more important functions to be parameterized for a particular site.

Two approaches are currently being investigated for determining the accuracy of BLOSSIM on a range of sites. In both approaches, BLOSSIM is used to generate branching patterns for known trees in permanent sample plots. BLOSSIM predictions are then compared with either subjective inventory data or PhotoMARVL data (Firth *et al.*, 2000) from the same trees. Both approaches have advantages and disadvantages. The advantage of inventory data is that they are quick to collect and would enable many sites to be covered in a short space of time. The disadvantage is the subjective nature of the data. The advantage of PhotoMARVL is that one can obtain quantitative measurements of tree dimensions from a photographic image. The disadvantage is the time required to take and measure the resulting image.

Acceptance by forest managers

As stated earlier, BLOSSIM is being developed with funding from a partnership between Forest Research and the forest companies who belong to the Stand Growth Modelling Cooperative. The industry representatives are kept up to date with progress through two meetings a year. However, it will still take time and effort for BLOSSIM to be accepted and used by the forest industry.

Industry representatives have been given a copy of a prototype for testing, but very limited use has been made of it. Possible reasons are that the prototype was not easy to use, the representatives did not have time to experiment with it and we did not provide training sessions for the prototype. Alternative approaches that allow industry representatives to be more involved with the model are needed. One approach is to involve them in model validation by asking them to collect inventory data for their permanent sample plots for comparison with BLOSSIM output.

Concurrently the model is being implemented within a software package being designed for the forest companies. Initial tests suggest that simulating the growth of all branches on trees may take too long for the system to be practical for a forest manager. One solution would be to simulate the growth of the largest branch in a cluster and then predict the diameter of the other branches at the end of the rotation.

Interacting with Forest Management

The modelling approach described above will ultimately provide a practical tool for forest managers. However, it will not be able to answer all their queries in respect of branching. This reinforces the point for models to be designed for specific end uses, for example research or forest management.

Predicting branch development following an inventory is only one concern forest managers have in respect of branching. During the development of BLOSSIM, some other management issues relating to branches were raised. I should like to discuss some of these as they illustrate some of the differences between scientists and forest managers. In particular, these examples illustrate the fact that managers need methods/tools that provide an immediate answer to a particular problem, whereas scientists are interested in why the problem occurs. Being able to understand why will ultimately benefit the forest managers.

Company A cuts a square piece of wood from the centre of the log. If bark is visible on the outside of the square then its value is downgraded. One cause is bark trapped on the upper side of a large branch. An immediate solution was needed to segregate logs into two categories: those with bark trapped and those without bark trapped. This was solved by the forest company observing that a hollow (fluting) below branches generally indicated that bark was trapped. This was a good solution for the forest managers. From a science perspective we need to determine why the bark is trapped in the first instance. If we can understand why it occurs, then we may be able to prevent it occurring in future crops. It appears to be a feature of young trees and one hypothesis is that a side bud expands before a terminal bud.

Company B has been telling me that branch diameters in their forest are larger than what I have measured in other forests with comparable silviculture. An analysis of diameter of the largest branch in the first pruning lift confirmed that branch diameters tend to be larger in this region of New Zealand. If we could understand why branches tend to be larger, then it might be feasible to suggest ways to minimize branch-diameter growth for a given silviculture regime. One possible reason that emerged after a visit to a few permanent sample plots in the forest was the number of branch cones formed – visually there appeared to be more cones than at other sites. From considering branches as beams, Cannell *et al.* (1988) hypothesized that an increase in weight would result in an increase in diameter (provided other variables do not change concurrently). Since cones increase the weight of the branch, the number of branch cones on a branch is a plausible cause.

In developing BLOSSIM, research has concentrated on understanding the general branching patterns of radiata pine, but it is the largest branches that cause timber to change from one grade to a lower one. Many of the large branches are not part of the regular branching pattern but are the result of lost leaders. A model function that will predict the likelihood of leader damage and its impact on branch diameter growth lower in the crown is likely to improve the model's performance.

Future Directions

It is becoming increasingly clear that there have been undesirable consequences from growing trees selected for improved growth and form under silviculture regimes designed for clearwood production. Wood density is lower and there are more resin pockets, more internal checking and more compression wood.

While the timber industry has to deal with the current resource, as scientists we need to understand why these problems have occurred and to ensure that they do not occur in successive rotations. It is considered that these interactions have occurred because we have an incomplete understanding of the growth of radiata pine. As Larson (1962) said: 'External factors of climate and environment exert their influence directly on the growth of the crown and only indirectly on the growth and quality of the wood.' Clearly, the modelling approach taken with branching needs to be extended to include wood properties. This has been started in a small way. To be able to understand the influence of tree lean on compression-wood development, we have been investigating whether a finite element model of the tree's righting mechanism (Fourcaud *et al.*, 2003) is applicable for radiata pine. To this end we have measured the crown architecture and selected wood properties by ring for two pairs of trees containing a straight and swept tree of the same clone. While this model predicts how a leaning stem will straighten over time, it is not yet capable of answering the question of interest to forest managers, 'What is the distribution of wood properties within a mature stem of a given shape?'

There are several ways in which the scope of this project could be expanded. For the second pair of trees, SilviScan (Evans *et al.*, 1995) was used to provide within-ring details of tracheid properties. A further extension would be to use carbon-isotope discrimination techniques on adjacent wood samples (e.g. Walcroft *et al.*, 1997) to determine how physiological processes within the crown have influenced the structure of cells. Measuring crown structure is a critical component since Walcroft *et al.* (1996) have shown that discrimination against ^{13}C in plant tissues varies in respect of crown structure, in particular branch length.

Since the late 1980s the Stand Growth Modelling Cooperative has been establishing and managing a series of trials designed to determine the impacts of tree breed, silviculture and site on tree growth. As these trials reach maturity, there is the opportunity to use the above techniques to determine retrospectively the interactions between crown development, growth and wood structure.

Conclusion

For science to provide an effective contribution to forest management, a thorough understanding of how trees function, from the growth through to the properties of the wood, is required. This information needs to be synthesized for forest managers and finally tools are needed so that foresters can measure the relevant parameters in the field. For this to be successful, it needs to be recognized that scientists work at different levels of complexity and that there should be collaboration rather than competition among scientists, since all are ultimately working towards the same goal.

Acknowledgements

This work has been supported by the Stand Growth Modelling Cooperative and the Foundation for Research, Science and Technology.

References

Bannister, H.M. (1962) Some variation in the growth pattern of *Pinus radiata* in New Zealand. *New Zealand Journal of Science* 5, 342–370.

Battaglia, M. and Sands, P.J. (1998) Process-based forest productivity models and their application in forest management. *Forest Ecology and Management* 102, 13–32.

Bollmann, M.P. and Sweet, G.B. (1976) Bud morphogenesis of *Pinus radiata* in New Zealand: I. The initiation and extension of the leading shoot of one clone at two sites. *New Zealand Journal of Forestry Science* 6, 376–392.

Bruce, D. (1990) Development of empirical forest growth models. In: Dixon, R.K., Meldahl, R.S., Ruark, G.A. and Warren, W.G. (eds) *Process Modelling of Forest Growth Responses to Environmental Stress*. Timber Press, Portland, Oregon, pp. 191–199.

Bruce, D. and Wensel, L.C. (1988) Modelling forest growth: approaches, definitions and problems. In: Ek, A.R., Shifley, S.R. and Burk, T.E. (eds) *Forest Growth Modelling and Prediction*. General Technical Report NC120, North Central Forest Experiment Station, Forest Service, United States Department of Agriculture, St Paul, Minnesota, pp. 1–8.

Burdon, R.D. (1994) Annual growth stages for height and diameter in *Pinus radiata*. *New Zealand Journal of Forestry Science* 24, 11–17.

Cannell, M.G.R., Morgan, J. and Murray, M.B. (1988) Diameters and dry weights of tree shoots: effects of Young's modulus, taper, deflection and angle. *Tree Physiology* 4, 219–232.

Deadman, M.W. and Goulding, C.J. (1979) A method for assessment of recoverable volume by log types. *New Zealand Journal of Forestry Science* 9, 225–239.

de Wit, C.T. (1970) Dynamic concepts in biology. In: Setlik, I. (ed.) *Prediction and Measurement of Photosynthetic Productivity. Proceedings of the IBP/PP Technical Meeting, Trebon, 14–21 September, 1969*, Pudoc, Wageningen pp. 17–23.

Dyck, B. and Thomson, P. (1999) Carter Holt Harvey's Millennium Forestry. *New Zealand Journal of Forestry* 43(4), 2–4.

Evans, R., Downes, G.M., Menz, D.N.J. and Stringer, S.L. (1995) Rapid measurement of tracheid transverse dimensions in a *P. radiata* tree. *Appita* 48, 134–138.

Firth, J.G., Brownlie, R.K. and Carson, W.W. (2000) Accurate stem measurements, key to new image-based system. *New Zealand Journal of Forestry* 45, 25–29.

Fourcaud, T., Blaise, F., Lac, P., Castéra, P. and de Reffye, P. (2003) Numerical modelling of shape regulation and growth stresses in trees. II. Implementation in the AMAPpara software and simulation of tree growth. *Trees Structure and Function* 17, 31–39.

Gordon, A.D., Lawrence, M.E. and Pont, D. (1995) Assessing the potential log yield of stands prior to harvesting. In: *Proceedings of the Institute of Foresters of Australia 16th Biennial conference 'Applications of New Technologies in Forestry', Ballarat, Victoria, 18–21 April, 1995*. Institute of Foresters of Australia, Canberra, pp. 193–200.

Goulding, C.J. (1994) Development of growth models for *Pinus radiata* in New Zealand – experience with management and process models. *Forest Ecology and Management* 69, 331–343.

Grace, J.C. (1990) Modelling the interception of solar radiant energy and net photosynthesis. In: Dixon, R.K., Meldahl, R.S., Ruark, G.A. and Warren, W.G. (eds) *Process Modelling of Forest Growth Responses to Environmental Stress*. Timber Press, Portland, Oregon, pp. 142–158.

Grace, J.C., Blundell, W. and Pont, D. (1998) Branch development in radiata pine – model outline and data collection. *New Zealand Journal of Forestry Science* 28, 182–194.

Grace, J.C., Pont, D., Goulding, C.J. and Rawley, B. (1999) Modelling branch development for forest management. *New Zealand Journal of Forestry Science* 29, 391–408.

James, R.N. (1990) Evolution of silvicultural practice towards wide spacing and heavy thinning in New Zealand. In: James, R.N. and Tarlton, G.L. (eds) *New Approaches to Spacing and Thinning in Plantation Forestry*. Bulletin 151, FRI, New Zealand Ministry of Forestry, Rotorua, pp. 13–20.

Landsberg, J.J. (2003) Modelling forest ecosystems: state-of-the-art, challenges and future directions. *Canadian Journal of Forest Research* 33, 385–397.

Landsberg, J.J. and Waring, R.H. (1997) A generalised model of forest productivity using simplified concepts of radiation-use efficiency, carbon balance and partitioning. *Forest Ecology and Management* 95, 209–228.

Larson, P.R. (1962) A biological approach to wood property. *Tappi* 45, 443–448.

Linder, S. (1981) *Understanding and Predicting Tree Growth*. No. 160, Studia Forestalia Suecica.

Ludlow, A.R., Randle, T.J. and Grace, J.C. (1990) Developing a process-based growth model for Sitka spruce. In: Dixon, R.K., Meldahl, R.S., Ruark, G.A. and Warren, W.G. (eds) *Process Modelling of Forest Growth Responses to Environmental Stress*. Timber Press, Portland, Oregon, pp. 249–262.

Ministry of Agriculture and Forestry (2001) *National Exotic Forest Description as at 1st April 2000*. Ministry of Agriculture and Forestry, Wellington, 63 pp.

Mohren, G.M.J. and Burkhart, H.E. (1994) Contrasts between biologically-based process models and management-oriented growth and yield models. *Forest Ecology and Management* 69, 1–5.

Munro, D. (1974) Forest growth models – a prognosis. In: Fries, J. (ed.) *Growth Models for Tree and Stand Simulation*. Research Notes No. 30, Royal College of Forestry, Stockholm, pp. 7–21.

Penning de Vries, F.W.T. (1983) Modelling of growth and production. In: Lange, O.L. and Aragno, M. (eds) *Encyclopedia of Plant Physiology*, New Series, Vol. 12D. Springer-Verlag, Berlin, pp. 117–150.

Pont, D. (2001) Use of phyllotaxis to predict arrangement and size of branches in *Pinus radiata*. *New Zealand Journal of Forestry Science* 31, 247–262.

Pont., D., Grace, J.C. and Todoroki, C. (1999) Modelling the influence of radiata pine branching characteristics on visual timber grade. In: Nepveu, G. (ed.) *Proceedings of Third Workshop 'Connection between Silviculture and Wood Quality through Modelling Approaches and Simulation Softwares'*. Publication Equipe de Recherches sur la Qualité des Bois, INRA, Nancy, France, pp. 63–71.

Robinson, A.P. and Ek, A.R. (2000) The consequence of hierarchy for modelling in forest ecosystems. *Canadian Journal of Forest Research* 30, 1837–1846.

Shelbourne, C.J.A., Burdon, R.D., Carson, S.D., Firth, A. and Vincent, T.G. (1986) *Development Plan for Radiata Pine Breeding*. Forest Research Institute, Ministry of Forestry, Rotorua, 142 pp.

Todoroki, C.L. (1991) Development of an automated sawing simulator. *New Zealand Journal of Forestry Science* 20, 332–343.

Todoroki, C.L. (1997) Developments of the sawing simulation software, AUTOSAW: linking wood properties, sawing, and lumber end-use. In: Nepveu, G. (ed.) *Proceedings of Second Workshop 'Connection between Silviculture and Wood Quality through Modelling Approaches and Simulation Softwares'*. Publication Equipe de Recherches sur la Qualité des Bois, INRA, Nancy, France, pp. 241–247.

Vincent, T.G. (1998) *GFPlus*^{TM}. Forest Research, New Zealand Radiata Pine Breeding Cooperative, Rororua, 4 pp.

Walcroft, A.S., Silvester, W.B., Grace, J.C., Carson, S.D. and Waring, R.H. (1996) Effect of branch length on carbon isotope discrimination in *Pinus radiata*. *Tree Physiology* 16, 281–286.

Walcroft, A.S., Silvester, W.B., Whitehead, D. and Kelliher, F.M. (1997) Seasonal changes in stable carbon isotope ratios within annual rings of *Pinus radiata* reflect environmental regulation of growth processes. *Australian Journal of Plant Physiology* 24, 57–68.

17 Thoughts on Forest Science

D.C. Malcolm
School of Forestry, Institute of Ecology and Resource Management, University of Edinburgh, Mayfield Road, Edinburgh EH9 3JU, UK

Introduction

This chapter is a brief attempt to consider some aspects of forest science in relation to its application to forest practice. It is not meant to be comprehensive in any sense but is a collection of subjective and eclectic, possibly random, thoughts on the history of the science and its impact on the management of forests. In the context of this volume, which is largely devoted to the interaction of forests and the atmosphere, it is necessary to say a few words about the scientific career and contribution of Paul Jarvis as well!

What is forest science? If one can distinguish forest science from its underpinning disciplines, it may be thought of as the description, analysis and prediction of forest growth. The underlying scientific disciplines are the plant sciences (genetics, plant physiology) and those that deal with the environment (geology, soil science, meteorology or climatology). These all come together in forest ecology, which considers environmental interactions with vegetation. The basic unit is the stand or population of trees or aggregations of these units in more extensive forests. However, an understanding of the processes involved in forest growth also demands study of the responses of the individual tree and its component parts – shoot, leaf and root.

History of Forest Science

Before forest science could really begin, it was necessary for individual plant function to be at least partially understood. For example, the dependence of growth on the uptake of carbon dioxide (de Saussure, quoted in Lieth, 1978) and the contribution made by mineral elements (Liebig, quoted in Lieth, 1978). Forest science, then, can be said to have begun in the latter part of the 19th century, when investigators started to consider the forest as an ecosystem, although

 Forests at the Land–Atmosphere Interface
(eds M. Mencuccini, J. Grace, J. Moncrieff and K.G. McNaughton)

initially with all the emphasis on the trees. There have been several main strands of development since then upon which modern forest science has been founded, of which some have been incorporated in practice. The following selective list highlights a few of the individuals who made fundamental contributions that have enabled many others to develop an understanding of forest function, processes and production. It is unlikely to match anyone else's selection.

Primary Production

In Bavaria, Ebermayer (1882) published values for the dry-matter production and nutrient budgets for beech, spruce and pine stands. The annual values for litter and wood increment were, for the three species on good sites, remarkably similar, at about 6500 kg/ha. This determination of primary productivity laid the foundation for a century's worth of similar studies, which still go on, as the nutrient, biomass and energy budgets of every conceivable forest type have been determined. The accuracy and precision of the estimates have been improved and the tedium of the techniques reduced by spawning along the way innumerable allometric equations – but essentially it has been good news for the manufacturers of ovens. Generally, these studies provide a 'snapshot' of existing conditions, rather than much insight into the productive processes involved. The impact of many of these studies on forest management has been slight as foresters have preferred to remain with their standard mensurational estimates of stem-volume production – the saleable material. However, some studies have estimated the effects of manipulations (e.g. addition of nutrients) and recently the calculation of renewable-energy supplies and the potential for forests to store carbon. A vast database now exists and, no doubt, may prove useful. M. Cannell (1982) collated many of these studies.

Adaptive Variation

Variation within species in adaptation to environmental conditions had been recognized for some time but began to be systematically studied by a number of research stations in the late 19th century. Cieslar (1907) published results from trials of Norway spruce at three elevations in the Tyrol (at 227, 795 and 1380 m), in which he measured height increment, needle length, needles/cm, colour, tree habit, root development, flushing times and frost damage. He established similar trials for larch and later for 21 provenances of oak. These studies coincided with the founding of IUFRO (in 1892), leading to international exchange of plant material. Adaptive variation was readily demonstrable at the population level, resulting in many similar studies for different species. Provenance trials became the basis of genecology, population genetics and tree-improvement programmes. The importance of matching the natural and exotic environments in planting programmes was so clear that it was immediately adopted by forestry practice, an irony being the failure (in Britain at least) to clarify the variation within native species. Recently, the advent of molecular techniques has enabled the history of species and their migration to be better understood but, as yet, there is no clear connection between this neutral variation and population adaptation to the environment. Controlling genetic variation in experiments on all aspects of forest tree growth should be a *sine qua non*.

Growth Analysis

Blackman's short paper (1919) on the applicability of compound interest to plant growth opened up an enormous field for analysing the comparative performance of plants of different species, different environmental conditions or imposed treatments. In particular, the concept of efficiency (interest rate) emphasized the role of leaf area in production and the partitioning of assimilates. Thus there has been a multitude of experiments that have investigated (and sometimes clarified) the effects of many variables (radiation, temperature, water availability and nutrients) in controlled environments or in field studies. These studies have always been easier to do in agricultural crops than in forest stands, where some ingenuity has had to be displayed to readily assess relative growth and net assimilation rates, leaf-area index and so on. A good example is the development of the 'pipe model theory' (Waring *et al.*, 1982) to relate sapwood area (conducting tissue) to leaf area. Also, the linking of relative growth rate and the supply of nutrients by Ingestad (1982) gave a notable insight into ecosystem functioning, with implications for fertilizer practice. Although frequently used in studies of seedlings in nurseries, growth-analysis techniques have not been used in the forest, even in silvicultural research, as much as might have been expected despite the difficulties.

Site Studies

The study of natural forest ecosystems could be said to have begun with Cajander's 'Ueber Waldtypen' ('Theory of forest types') in 1909. The classification of forest by reference to readily identifiable site type, originally based on ground vegetation alone, has stood the test of time as a valuable management tool. It allows the manager to rationalize species distribution and the application of silvicultural systems, as well as estimating site potential for wood production, wildlife habitat, conservation and so on. The concept has been applied worldwide and is still being developed; for example, a new ecological site classification for British forests appeared in 2001 (Pyatt *et al.*, 2001). This, like most other recent classifications, now links measures of climate and nutrient supply to the vegetation in separating sites and predicting productivity. Although there have been many scientists involved in developing site studies, it is worth mentioning Krajina (1965) and his colleagues in British Columbia, who delineated climatic zones as an initial stratification in systematizing the vegetation. In a long series of experiments in Sweden, Tamm (1979) investigated the role of soil nutrients, particularly nitrogen, and the effects of their manipulation.

Statistics and Computation

The introduction of statistical tests and especially the analysis of variance, by Fisher (1925), transformed the design of experiments. Replication of treatments and factorial trials became the norm, with later 'pseudoreplication' coming to haunt some field ecologists. Sampling schemes made all sorts of investigations possible, whether for inventory of standing volumes or detecting small differences between populations or treatment effects. Multivariate statistics expanded the possibilities further. The study of site factors and the growth of different species using multiple regression launched many studies where large numbers of variables were assessed for their combined effects on forest growth. Sometimes the collection of data seemed to replace ecological thought processes – a situation exacerbated by the ease of analysis after computers became readily available in the 1960s. Nevertheless, the suite of multivariate analyses (PCA, CCA, etc.) now in common use have advanced understanding in forest ecology. The advent of computing has also made it possible to model complex systems, so that the response of forest stands to environmental variables can be integrated and predicted. The requirement for satisfactory data to run models based on physiological processes has not encouraged their use in forest practice. Managers have been more ready to apply the results of empirical models based on mensurational databases. The recent development of predictive models of forest growth (e.g. Landsberg and Waring's 3-PG, 1997), which utilize more readily obtained driving variables, or indices of them, may change this attitude.

Stand Tending

Once a stand is established, almost the only way forest managers can influence the economic outcome is by controlling stand density. Following early investigations on thinning by Hartig (1813) in German forests, there have been innumerable experiments to gauge the effects of stand density on the yield and the timber assortments obtainable from a stand rotation. Many of these studies depended on the repeated measurements of individual trees over long periods. Before 1930 there were no properly replicated trials (the first was at Bowmont in Scotland) but, if they were based on a sufficient number of sample plots, it was possible to devise generalized yield tables for uniform stands, which could be applied in forest planning. The yield tables were based on the application of a particular thinning grade, usually set to maintain yield close to the maximum attainable by an unthinned stand. Thinning, or the selective reduction in competition between individual trees, transfers leaf-area development between trees and, *inter alia*, alters the distribution of light in the canopy. Based on many studies, Assman (1970) showed that reductions in basal area down to about 70% of full stocking (i.e. a self-thinning stand) did not reduce overall production of measurable volume, thus allowing production to be concentrated on the most valuable trees. With some economic objectives, reductions in total yield are accepted to capitalize on the increased growth of fewer stems, although a price may have to be paid in stem form, wood density and branchiness. The essential point is that the variables involved in marking a thinning are so many (management objectives, site variation, stand history) that it remains a highly subjective operation and only subject to generalized rules of thumb. With a few exceptions, forest science has had little impact on practice, although there are published guidelines for controlling the proportion of standing volume removed in thinning.

Stand-density Relations

Competition between individual trees and the attendant mortality have attracted attention in attempts to optimize stand treatments. Reineke (1933) first drew attention to the generality, across species, of stand-density relationships in proposing a stand-density index to which different treatments could be related. In South Africa, Craib (1939) installed his CCT plots to explore the effects of competition on pine, which demonstrated the inexorable increase to a maximum basal area of stands of initially very different densities. This trend was substantiated by Kira *et al.* (1953) in Japan, who showed, for a number of plant species, that yields were uniform for different starting densities. Finally, Yoda *et al.* (1963) introduced the so-called '$-3/2$ power law' relating the mean weight (size) of individuals to density in self-thinning stands of a wide range of species. After a lag of about a decade, this useful rule was noticed in the literature, with a small flurry of papers challenging its universality. Nevertheless, the relationship has been shown to be robust but has only infrequently been used in studies, or the modelling, of stand development.

Application of Forest Science

These rather few examples of developments in forest science have deliberately omitted ecophysiological topics (the subject of this meeting) and swathes of research in many areas, such as hydrological effects of forests or, say, mycorrhizal relationships. They do, however, include some of the major themes that have been developed in the last century or so. The interesting question is why the very considerable knowledge and literature describing the processes and functions that underlie forest growth have not always been incorporated or made use of in general forest practice. Some developments have, of course, been utilized: for example, the adoption of site classifications generally, the recognition of differences in adaptive variation as a basis for tree breeding and research on nutrition in fertilizer practice. But even these have depended on clear economic/financial benefits being discernible to the manager, who has tended to rely on the results of empirical trials and field experiments. Reliance on the early results of field experiments, has, on occasion, been illusory. For example, in some afforestation schemes, the demonstrable boost to early growth obtained from intensive cultivation has not been maintained, as nutrient capital has been exploited faster than the site could continue to provide it. Some of these sorts of error could have been avoided by the application of existing knowledge or by conducting research into the processes involved.

There is a sense in which forest science can be thought of as following forest practice: that is, it has been explicative of observed phenomena that had already been made use of by foresters aiming for desired end results. An example might be the development of the shelter-wood silvicultural system in the oak/beech forests in France.

In 1666 Colbert, Chancellor to Louis XIV, devised rules for the upgrading of derelict stands of oak for the production of naval timber. Sustained yields are now obtained over quite large areas (~30,000 ha at Troncais) by allocating equal areas to 25-year age classes throughout the forest. The silvicultural technique requires the establishment, by natural regeneration, of very dense carpets of oak seedlings (~100,000/ha) under an existing canopy, which has been slightly opened in a seeding felling. Once obtained, the regeneration is released in a series of secondary fellings, which provide progressively more radiation at ground level, while providing protection against spring frosts. The ensuing intense competition in the thicket and sapling phases leads to strong vertical growth and early self-pruning. The beech component is actively kept in the understorey to maintain shaded ground cover and prevent epicormic growth in the oak, as it is slowly and steadily thinned, to ensure a radial increment of less than 3 mm/year. After a rotation of 225 years, the yield is taken of very valuable veneer-quality stems and barrel-stave wood and the process starts again.

The timing and intensity of the stand treatments are governed by the observed responses of the desired trees (oak) and their perceived need for release from competition. This is silvicultural skill (art) rather than forest science. The physiological basis for these treatments came much later, with, for example, studies in the 1960s on oak-seedling physiology (Jarvis, 1964) or the effect of stem shading on height growth in young oak (Roussel, 1963). (The latter gave rise to the rash of plastic-tube cemeteries around the country – another triumph for empiricism?)

Yield regulation in the uniform stands of French oak forests is generally simple to achieve, with only occasional problems from climatic perturbations (late frosts or storm damage). Instituting effective yield regulation is another matter in tropical forests, where utilization is based on logging a few so-called ‘commercial species’ from the ecologically complex forest. Harvesting disrupts the ecosystem, damages many of the residual trees, destroys seedling banks of desirable species and removes or compacts the topsoil. Attempts to regulate felling practices to ensure continued productivity have set diameter cutting limits (e.g. > 50 cm) and, recently, reduced impact logging with directional felling and planned skid-lines. However, little is known of the effects of removing dominant trees on the regeneration and subsequent growth patterns of the remaining stand.

Recent physiological studies of seedlings in dipterocarp forests in Kalimantan, Borneo, demonstrate how the application of forest science may assist in the design of sustainable-management systems. Some desirable species of the ‘light red meranti’ group were shown to be light-saturated at an irradiance (PPFD) of 5–10 mol/m^2/day, whereas less desirable pioneer species appear to need at least 8 mol/m^2/day for establishment. Thus the maximum size of canopy gap to promote the more valuable trees was calculated to be 650 m^2, equivalent to the crown area of two adjacent trees (van Gardingen *et al.*, 1998). Larger gaps encourage pioneer species and lead to undesirable ecological effects on soil-moisture availability and, importantly, the viability of the host-specific mycorrhizas upon which seedling growth is dependent.

Based on these studies and the data accumulated from logged and unlogged sample plots, it was possible to test the effects of different harvesting regimes on the future yield, growth and economic out-turn of managed forest (P.R. van Gardingen *et al.*, unpublished observations). The treatments simulated in a suite of linked biological and economic models ranged from the current > 50 cm dbh cutting limit, through felling eight stems/ha to volume limits of 50 m^3/ha, all in felling cycles of 25–45 years. The economic analysis (cost/benefit) set a goal of 16% internal rate of return (IRR). The current diameter limit and the eight stem/ha regime proved unsustainable both in yield and financially, while the best option appears to be harvesting 50 m^3/ha at 35-year intervals (or 60 m^3 at 45 years). All regimes tested showed that it would take about 150 years for the forest to attain a new equilibrium in stand structure. The high IRR sought was not reached by any treatment, but 'volume-controlled' harvesting could achieve it if the current utilization efficiency of 52% recovery of felled timber were improved by about 15%.

This analysis illustrates the way basic studies and modelling techniques can be used to estimate the effects of possible treatments but it leaves out an important aspect of current thinking in forest management. Increasingly, emphasis is being placed on the social implications of management decisions for local populations. Although not a new concept, funding bodies are increasingly reluctant to fund basic studies in forest science that do not specifically address the needs of the local inhabitants. In the case of the Kalimantan forests, the economics might look different if values for non-timber products and environmental conservation had been able to be included. Unfortunately, there are often also political constraints on the application of scientifically based solutions to management problems.

These two examples of forest-management regimes illustrate some of the difficulties for forest scientists in transferring the results of their efforts into practice. In France, silviculturists had already developed, through close observation of the effects of treatments, a sophisticated set of techniques to sustainably achieve a highly valued product, without needing to understand the processes involved. In the Kalimantan case, forest science can indicate the ecologically desirable regime for management to follow but conservation of the forest is likely to fail because of the economic and political environment.

Paul Jarvis – Forest Scientist

We are here to mark Paul's official retirement and to celebrate an outstanding scientific career. It therefore seems appropriate to comment briefly on some aspects of that career. He will give his recollection of it shortly!

I believe that Paul went up to Oxford intending to read forestry but maybe wisely switched to botany, graduating in 1957. His Sheffield PhD was awarded in 1961 for a thesis on 'Factors affecting the growth and regeneration of

Quercus petraea in the Sheffield region'. This was followed by a spell in Sweden and then Australia before he moved to Aberdeen, where he initiated work on the physiology of Sitka spruce canopies. In 1975 he came to Edinburgh as Professor of Forestry and Natural Resources, where he has continued to pursue many aspects of the environmental relations of forest plants.

In terms of publications, he (and Margaret) burst on to the scene with nine papers in 1963, including six papers on water relations of tree seedlings in one volume (16) of *Physiologia Plantarum*, which are still quoted. Gas exchange began to appear in titles from 1965, followed by the results from the Fetteresso studies in Sitka spruce. This series of 20 or so co-authored papers, mainly in the *Journal of Applied Ecology*, ran from 1971 to about 1982 as a string of postgraduates published. Two notable books appeared in this period as well: a *Manual of Photosynthetic Methods*, with co-editors Sestak and Catsky, to which Paul contributed 330 pages out of the 820 total; the other very different book was a translation from Swedish (with Margaret) of *Stalfelt's Plant Ecology* – a delightful book.

The work at Fetteresso, detailing canopy interactions with the atmosphere, seemed at the time to have little to say to the forest manager – other than that they should pay attention to the water relations of the species. However, that work provided an excellent platform on which to develop thinking about larger-scale canopy processes and forest and regional uptake of carbon dioxide. With that, Paul and his colleagues were well placed to be involved on a global scale with the concerns about climate change and the role of forests as sinks or sources for increasing levels of CO_2 in the atmosphere. This was a contribution to global policy-making, way beyond simple forest management. Beginning with the individual plant, the work has gone to the stand, forest, region and Earth.

To iterate all the publications would be tedious; suffice it to say that, in the eight quinquennia since 1963, his publication rate has never fallen below 300 pages and reached 830 pages in the mid-1990s (i.e. an average > 100 pages/year over 38 years). Included in this output are a number of important reviews of forest : atmosphere relations. He has always been a fierce editor and referee and woe betide anyone failing to use SI units properly or indulging in split infinitives.

The academic world places too much emphasis on output and not enough on other aspects of university life. Some second-year students used to be stunned by Paul's demanding lectures but he was always willing to help them understand. His ability to develop honours students' thinking and expand his postgraduates' abilities was sometimes little short of miraculous. On field courses he enjoyed sharing his expertise and getting to know the students. On the other hand, administration, budgeting and dealing with some of the bureaucratic 'nonsenses' in the university were not so interesting!

Outside the academic scene, Paul has never lost his basic interest in plants and natural history, whether as a gardener or field ecologist. His contribution to conservation bodies, such as the John Muir Trust and the National Trust for Scotland, is not well enough known.

With all this dynamism and a massive contribution to forest science, he remains a strongly committed family man. We all salute his achievements – and do not expect him to stop yet!

References

Assman, E. (1970) *The Principles of Forest Yield Study*. Translated by S.H. Gardiner and P.W. Davis. Pergamon Press, Oxford, 506 pp.

Blackman, V.H. (1919) The compound interest law and plant growth. *Annals of Botany* 33, 353–360.

Cajander, A.K. (1909) Ueber Waldtypen. *Acta Forestalia Fennica* 1 and *Fennia* 28 (quoted in Cajander, A.K. (1926) The theory of forest types. *Acta Forestalia Fennica* 31, 1–108).

Cannell, M.G.R. (1982) *World Forest Biomass and Primary Production Data*. Academic Press, London, 391 pp.

Cieslar, A. (1907) Die Bedeutung klimatischer Varietaten unserer Holzarten für den Waldbau. *Centralblatt für des geschichte Forstwesen* 33 (quoted by Langlet O. (1972) *Taxon* 20).

Craib, I.J. (1939) *Thinning, Pruning and Management Studies on the Main Exotic Conifers Grown in South Africa*. Bulletin 196, Department of Agriculture and Forestry Science, Government Printer, Pretoria.

Ebermayer, E. (1882) *Die gesamte Lehre der Waldstreu mit Rucksicht auf die chemische Statik des Waldbaues*. Springer-Verlag, Berlin, Heidelberg (quoted in Lieth, H.F.H. (1978) *Patterns of Primary Production in the Biosphere*. Dowden, Hutchinson and Ross, Stroudsberg, Pennsylvania).

Fisher, R.A. (1925) *Statistical Methods for Research Workers*. Oliver and Boyd, Edinburgh.

Hartig, G.L. (1813) *Anweisung zur Taxation und Beschreibung den Forsten*, 3rd edn. Giessen (quoted in Kostler, J. (1956) *Silviculture*. Oliver and Boyd, Edinburgh).

Ingestad, T. (1982) Relative addition rate and external concentration: driving variables used in plant nutrition research. *Plant, Cell and Environment* 5, 443–453.

Jarvis, P.G. (1964) The adaptability to light intensity of seedlings of *Quercus petraea* (Matt.) Liebl. *Journal of Ecology* 52, 545–571.

Kira, T., Ogawa, H. and Shinozaki, K. (1953) Intraspecific competition among higher plants. I. Competition–density–yield interrelationships in regularly dispersed populations. *Journal of the Institute of Polytechnics, Osaka City University* D7, 73–94.

Krajina, V.J. (1965) Biogeoclimatic zones and classification of British Columbia. In: Krajina, V.J. (ed.) *Ecology of Western North America*, Vol. 1. Department of Botany, University of British Columbia, Vancouver, pp. 1–17.

Landsberg, J.J. and Waring, R.H. (1997) A generalised model of forest productivity using simplified concepts of radiation-use efficiency, carbon balance and partitioning. *Forest Ecology and Management* 95, 209–228.

Lieth, H.F.H. (ed.) (1987) *Patterns of Primary Production in the Biosphere*. Benchmark Papers in Ecology V.8 Dowhen, Hutchinson and Ross, Strondsburg, Pennsylvania, 342 pp.

Pyatt, D.G., Ray, D. and Fletcher, J. (2001) *An Ecological Site Classification for Forestry in Great Britain*. Bulletin 124, Forestry Commission, Edinburgh, 74 pp.

Reineke, L.H. (1933) Perfecting a stand density index for even-aged forests. *Journal of Agricultural Research* 46, 627–638.

Roussel, L. (1963) Effets obtenus en ombrageant la tige du chêne pédoncule. *Annales scientifique de l'Université de Besançon*, Série Botanique 19.

Tamm, C.O. (1979) Nutrient cycling and productivity of forest ecosystems. In: Leaf, A.L. (ed.) *Impact of Intensive Harvesting on Forest Nutrient Cycling*. State University of New York, Syracuse, pp. 2–21.

van Gardingen, P.R., Clearwater, M.J., Nifinluri, T., Effendi, R., Rusmantoro, W., Noor, M., Mason, P.A., Ingleby, K. and Munro, R.C. (1998) Impacts of logging on the regeneration of lowland dipterocarp forest in Indonesia. *Commonwealth Forestry Review* 77, 71–82.

Waring, R.H., Schroeder, P.E. and Oren, R. (1982) Application of the pipe model theory to predict canopy leaf area. *Canadian Journal of Forest Research* 12, 556–560.

Yoda, K., Kira, T., Ogawa, H. and Hozumi, H. (1963) Self-thinning in overcrowded pure stands under cultivated and natural conditions. *Journal of Biology, Osaka City University* 14, 107–129.

Index

cabi-publishing.org/bookshop
Browse Read and Buy
ANIMAL & VETERINARY SCIENCES
BIODIVERSITY CROP PROTECTION
HUMAN HEALTH NATURAL RESOURCES
ENVIRONMENT PLANT SCIENCES
SOCIAL SCIENCES
CABI Publishing
A division of CAB International
Online BOOK SHOP
Search
Reading Room
Bargains
New Titles
Forthcoming
Order & Pay Online!
MasterCard
VISA
AMERICAN EXPRESS
Crop Pollination by Bees
Keith S. Delaplane and Daniel F. Mayer
A DICTIONARY OF Entomology
G Gordh and D H Headrick
Principles of CATTLE PRODUCTION
C.J.C. Phillips
Seeds
THE ECOLOGY OF REGENERATION IN PLANT COMMUNITIES
2nd EDITION
Edited by Michael Fenner
FULL DESCRIPTION
BUY THIS BOOK
BOOK OF THE MONTH
Tel: +44 (0)1491 832111 Fax: +44 (0)1491 829292